Designing Embedded Systems and the Internet of Things (IoT) with the ARM® Mbed™

Designing Embedded Systems and the Internet of Things (IoT) with the ARM® Mbed™

Perry Xiao
London South Bank University
UK

WILEY

This edition first published 2018

Registered Offices
John Wiley & Sons, Inc., 111 River Street, Hoboken, NJ 07030, USA
John Wiley & Sons Ltd, The Atrium, Southern Gate, Chichester, West Sussex, PO19 8SQ, UK

Editorial Office
The Atrium, Southern Gate, Chichester, West Sussex, PO19 8SQ, UK

For details of our global editorial offices, customer services, and more information about Wiley products visit us at www.wiley.com.

Wiley also publishes its books in a variety of electronic formats and by print-on-demand. Some content that appears in standard print versions of this book may not be available in other formats.

Library of Congress Cataloging-in-Publication Data

Names: Xiao, Perry, author.
Title: Designing embedded systems and the internet of things (IoT) with the ARM Mbed / by Perry Xiao.
Description: First edition. | Hoboken, NJ : John Wiley & Sons, Inc., [2018] | Includes bibliographical references and index. |
Identifiers: LCCN 2018008687 (print) | LCCN 2018015034 (ebook) | ISBN 9781119364016 (pdf) | ISBN 9781119364047 (epub) | ISBN 9781119363996 (cloth)
Subjects: LCSH: Embedded computer systems–Design and construction | Internet of things–Equipment and supplies. | Microcontrollers.
Classification: LCC TK7895.E42 (ebook) | LCC TK7895.E42 X56 2018 (print) | DDC 006.2/2–dc23
LC record available at https://lccn.loc.gov/2018008687

Cover design by Wiley
Cover image: © matejmo/Getty Images; © Raimundas/Shutterstock

Set in 10/12pt WarnockPro by SPi Global, Chennai, India

10 9 8 7 6 5 4 3 2 1

This book is dedicated to my family. To my wife, May, my son, Zieger, and my daughter, Jessica, who make my life complete—without them, life would be meaningless. To my parents and my brother, who have shared their life and love with me that ultimately made me what I am today. To my friends and colleagues, who supported me throughout my career.

Contents

About the Author

Dr. Perry Xiao is an associate professor and course director at the School of Engineering, London South Bank University, London, United Kingdom. He got his BEng degree in Opto-Electronics, MSc degree in Solid State Physics, and PhD degree in Photophysics. He is a chartered engineer (CEng), a Fellow (FIET) from Institution of Engineering and Technology (IET) and a Senior Fellow (SFHEA) from Higher Education Academy (HEA). He has been teaching electronics, software, computer networks, and telecommunication subjects at both undergraduate level and postgraduate level for nearly two decades. He is also supervising BEng final project students and MSc project students every year. His main research interest is to develop novel infrared and electronic sensing technologies for skin bioengineering applications and industrial nondestructive testing (NDT). To date, he has finished seven PhD student supervisions, obtained two UK patent applications, published more than 100 scientific papers, been editorial reviewer for nine journals, and generated nearly £1 million in research grants.

He is also a director and co-founder of Biox Systems Ltd., UK—a university spin-off company that designs and manufactures state-of-the-art skin measurement instruments, AquaFlux and Epsilon, which have been used in more than 70 organizations worldwide, including leading cosmetic companies, universities, research institutes, and hospitals.

Preface

When I first got the Arm® Mbed™ Lab-in-a-Box (LiB) kits from a colleague a few years ago, I could not hide my excitement. It was a box of mbed NXP LPC1768 development boards donated through the ARM University Program. One of the key features of the Arm® Mbed™ system is that you can write and compile your code online through a web browser. This was completely new to me. I have been using various microcontrollers throughout my life. I did my BEng final year project on laser energy control using Intel's 8051 single-chip microcontroller back in the 1980s. The concept was very simple: read the voltage value from the laser power monitor, compare it with a desired value, and calculate the required adjustment to feed back to the laser to increase or decrease the laser output. But we had to design and make our own printed circuit boards (PCB) and to write our own code that would run on 8051. At that time, programming microcontrollers was not a trivial task. You needed to write the program in assembly language and punch in the corresponding hexadecimal code into the microcontroller. We spent many sleepless nights in the lab, mainly for debugging the code. I have since worked with many other microcontroller-based embedded systems, and the experiences were very mixed. Some of the embedded systems were so difficult to use that you would need to download this software, download that toolchain, etc. Using my students' words, you needed a PhD just to get the compiler software running. The code was also sophisticated—you would need to configure this register, and configure that port. You could produce lines of lines of code, which did not even do much!

The two embedded systems that impressed me most were Raspberry Pi and Arduino. Raspberry Pi is very attractive for its price and its compact, credit-card size. With a full Debian-based Linux operating system and graphical user interface, it is a great kit for people to learn computing and coding. But for many of our student projects, we don't need a full operating system, and a lack of analog to digital converter (ADC) and digital to analog converter (DAC) are also major drawbacks. Arduino is also attractive for its price and size, but what impressed me most is its simplicity, both in hardware and software. I have read many "24 hours" books, and Arduino is genuinely one of the things that you can truly learn in 24 hours. It is just that simple. However, the limited memory size means you cannot write too-long programs, and 10-bit ADC is often proven to be inadequate in many applications.

So when I introduced the Arm® Mbed™ NXP LPC1768 development boards to my students, they loved them. They liked the web-based compiler. The very fact that you don't need to download and install any software on your computer in order to run it is amazing. It makes life so much easier. The code was also much simpler, much more understandable. As it is claimed on the Arm® Mbed™ website (https://os.mbed.com/platforms/FRDM-K64F/), it can really just take 30 seconds to get the development board out of the box, and run an application without installing any software!

The Arm® Mbed™ NXP LPC1768 is one of the most popular microcontroller development boards, widely used among students and electronic hobbyists. It is based on 32-bit ARM® Cortex™-M3 microcontroller with 96 MHz clock speed, 512 KB flash, 32 KB RAM, and, most importantly, 12-bit ADCs. It is more powerful and runs much faster than Arduino. It also has lots of interfaces, including Ethernet, USB, CAN, SPI, I2C, DAC, PWM, and other I/O interfaces.

However, the 32-bit ARM® Cortex™-M3 microcontroller is gradually reaching its shelf life; its replacement is the 32-bit ARM® Cortex™-M4 microcontroller. So this book will be focusing on the new, exciting Arm® Mbed™ Ethernet Internet of Things (IoT) Starter Kit, which includes an Arm® Mbed™ NXP FRDM-K64F development board and an Arm® Mbed™ application shield. The Arm® Mbed™ NXP FRDM-K64F is the next generation, flagship development board, which is based on ARM® Cortex™-M4 microcontroller with a CPU frequency up to 120 MHz, 1024 KB Flash, 256 KB RAM, and astonishing two 16-bit ADCs. It is much faster and more powerful than NXP LPC1768. It also has DACs and Timers, as well as other interfaces such as Ethernet, USB device crystal-less, and Serial. The Arm® Mbed™ Ethernet IoT Starter Kit is a cloud-based development kit jointly developed by ARM and IBM. It provides the user with a slick experience, sending data from the onboard sensors into the IBM cloud. It allows you to access the IBM cloud applications through IBM's BlueMix platform. It is particularly suitable for developers with no specific experience in embedded or web development, as it provides a platform for learning new concepts and creating working prototypes. The starter kit hardware can also be modified and extended to satisfy your specific requirements.

For backward compatibility reasons, many example codes also work for NXP LPC1768 development board and its mbed application board.

As of the time this book was written, the Arm® Mbed™ had just released its latest Arm® Mbed™ OS (operating systems) version 5.7, which has quite a few changes compared with the previous version mbed OS 3.0 and 2.0. This book is mainly based on the Arm® Mbed™ OS 5.7, and more details about the new OS are available at the Arm® Mbed™ documentation website (https://os.mbed.com/docs).

I have thoroughly enjoyed working with the Arm® Mbed™ development boards, and I hope you will enjoy it too.

How This Book Is Organized

This book aims to teach students how to design and develop embedded systems as well as Internet of Things (IoT) applications using Arm® Mbed™ development boards. It is divided into three parts.

Part I: Introduction to Arm® Mbed™ and IoT (Chapters 1–3) gives an introduction of embedded systems, microcontrollers and microprocessors, Arm® architecture and

Arm® Mbed™ system. It also provides an overview of the Internet of Things (IoT), including IoT applications and IoT enabling technologies.

Part II: Arm® Mbed™ Development (Chapters 4–10) illustrates how to get started with Arm® Mbed™ development, as well as how to work with analog inputs/outputs, digital inputs/outputs, communication interfaces, debugging, online libraries, and project managements.

Part III: The IoT Starter Kit and The IoT Projects (Chapters 11–12) introduces the Arm® Mbed™ Ethernet IoT Starter Kit and provides some example IoT projects.

Part IV: Appendices

Appendix A: Example Codes
Appendix B: HiveMQ MQTT Broker
Appendix C: Node-RED on Raspberry Pi
Appendix D: String and Array Operations
Appendix E: Useful Resources

Example Codes

All the example source codes are available on the website that accompanies this book. Appendix A has more details on how to use the codes.

Who This Book Is For

This book is intended for university/college students as well as amateur electronic hobbyists. It assumes readers have a basic concept of how computers work and can competently use a computer, i.e., can switch the computer on, log in, run some programs, and copy files to and from a USB memory stick (without losing their temper!).

It assumes that readers have some electronics experience, such as handling a breadboard, wires, resisters, power supply, and LEDs. It also assumes readers have some basic programming experiences (ideally in C/C++, but other languages are also fine), and know the basic syntax, the different types of variables, the conditional selections, and the loops and subroutines. Prior knowledge and experiences of microcontrollers are desirable, but not necessary.

Finally, it assumes readers have a basic concept of computer networks and the Internet, i.e., understand the concept of IP addresses and port numbers, know how to find out the IP address of a computer, and can use some of the most commonly used Internet services such as the World Wide Web, email, file download/upload, online audio, online video, and even some cloud-based services.

This book can be used as a core textbook as well as a background-reading textbook.

Suggested Prerequisite Readings

Electronics:

Electronics All-in-One for Dummies, 2nd edition, Doug Lowe, ISBN: 978-1-119-32079-1, March 2017.

C/C++ Programming:

Beginning Programming with C for Dummies, Dan Gookin, ISBN: 978-1-118-73763-7, November 2013.

C++ Primer, 5th edition, Stanley B. Lippman, Josée Lajoie, Barbara E. Moo, Addison Wesley, ISBN: 978-0-321-71411-4, August 2012.

Computer Networks and Internet:

Computing Fundamentals: Digital Literacy Edition, Faithe Wempen with Rosemary Hattersley, Richard Millett, Kate Shoup, ISBN: 978-1-118-97474-2, August 2014.

Understanding Data Communications: From Fundamentals to Networking, 3rd edition, Gilbert Held, ISBN: 978-0-471-62745-6, October 2000.

What You Need

In this book, you will need:

- Arm® Mbed™ Ethernet IoT Starter Kit
  - NXP FRDM-K64F development board
  - Mbed application shield
- Breadboard with jump wires
- Various sensors
- A digital or analog oscilloscope (optional)
- NXP LPC1768 development board and its Application board (optional)
- Raspberry Pi (https://www.raspberrypi.org/) (optional)
- Java JDK software (http://www.oracle.com/technetwork/java/javase/downloads/index.html)
- Python software (https://www.python.org/downloads/) (optional)

Author's Acknowledgments

I would like to express my sincere gratitude to Wiley Publishing for giving me this opportunity. I would also like to thank Ella Mitchell for her persistence and patience. Without it, this book would be not possible.

About the companion website

Don't forget to visit the companion website for this book:

www.wiley.com\go\xiao\designingembeddedsystemandIoTwitharmmbed

There you will find valuable material designed to enhance your learning, including:

- Examples

Scan this QR code to visit the companion website

Part I

Introduction to Arm® Mbed™ and IoT

In this part:

1

Introduction to Arm® Mbed™

Isn't it a pleasure to study and practice what you have learned?

- Confucius

1.1 What is an Embedded System?

An embedded system is a small-scale computer system that is part of a machine or a larger electrical/mechanical system. It is often designed to perform certain dedicated tasks and often a real-time system. It is called *embedded* because the computer system is embedded within a hardware device. Embedded systems are important, as they are getting increasingly used in many daily appliances, such as digital watches, cameras, microwave ovens, washing machines, boilers, fridges, smart TVs, and cars. Embedded systems also often need to be small in size, low in cost, and have low power consumption.

Figure 1.1 shows the schematic diagram of a typical embedded system that includes a microcontroller, inputs/outputs, and communication interfaces.

Microcontroller

Microcontroller is the brain of an embedded system, which orchestrates all the operations. A microcontroller is a computer processor with memory and all input/output peripherals on it. More details about microcontrollers will be illustrated in the next section.

Inputs

An embedded system interacts with the outside world through its inputs and outputs. Inputs can be digital inputs or analog inputs. Inputs are typically used for reading data from sensors (temperature sensor, light sensor, ultrasound sensor, etc.) or other types of input devices (keys, buttons, etc.).

Outputs

Outputs can also be digital outputs or analog outputs. Outputs are typically used for display, driving motors, or other devices (actuators).

Designing Embedded Systems and the Internet of Things (IoT) with the ARM® Mbed™, First Edition. Perry Xiao

Companion website: www.wiley.com\go\xiao\designingembeddedsystemandIoTwitharmmbed

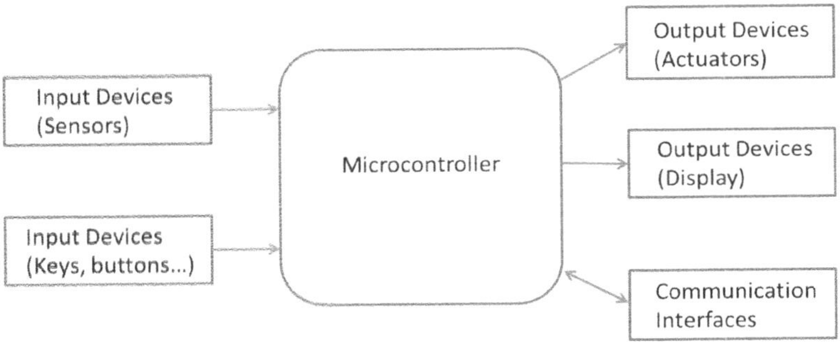

Figure 1.1 Schematic diagram of a typical embedded system.

Communication Interfaces
An embedded system communicates with other devices using communication interfaces, which includes Ethernet, USB (Universal Serial Bus), CAN (Controller Area Network), Infrared, ZigBee, WiFi and Bluetooth, for example.

1.2 Microcontrollers and Microprocessors

At the heart of embedded systems are microcontrollers. Although there are embedded systems built on microprocessors, modern embedded systems are largely based on microcontrollers. A typical microcontroller contains a central processing unit (CPU), interrupts, timer/counter, memory, and other peripherals, all in a single integrated circuit (IC). A microcontroller is a true computer on a chip or system-on-a-chip (SoC). Microcontrollers are ideal for control applications because you can use them to build an embedded system with little additional circuitry.

Microcontrollers (MCU or μC) are different from microprocessors (MPU). A microprocessor is a single IC with only a central processing unit (CPU) on it. In order to make it functional, you will need to add external memory and other peripheral devices. Figure 1.2 shows the main differences between a microprocessor and a microcontroller. To put it simply, you can imagine that a microprocessor is just a CPU on a single IC, while a microcontroller is a small computer with CPU, memory, and other peripherals.

Microprocessors are mainly used in general-purpose systems like personal computers. They have relatively high computational capacity and can perform numerous tasks. Microprocessors have relatively high clock frequency, usually in the order of gigahertz. Microprocessors generally consume more power and often require external cooling system.

Microcontrollers are designed for control applications and are generally used in embedded systems. They have relatively low computational capacity and can perform single or very few tasks. Microcontrollers have relatively low clock frequency, usually in the order of megahertz. Microcontrollers consume less power and have no need for a cooling system.

Figure 1.3 shows a more detailed schematic diagram of a microcontroller. Following are its key components.

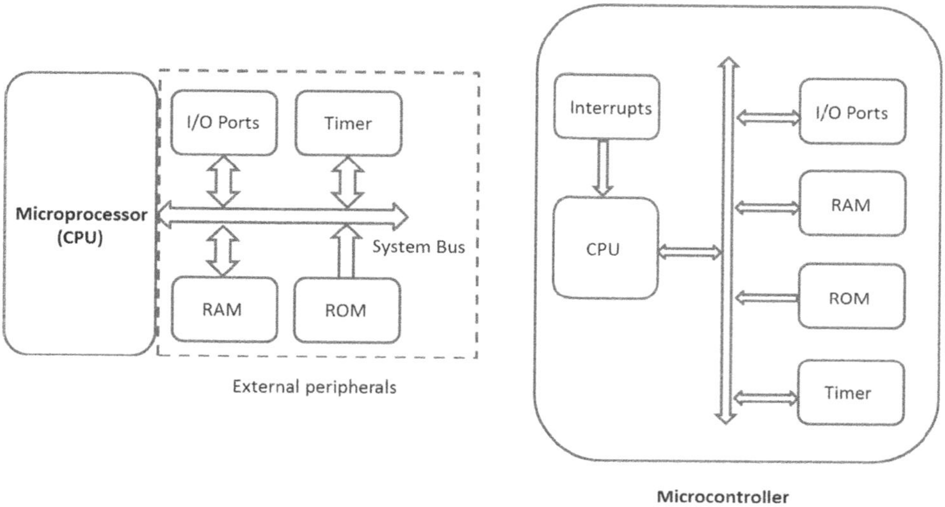

Figure 1.2 Comparison between a microprocessor and a microcontroller.

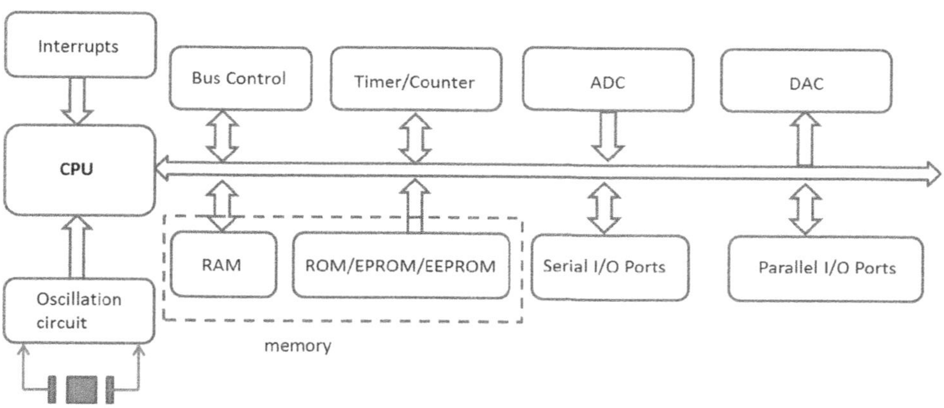

Figure 1.3 Detailed schematic diagram of a microcontroller.

CPU

CPU, often referred to as a processor or central processor, is the brain of a microcontroller. See Figure 1.4 for details. It contains three main components: the arithmetic logic unit (ALU), the control unit, and registers. ALU performs arithmetic and logical operations, registers provide operands to the ALU and store the results of ALU operations, and the control unit controls the overall operations and communicates with both ALU and registers. The operation cycle of CPU can be described as fetch, decode, and execute.

CPU communicates with its external peripherals such as memory and input/output through system bus, which includes a data bus, an address bus, and a control bus. The data bus is for carrying information, the address bus is for determining where the information should be sent, and the control bus is for determining the operation. The address

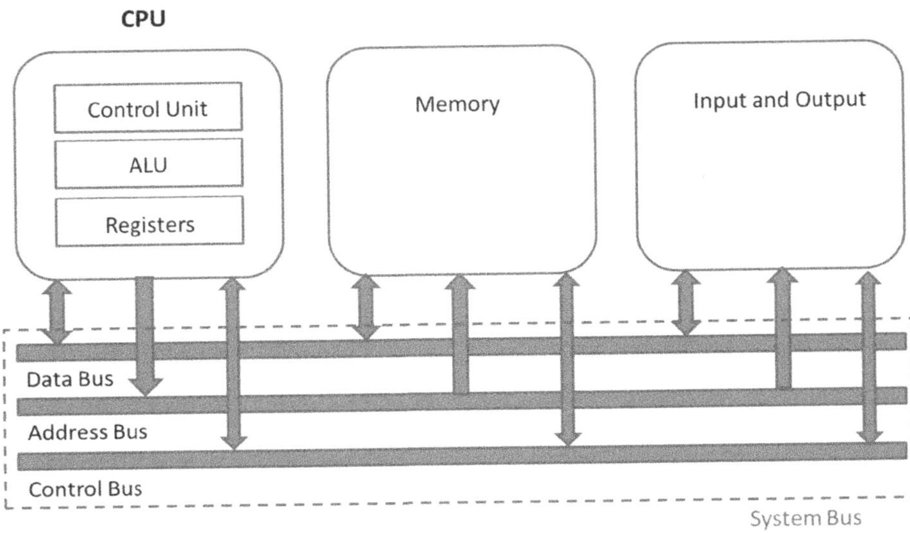

Figure 1.4 Detailed schematic diagram of a CPU.

bus is unidirectional, from CPU to peripherals, while the data bus and control bus are bidirectional.

CPU can be divided into different types, depending on the instruction set implemented. The instruction set, also called instruction set architecture (ISA), is a set of basic operations that CPU can perform. Two main types are complex instruction set computing (CISC) and reduced instruction set computing (RISC). A CISC CPU has very large instruction sets (300 and more) and more complex hardware, but has more compact software code. It takes more cycles per instruction. It also uses less RAM, as there is no need to store intermediate results. A RISC CPU has small instruction sets (100 and less) and simpler hardware, but has more complicated software code. It takes one cycle per instruction and uses more RAM to handle intermediate results. Typical examples of CISC CPUs are AMD and Intel x86, which are mainly used in personal computers, workstations, and servers, as they are capable of more sophisticated tasks. Typical examples of RISC CPUs are Atmel AVR, PIC, and ARM®, which are mainly used in microcontrollers because they consume less power.

Memory

Microcontrollers use memories for storing programs and data. There two types of memory, internal and external. Internal memory is limited in size but is fast. For applications in which internal memory is not enough, external memory is then needed. Traditionally, there are two types of external memory, random access memory (RAM) and read-only memory (ROM). RAM can be accessed randomly and you can perform both read and write operations to RAM. RAM will lose all its contents when power is switched off. ROM is read-only memory, which means you can read data from it but cannot write data to it. ROM does not lose its contents even if power is switched off; therefore, it is used to store programs and data permanently.

However, there are new types of memories, such as electrically erasable programmable ROM (EEPROM) and non-volatile RAM (NVRAM). Both can be read and write and do not lose its contents even if power is switched off. Flash memory is the best example of NVRAM. It is high-density, low-cost, fast, and electrically programmable. Flash memory is being extensively used for embedded systems that contain embedded operating systems and the application program.

Parallel Input/Output Ports

Parallel input/output ports have multiple wires (or pins) running parallel to each other. It is called *parallel* because multiple signals can be accessed all at once. Parallel input/output ports are mainly used to drive/interface various devices such as LCDs, LEDs, printers, memories, and so on to a microcontroller. Parallel ports can transfer data much faster than serial ports, but only suitable for short distance communications due to interference and noise.

Serial Input/Output Ports

Serial input/output ports use a single data wire to transfer data. Serial ports therefore are much slower than parallel input/output ports. However, serial ports can have higher bandwidth, and can be used over longer distances. Universal Asynchronous Receiver/Transmitter (UART) peripheral is a commonly used serial input/output port in embedded systems. It uses one wire for receiving data (Rx) and one wire for transmitting data (Tx).

Timers/Counters

Timers and counters are useful functions for a microcontroller. A microcontroller may have more than one timer and counters. The timers and counters provide all timing and counting functions inside the microcontroller, including clock functions, modulations, pulse generations, frequency measuring, and making oscillations.

Analog to Digital Converter (ADC)

ADC converts analog signals to digital signals. It is mainly used for reading voltage outputs of sensors. ADC can be 8 bits, 10 bits, 12 bits, 16 bits, 24 bits, and even 32 bits. The higher the number of bits means the higher conversion resolution. The bandwidth of an ADC (i.e., the range of frequencies it can measure) is determined by its sampling rate or sampling frequency. According to Nyquist–Shannon sampling theorem, the highest frequency that an ADC can measure is less than half of its sampling rate. The typical ADC sampling rate of mbed boards is about a few hundreds kilohertz.

Digital to Analog Converter (DAC)

DAC is the opposite of ADC. DAC converts the digital signals into analog signals. It usually used for controlling analog devices such as audio speakers, DC motors, and various drives.

Interrupt Control

Interrupt is one of the most important and powerful features in microcontroller applications. The interrupt control is used to interrupt a working program. The interrupts can be either hardware interrupts (external, activated by using interrupt pin) or software interrupts (internal, by using interrupt instruction during programming).

Reset
Reset is an important function that exists in all microcontrollers. Reset can make sure microcontrollers go back to its original state. This is important, especially when things go wrong.

Watchdog
A watchdog, or watchdog timer, is a piece of electronic hardware that is commonly used in embedded systems to automatically detect software malfunctions and to reset the processor. A watchdog timer basically counts down from some initial value to zero. The embedded software selects the counter's initial value and periodically restarts it. If the counter ever reaches zero before the software restarts it, the software is presumed to be malfunctioning and the processor will be reset.

1.3 ARM® Processor Architecture

ARM® (Advanced RISC Machine) architecture is a computer processor architecture based on reduced instruction set computing (RISC). ARM® architecture was originally developed by British company Acorn Computers based in Cambridge, United Kingdom in the 1980s. ARM® originally stood for Arcon RISC Machine. The first ARM processors were used in BBC Microcomputers. Acorn started working with Apple Computer and VLSI Technology in the late 1980s. In 1990, Acorn spun off the design team into a new company named Advanced RISC Machines (ARM®) Ltd. The company name was later changed to ARM® Holdings plc. ARM® Holdings plc floated on the London Stock Exchange and NASDAQ in 1998. It became a member of the FTSE 100 in 1999.

ARM® processors become increasingly popular after being used on Apple's iPhone and iPad since 2007. To date, ARM® processors are widely used in smartphones, tablets, and smart TVs. Over 50 billion ARM® processors were produced as of 2014. In July 2016, ARM® Holdings has an annual turnover about £1 billion and agreed to a £24.3 billion takeover by Japan's Softbank company. The takeover is largely seen as an investment for the Internet of Things (IoT), in which ARM® processors will be likely taking a dominant role.

To date, ARM® processors can be generally divided into three categories: Application, Real-time, and Microcontroller, as shown in Table 1.1. The ARM® application processors (Cortex-A series) are the most powerful processors, optimized for higher performance, and can be typically used in phones, pads, tablets, and computers. The ARM® Real-time processors (Cortex-R series), optimized for faster response, can be typically used in industrial, home, and automotive applications. The ARM® Microcontroller processors (Cortex-M series), optimized for smaller size, and lower power consumption, can be typically used in embedded systems, and, of course, IoT applications!

Figure 1.5 shows the performance functionality and capacity of the ARM® Cortex-A, Cortex-R, and Cortex-M processors.

Table 1.1 ARM® architecture categories

Application	Real-time	Microcontroller
32-bit and 64-bit A32, T32, and A64 instruction sets Virtual memory system Supporting rich operating systems	32-bit A32 and T32 instruction sets Protected memory system (optional virtual memory) Optimized for real-time systems	32-bit T32 / Thumb® instruction set only Protected memory system Optimized for microcontroller applications

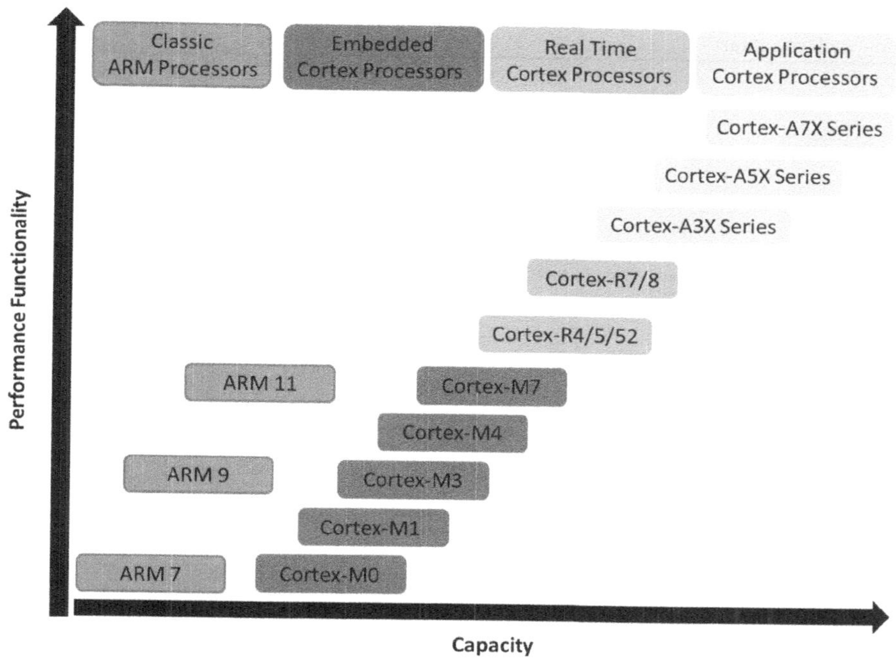

Figure 1.5 The performance functionality and capacity of ARM® processors (reproduced according to: https://www.arm.com/products/processors/).

Table 1.2 shows the different Cortex-M microcontrollers. The Cortex-M0, Cortex-M0+, and Cortex-M23 controllers are designed for the lowest power consumptions. The Cortex-M3, Cortex-M4, and Cortex-M33 controllers are designed for the highest efficiency. The Cortex-M7 controllers are designed for the highest performance. In this book, we will focus only on the ARM® Microcontroller processors, specifically the Cortex-M4 series.

Figure 1.6 shows the features and functions of ARM® Cortex-M series processors.

Further Information about ARM® Processor Architecture

https://www.arm.com/products/processors/instruction-set-architectures/index.php
https://en.wikipedia.org/wiki/ARM_architecture

Table 1.2 The Cortex-M Series Microcontrollers

Lowest Power and Area	Performance Efficiency	Highest Performance
Cortex-M23 TrustZone in smallest area, lowest power **Cortex-M0+** Highest energy efficiency **Cortex-M0** Lowest cost, low power Freely available for design and simulation via DesignStart	**Cortex-M33** Flexibility, control and DSP with TrustZone **Cortex-M4** Mainstream control and DSP **Cortex-M3** Performance efficiency	**Cortex-M7** Maximum performance, control and DSP

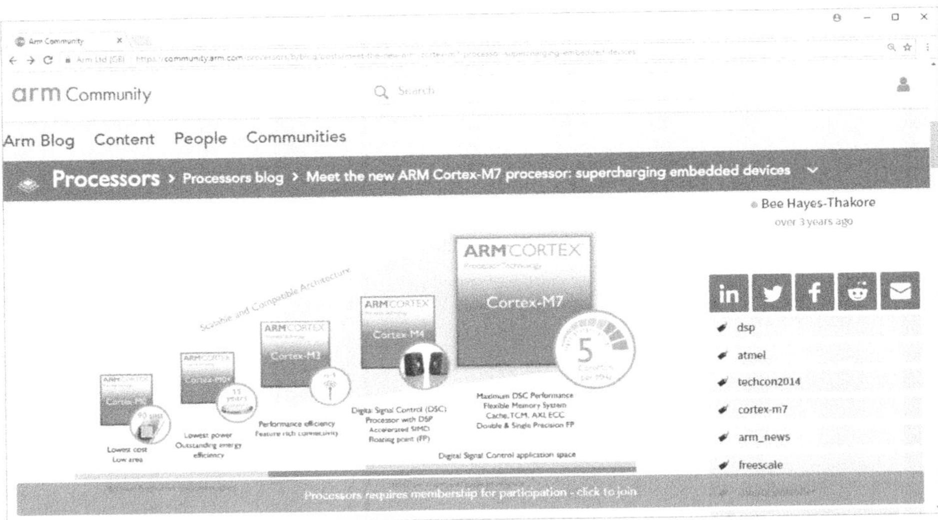

Figure 1.6 The features and functions of ARM® Cortex-M series processors from the ARM® website. (*Source:* https://community.arm.com/processors/b/blog/posts/meet-the-new-arm-cortex-m7-processor-supercharging-embedded-devices)

1.4 The Arm® Mbed™ Systems

The Arm® Mbed™ is a platform and operating system based on 32-bit ARM® Cortex-M microcontrollers. It is collaboratively developed by ARM® and its technical partners, and is designed for Internet of Things (IoT) devices. It provides the operating system, cloud services, tools, and developer ecosystem to make the creation and deployment of IoT solutions possible.

One of the major features of the Arm® Mbed™ systems is its web-based development environment. Just plug the device into computer using a USB cable, which will appear on your computer as a USB memory stick. Write and compile your software code using the Arm® Mbed™ Online Compiler, download the compiled code into the device, and press the onboard reset button to run!

Arm® Mbed™ provides all you need to develop IoT and embedded devices. It has a full support for over 100 mbed-enabled boards and more than 400 components. It also has tools for writing, building, and testing applications, and server and client-side tools to communicate with your devices.

The mbed microcontrollers provide experienced embedded developers a powerful and productive platform for building proof-of-concepts. For developers new to 32-bit microcontrollers, mbed provides an accessible prototyping solution to get projects built with the backing of libraries, resources, and support shared in the mbed community.

Figure 1.7 and Figure 1.8 show the Arm® Mbed™ home page, and the corresponding developer website. Figure 1.9 shows a list of development boards that is supported by the Arm® Mbed™. There are several development boards worth mentioning:

1.4.1 NXP LPC1768

This is one of the most popular development boards. It is based on the NXP LPC1768 microcontroller, with a 32-bit ARM® Cortex-M3 core running at 96 MHz. It has 512 KB

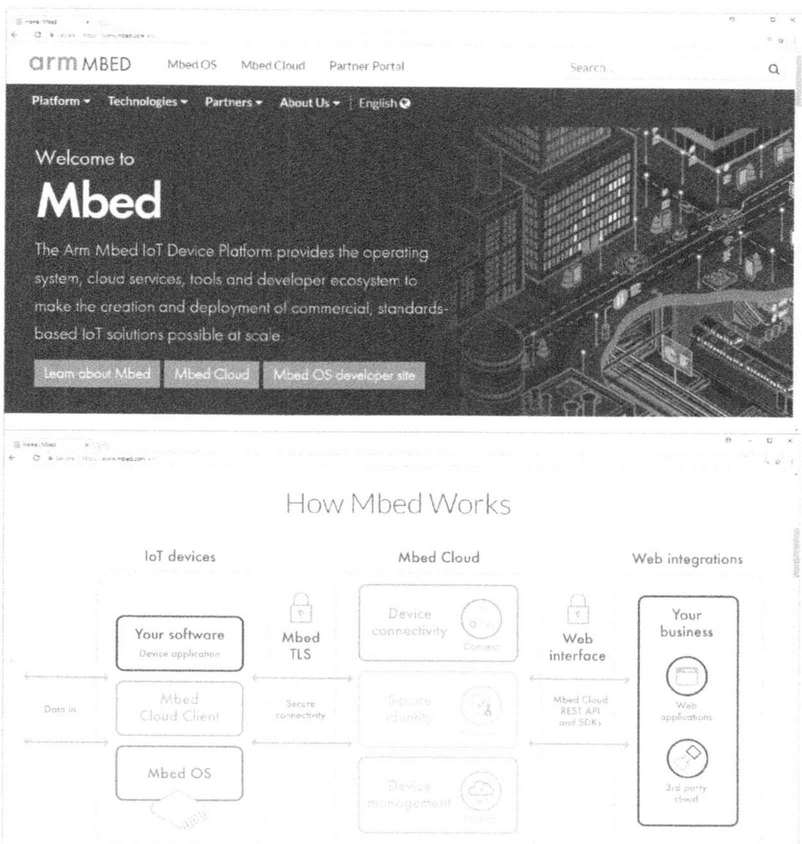

Figure 1.7 The Arm® Mbed™ website (top) and the schematic diagram of mbed systems (bottom).

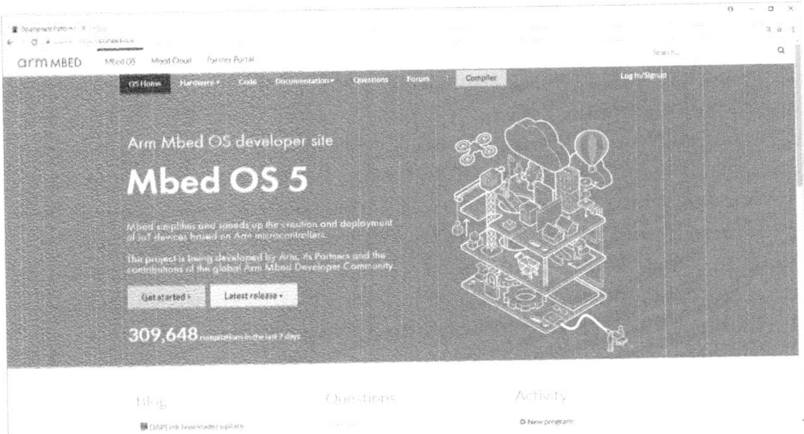

Figure 1.8 The Arm® Mbed™ developed website. The URL used to be https//developer.mbed.org, but has changed to https//os.mbed.com.

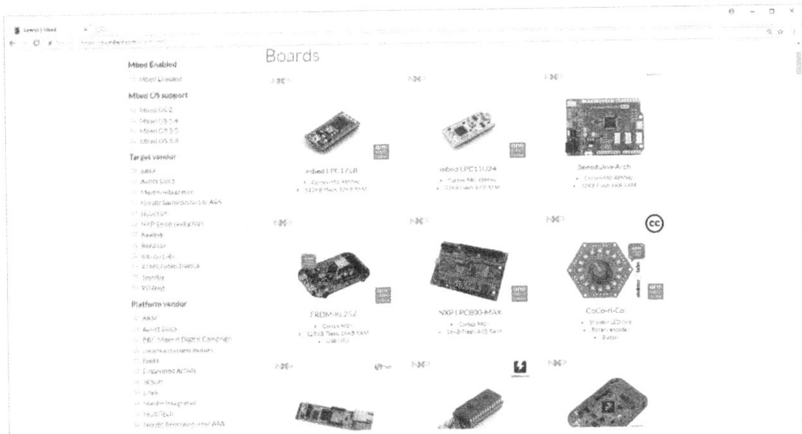

Figure 1.9 The Arm® Mbed™ development boards. The URL used to https//developer.mbed.org/products, but has changed to https//os.mbed.com/products.

flash, 32 KB RAM and lots of interfaces, including built-in Ethernet, USB host and device, CAN, SPI, I2C, ADC, DAC, PWM, and other I/O interfaces. The 12 bits of ADC are particularly useful. Figure 1.10 shows the board and its pinouts, including commonly used interfaces and their locations. The pins P5-P30 can also be used as DigitalIn and DigitalOut interfaces.

Features

- NXP LPC1768 MCU
  - High-performance ARM® Cortex™-M3 Core
  - 96 MHz, 32 KB RAM, 512 KB flash
  - Ethernet, USB host/device, 2×SPI, 2×I2C, 3×UART, CAN, 6×PWM, 6×ADC (12 bits), GPIO

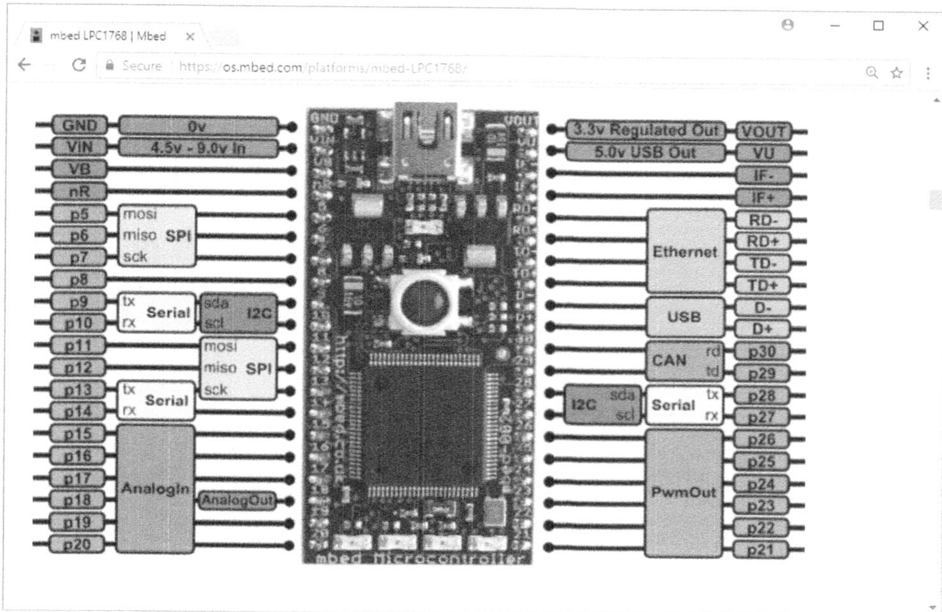

Figure 1.10 The NXP LPC1768 development board and its pinout from the Arm® Mbed™ website. (*Source:* https://os.mbed.com/platforms/mbed-LPC1768/)

- Prototyping form-factor
  - 40-pin 0.1" pitch DIP package, 54×26mm
  - 5V USB or 4.5-9V supply
  - Built-in USB drag 'n' drop flash programmer
- mbed.org developer website
  - Lightweight online compiler
  - High level C/C++ SDK
  - Cookbook of published libraries and projects

There is also an mbed Application Board for NXP LPC1768 (Figure 1.11. The NXP LPC1768 and its mbed application board make a great learning kit.

Features

- 128 × 32 graphics LCD
- 5 way joystick
- 2 × potentiometers
- 3.5 mm audio jack (analog out)
- Speaker, PWM connected
- 3 Axis +/1 1.5g accelerometer
- 3.5mm audio jack (analog in)
- 2 × Servo motor headers
- RGB LED, PWM connected
- USB-mini-B connector
- Temperature sensor

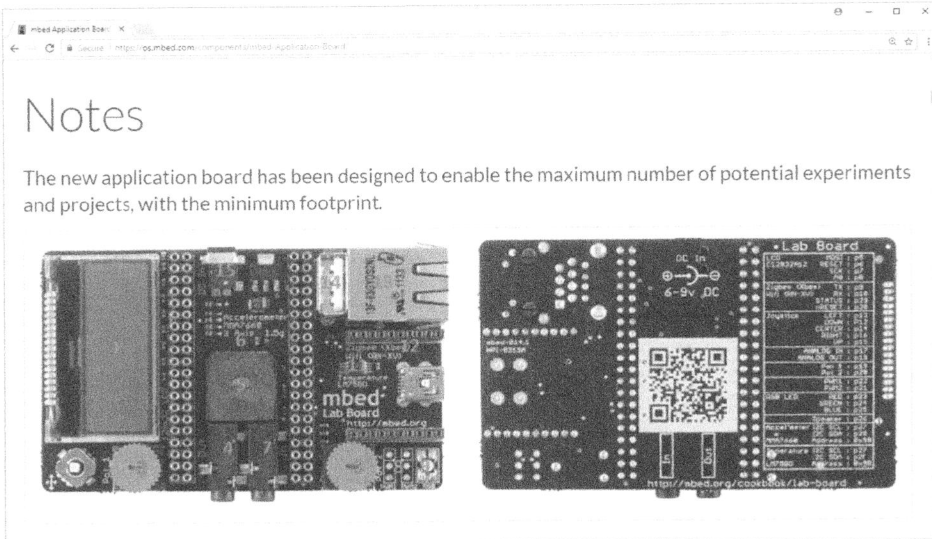

Figure 1.11 The mbed application board for NXP LPC1768 development board, front (left) and back (right), from the Arm® Mbed™ website. (*Source:* https://os.mbed.com/components/mbed-Application-Board/)

- Socket for Xbee (Zigbee) or RN-XV (WiFi)
- RJ45 Ethernet connector
- USB-A connector
- 1.3 mm DC jack input

Further Information about LPC1768

https://os.mbed.com/platforms/mbed-LPC1768/
https://os.mbed.com/components/mbed-Application-Board/
http://www.nxp.com/products/microcontrollers-and-processors/arm-processors/lpc-cortex-m-mcus/lpc1700-cortex-m3/arm-mbed-lpc1768-board:OM11043

1.4.2 NXP LPC11U24

This is another interesting development board. It is based on the NXP LPC11U24, with a 32-bit ARM® Cortex-M0 core running at 48 MHz. It includes 32 KB flash, 8 KB RAM, and lots of interfaces, including USB device, SPI, I2C, ADC, and other I/O interfaces. Figure 1.12 shows the board and its printout, including the commonly used interfaces and their locations. The pins P5–P30 can also be used as DigitalIn and DigitalOut interfaces.

Different from NXP LPC1768, NXP LPC11U24 is much slower and less powerful, but it uses less power and is much cheaper, so it is mainly designed for low-cost USB devices and battery-powered applications.

Features

- NXP LPC11U24 MCU
- Low power ARM® Cortex™-M0 core

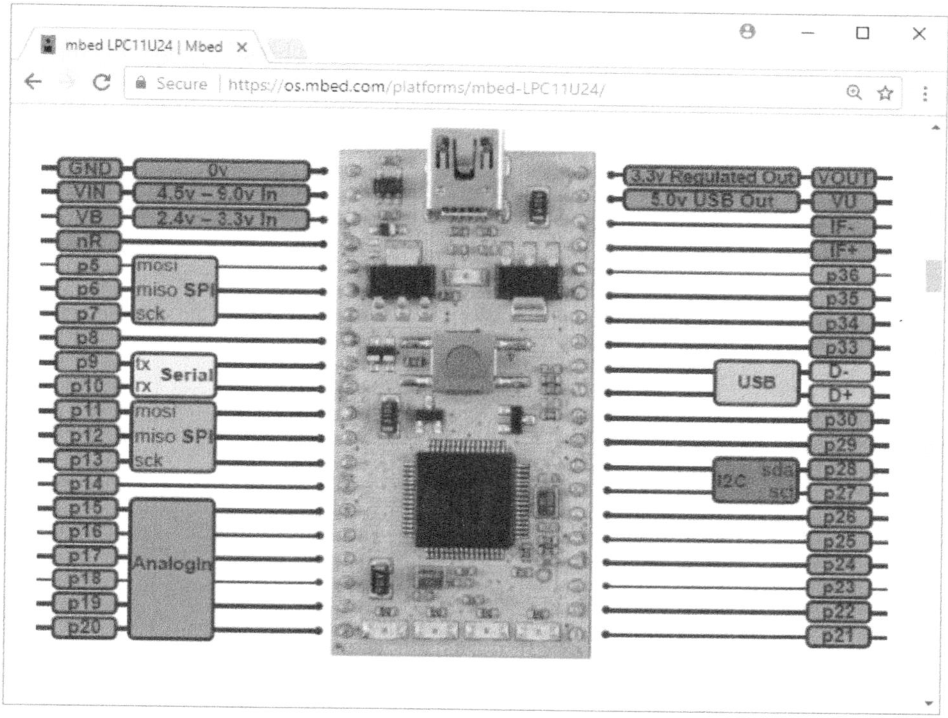

Figure 1.12 The NXP LPC11U24 development board and its pinout from the Arm® Mbed™ website. (*Source:* https://os.mbed.com/platforms/mbed-LPC11U24/)

- 48 MHz, 8 KB RAM, 32 KB flash
- USB device, 2×SPI, I2C, UART, 6×ADC, GPIO
- Prototyping form-factor
- 40-pin 0.1" pitch DIP package, 54×26mm
- 5V USB, 4.5–9V supply or 2.4–3.3V battery
- Built-in USB drag 'n' drop flash programmer
- mbed.org developer website
- Lightweight online compiler
- High-level C/C++ SDK
- Cookbook of published libraries and projects

Further Information about LPC11U24

https://os.mbed.com/platforms/mbed-LPC11U24/
http://www.nxp.com/products/microcontrollers-and-processors/arm-processors/lpc-cortex-m-mcus/lpc1100-cortex-m0-plus-m0/arm-mbed-lpc11u24-board:OM13032

1.4.3 BBC Micro:bit

The BBC micro:bit is a pocket-sized, codable computer, developed by BBC through a major partnership with 31 organizations, including ARM®, NXP, element14, Microsoft, and Cisco. Figure 1.13 shows the board and its printout. It allows anyone to get creative with technology. BBC has donated a free micro:bit to every 11- or 12-year-old child in year 7 or equivalent across the United Kingdom.

The BBC micro:bit is based on a Nordic nRF51822 MCU with 16K RAM and 256K Flash. There's also an onboard accelerometer and magnetometer from Freescale.

Features

- Can be programmed with high-level online IDEs using the BBC's website at http://www.microbit.co.uk/create-code including:
  - Microsoft TouchDevelop IDE
  - Microsoft Blocks
  - CodeKingdoms Javascript
  - MicroPython
- mbed enabled
  - Online IDE at developer.mbed.org
  - Easy to use C/C++ SDK
  - Dedicated micro:bit runtime libraries for rapid development (developed by Lancaster University)

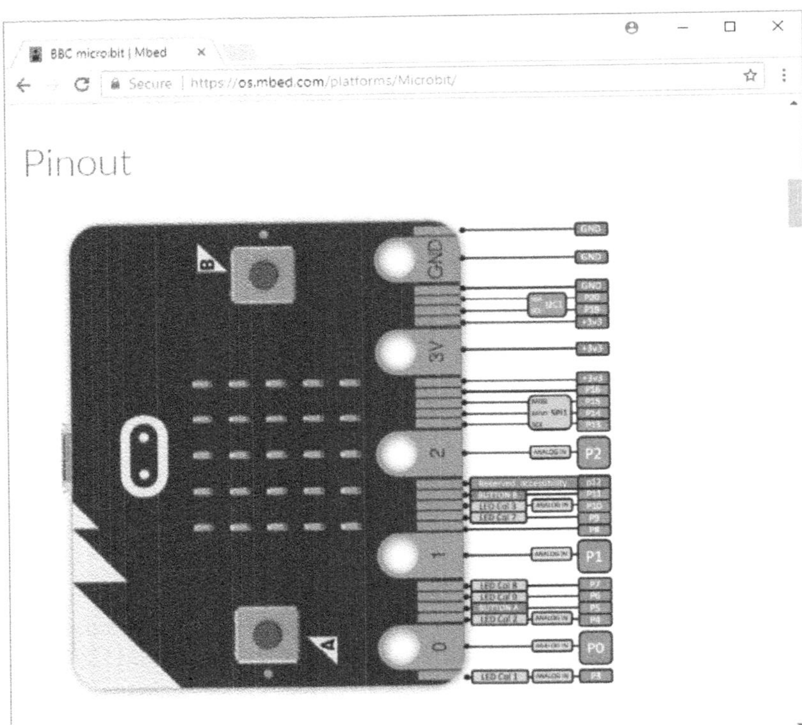

Figure 1.13 The BBC Micro:bit development board and its pinout from the Arm® Mbed™ website. (*Source:* https://os.mbed.com/platforms/Microbit/)

- Nordic nRF51822 multi-protocol Bluetooth® 4.0 low energy/2.4GHz RF SoC
  - 32-bit ARM® Cortex M0 processor (16MHz)
  - 16 kB RAM
  - 256 kB Flash
  - Bluetooth Low Energy Master/Slave capable
- Input/Output
  - 25 LED matrix
  - Freescale MMA8652 3-axis accelerometer
  - Freescale MAG3110 3-axis magnetometer (e-compass)
  - Push Button ×2
  - USB and Edge connector serial I/O
  - 2/3 reconfigurable PWM outputs
  - 5 × banana/croc-clip connectors
  - Edge connector
  - 6 × analog in
  - 6-17 GPIO (configuration dependent)
  - SPI
  - i2c
- USB Micro B connector
- JST power connector (3v)

Further Information about Micro:bit

https://www.microbit.co.uk/
https://os.mbed.com/platforms/Microbit/

1.4.4 The Arm® Mbed™ Ethernet Internet of Things (IoT) Starter Kit

This Ethernet IoT Starter Kit includes an Arm® Mbed™ Freedom FRDM-K64F development board and mbed application shield (Figure 1.14). It is designed for the IBM IoT Foundation and is aimed to provide the user with a slick experience. It allows the user to send data from the onboard sensors into the IBM cloud easily. It is particularly suitable for developers with no specific experience in embedded or web development, as it provides a platform for learning new concepts and creating working prototypes. It allows the user to access to IBM cloud applications through IBM's BlueMix platform, in which deployment and device management are very simple. The starter kit hardware can also be modified and extended to suit specific needs.

The FRDM-K64F development board is the next-generation development board. It uses a power-efficient Kinetis K64F MCU featuring an ARM® Cortex®-M4 core running up to 120 MHz and embedding 1024 KB Flash, 256 KB RAM, and lots of peripherals (16-bit ADCs, DAC, timers) and interfaces (Ethernet, USB device crystal-less, and serial). The new mbed application shield has been designed to enable the maximum number of potential experiments with Arduino form factor development boards, keeping as much in common with the mbed application board as possible.

This book focuses on the Arm® Mbed™ IBM Ethernet IoT Starter Kit.

The Arm® Mbed™ Ethernet IoT Kit contents:

Figure 1.14 The Arm® Mbed™ Ethernet Internet of Things (IoT) Starter Kit, which includes a FRDM-K64F development board (left) and its application shield (right).

Mbed Enabled NXP K64F Development Board

- NXP K64F Kinetis K64 MCU (MK64FN1M0VLL12)
- High-performance ARM® Cortex™-M4 Core with floating point unit and DSP
- 120 MHz, 256 KB RAM, 1 MB flash

mbed Application Shield

- 128×32 graphics LCD
- 5-way joystick
- 2 × potentiometers
- Speaker, PWM connected
- 3-axis +/1 1.5 g accelerometer
- RGB LED, PWM connected
- Temperature sensor
- Socket for XBee (ZigBee) or RN-XV (WiFi)

MCU Features

- Kinetis MK64FN1M0VLL12 in 100LQFP
- Performance
  - ARM® Cortex™-M4 32-bit core with DSP instructions and floating point unit (FPU)
  - 120 MHz max CPU frequency
- Memories and memory interfaces
  - 1024 KB program flash memory

  - 256 KB RAM
  - FlexBus external bus interface
- System peripherals
  - Multiple low-power modes, low-leakage wake-up unit
  - 16-channel DMA controller
- Clocks
  - 3× internal reference clocks: 32 KHz, 4 MHz, and 48 MHz
  - 2× crystal inputs: 3–32 MHz (XTAL0) and 32 kHz (XTAL32/RTC)
  - PLL and FL
- Analog modules
  - 2× 16-bit SAR ADCs up 800 ksps (12-bit mode)
  - 2× 12-bit DACs
  - 3× analog comparators
  - Voltage reference 1.13 V
- Communication interfaces
  - 1× 10/100 Mbit/s Ethernet MAC controller with MII/RMII interface IEEE1588 capable
  - 1× USB 2.0 full-/low-speed device/host/OTG controller with embedded 3.3V/120mA Vreg, and USB device crystal-less operation
  - 1× Controller area network (CAN) module
  - 3× SPI modules
  - 3× I2C modules. Support for up to 1 Mbit/s
  - 6× UART modules
  - 1× Secure dgital host controller (SDHC)
  - 1× I2S module
- Timers
  - 2× 8-channel Flex-Timers (PWM/Motor control)
  - 2× 2-channel Flex-Timers (PWM/Quad decoder)
  - 32-bit PITs and 16-bit low-power timers
  - Real-time clock (RTC)
  - Programmable delay block
- Security and integrity modules
  - Hardware CRC and random-number generator modules
  - Hardware encryption supporting DES, 3DES, AES, MD5, SHA-1, and SHA-256 algorithms
- Operating characteristics
  - Voltage range: 1.71 to 3.6 V
  - Flash write voltage range: 1.71 to 3.6 V

Board Features

- Onboard components
  - FXOS8700CQ—6-axis combo sensor accelerometer and magnetometer
  - 2 user push buttons
  - RGB LED
- Connectivity
  - USB full-/low-speed on-the-go/host/device controller with on-chip transceiver, 5 V to 3.3 V regulator and micro-USB connector

  - Ethernet 10/100 controller with onboard transceiver and RJ45 connector
  - Up to 5× UARTs, 2× SPIs, 2× I2Cs and 1× CAN connected to headers (multiplexed peripherals)
- Extensions
  - Micro SD-card socket
  - Headers compatible with Arduino R3 shields (32-pins / outer row)
  - Headers for proprietary shields (32-pins / inner row)
- Analog and digital IOs (multiplexed peripherals)
  - Up to two ADC 16-bit resolution with 28 analog I/O pins connected to headers
  - Up to three timers with 18 PWM signals accessible from headers
  - Up to six comparator inputs or one DAC output
  - Up to 40 MCU I/O pins connected to headers (3.3 V, 4 mA each, 400 mA max total)
- Board power-supply options (onboard 5 to 3.3 V regulator)
  - USB debug 5 V
  - USB target 5 V
  - 5–9 V Vin on Arduino headers
  - 5 V PWR input
  - Coin-cell 3.3 V
- Integrated open SDA USB debug and programming adapter
  - Several industry-standard debug interfaces (PEmicro, CMSIS-DAP, JLink)
  - Drag-n-drop MSD flash programming
  - Virtual USB to serial port
- Form factor: 3.2" × 2.1" / 81 mm × 53 mm
- Software development tools
  - mbed HDK & SDK enabled
  - Online development tools
  - Easy-to-use C/C++ SDK
  - Lots of published libraries and projects
  - Alternate offline options NXP free KDS (compiler toolchain) and KSDK library/ examples
- Supplier website: http://www.nxp.com/frdm-k64F

Figure 1.15 shows the FRDM-K64F development board's component layout and pinout. Following are the most used pins:
RGB LED **LED1 (LED_RED), LED2(LED_GREEN), LED3 (LED_BLUE), LED4 (LED_RED)**
Digital inputs/outputs **D0, D1, D2, ..., D15**
Analog inputs **A0, A1, A2, A3, A4, A5**
Analog outputs **DAC0_OUT**
PWM (pulse width modulation) **A4, A5, D3, D5, D6, ..., D13**

Further Information about FRDM-K64F

https://os.mbed.com/platforms/IBMEthernetKit/
https://os.mbed.com/platforms/FRDM-K64F/
https://os.mbed.com/components/mbed-Application-Shield/
http://www.nxp.com/frdm-k64f

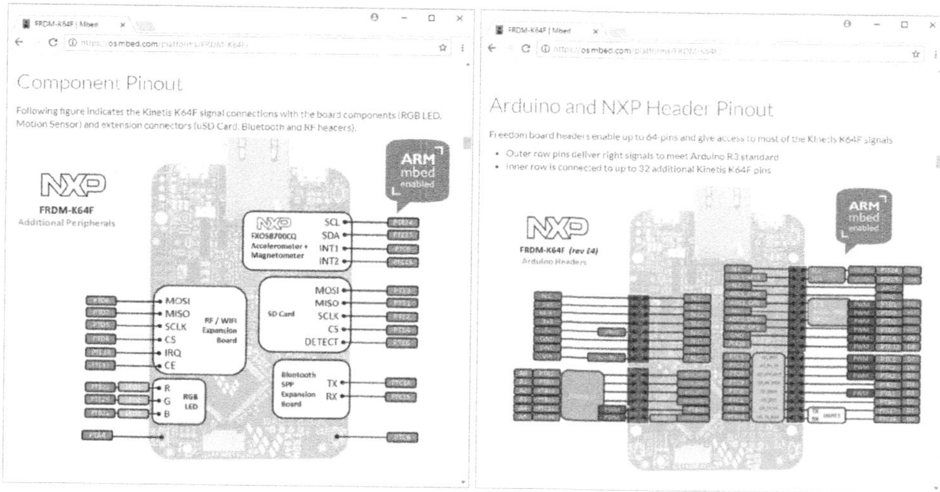

Figure 1.15 FRDM-K64F development board's component layout (left) and the Arduino and NXP header pinout from the Arm® Mbed™ website. (*Source:* https://os.mbed.com/platforms/FRDM-K64F/)

1.5 Summary

This chapter first explains what an embedded system is and discusses the difference between microcontrollers and microprocessors. It then introduces the ARM® architecture and Arm® Mbed™ systems.

1.6 Chapter Review Questions

Q1.1 What is an embedded system?

Q1.2 What is the difference between microcontrollers and microprocessors?

Q1.3 How does CPU work?

Q1.4 Use a suitable diagram to describe ARM® Processor Architecture.

Q1.5 What is Arm® Mbed™? Describe the concepts of mbed cloud services, clients, and mbed OS.

Q1.6 Use a table to compare the key features between LPC1768 and FRDM-K64F.

2

Introduction to the Internet of Things (IoT)

Genius is 1 percent inspiration and 99 percent perspiration.

- Thomas A. Edison

2.1 What is the Internet of Things (IoT)?

The Internet of Things (IoT) refers to the network of physical objects, It is fast growing and already has billions of devices connected (Figure 2.1). This is different from the current Internet, which is largely a network of computers, including phones and tablets. The "things" in the IoT can be anything from household appliances, machines, goods, buildings, and vehicles to people, animals, and plants. With the IoT, all the physical objects are interconnected, capable of exchanging data with each other without human intervention. They can be accessed and controlled remotely. This is going to completely transform our lives—it will be truly revolutionary.

The concept of connecting devices together is not new. In 1982, a Coke machine at Carnegie Mellon University became the first appliance connected to the Internet. It could keep track of inventory and whether drinks were cold. Since then, connectedness has greatly expanded, in the areas of ubiquitous computing, machine-to-machine (M2M) communications, and device-to-device (D2D) communications. But the term IoT was invented by British entrepreneur Kevin Ashton in 1999, in a presentation he made to Procter & Gamble. As that time, he was the cofounder and executive director of the Auto-ID Center at MIT, and the vision of IoT was based on radio-frequency identification, or RFID (radio-frequency identification). IoT has evolved since, and became increasingly popular in recent years, due to the convergence of several enabling technologies, such as microcontrollers, sensors, wireless communications, embedded systems, and micro-electromechanical systems (MEMS).

Today, the IoT is largely seen as the next big thing, the future of the Internet. According to *Internet Society,* there will be about 100 billion IoT devices and a global market of more than $11 trillion by 2025. IoT will grow exponentially just like what the Internet did about two decades ago.

Further Information about IoT

http://internetofthingswiki.com/

http://www.theinternetofthings.eu/

Designing Embedded Systems and the Internet of Things (IoT) with the ARM® Mbed™, First Edition. Perry Xiao

Companion website: www.wiley.com\go\xiao\designingembeddedsystemandIoTwitharmmbed

Figure 2.1 The symbolic view of the Internet of Things (IoT). (*Source:* https://pixabay.com/en/network-iot-internet-of-things-782707/)

http://www.computerweekly.com/resources/Internet-of-Things-IoT
http://www.ibm.com/Watson/IoT
https://www.microsoft.com/en-gb/internet-of-things/
http://www.cisco.com/c/en_uk/solutions/internet-of-things/overview.html

2.2 How Does IoT Work?

There are several steps in order to make the Internet of Things (IoT) work.

First, each "thing" on the Internet of Things must have a unique identity. Thanks to Internet Protocol Version 6 (IPv6) address, the 128-bit next-generation Internet Protocol (IP) address can provide 2^{128} different addresses—that is about 6.7×10^{23} addresses per square meter. We should be able to assign a unique ID to every physical object on the planet.

Second, each "thing" must be able to communicate. There are number of modern wireless technologies which make communications possible, such as WiFi, Bluetooth Low Energy, Near-field communication (NFC), RFID, as well as ZigBee, Z-Wave, and 6LoWPAN (IPv6 over Low power Wireless Personal Area Networks), etc.

Third, each "thing" needs to have sensors so that we can get information about it. Sensors can be temperature, humidity, light, motion, pressure, infrared, ultrasound sensors, etc. The new sensors are increasingly getting smaller, cheaper, and more durable.

Fourth, each "thing" needs to have a microcontroller (or microprocessor) to manage the sensors and communications, and to perform the tasks. There are many microcontrollers exist that could be used for IoT, but the ARM® based microcontrollers are no doubt one of the most influential ones. This book is focused on the Arm® Mbed™ microcontrollers.

Finally, we will need cloud services to store, analyze, and display data so that we can see what's going on and take action via phone apps. There are already a lot of big companies working on this, such as IBM's IBM Watson, Google's Google Cloud Platform, Microsoft's Azure, and Oracle's Oracle Cloud etc. The Arm® Mbed™ is also developing its own cloud (https://cloud.mbed.com/), but as of this writing, it is still on its first release, only available to a select group of industrial lead partners.

2.3 How Will IoT Change Our Lives?

The Internet of Things will fundamentally change the way we live and change the way we interact with world.

We all had this "*where are my keys*" experience before. Well, in the world of IoT, we probably won't need our keys anymore! Our phones are the keys, we are the keys. We could open the doors using phones, or using our biometric information, such as fingerprints, palm prints, palm veins, iris, retina, face, and voice. For example, just like in the folk tale *Ali Baba and the Forty Thieves,* you could open your home door by saying "Open Sesame." But this time is different—only you and your family can open the door, while others cannot, thanks to voice recognition, which can uniquely identify you and your family.

IoT can also make our homes smarter. This is already happening, with all these smart locks, smart meters, smart thermostats, smart lighting, smart grid, and smart cars, etc. Figure 2.2 shows an example of smart meters and thermostat. The smart home can wake you up in the morning and start the coffee machine while you are in still the shower. It can switch on the lights just before you enter the room and switch the lights off immediately after you leave. As a father of two teenager kids, it will save me the hassle of running upstairs and downstairs to switch off the lights after they leave the rooms. It can allow you to switch on the TV and change channels using your voice commands.

Figure 2.2 Landis Smart Electricity Meter (left) and Nest Learning Thermostat (right). (*Source:* https://commons.wikimedia.org)

It can also sense you are coming home and adjust the thermostat or even preheat the oven. With the advances of artificial intelligence (AI), it also possible to analyze or learn your living pattern, to turn the heating or air conditioning on when you are at home and turn the heating or air conditioning off when you are not. This is even more significant for public buildings, like offices, theaters, hospitals, and museums, where the utility bill is always a large chunk of the monthly spending.

Both my father and my wife's father suffered a stroke, which paralyzed them and confined them to beds. Wouldn't it be nice if we all could have wearable or in-plant sensors, that could monitor our heart rate, blood pressure, body temperature, body mass index (BMI), maybe even blood sugar levels or cholesterol levels, 24 hours a day and 7 days a week? We could then predict and prevent the stroke even before it happens. We might also be able to predict many other deadly diseases such as cancers by using big data analysis and machine learning, so that we could get treated in the earlier stages, and significantly reduce the needs for hospital admissions. We are going to live healthier and longer, much, much longer.

Many people around the world suffer food allergies or intolerances. The most common are to milk, eggs, peanuts, shellfish, tree nuts, soy, and wheat. This can make routine food shopping a daunting tasking. You must carefully read through the small prints to figure out whether this product contains the ingredient you would want to avoid. This will change, of course. Your phone, or any other devices, will tell you which food product is right for you. This will also help people on a diet and the people who have special nutrition requirements, such as athletes. After finishing shopping, there will be no need to check out. By the time you walk out of the supermarket, your purchases will already have been tallied and the total charged to your credit card. The receipt will be sent to your email or your phone. Just think of the time you could save in the queue for checking out, especially during peak hours—imagine no longer waiting while the person in front of you gets a price check or tries to pay in coins! If you don't like something you bought, or simply just changed your mind, you can simply return it to a designated area, where the item will be automatically examined. If it is in satisfactory condition, a refunded will be issued. No questions asked, no signature required. I like that!

Bullying is a serious issue in many schools. In countless occasions, the victims, or the parents of the victims, feel powerless, as they can neither prevent, nor prove, what has happened. With IoT, this will be completely different. Victims could be wired up with sensors that can automatically logged voice, video, as well as other information in the cloud storage, and warning messages could be sent to parents and the school. The school could immediately know at exactly which place, at exactly what time, exactly who were involved and exactly what has happened. Bullies could be punished swiftly and fairly. Common sense tells us, if you know that you are definitely going to be caught for what you are going to do, you probably would not do it in the first place. Bullying could be a thing of past!

This might also apply to crimes. Many big criminals start with small crimes, and they often commit the crime because they thought they could get away with it. After successfully getting away with a series of small petty crimes, they start to commit more serious crimes. This escalating cycle continues until one day they caught by police and sent to prison. After release from prison, with a criminal record, it is very difficult to get any decent job. So many of them return to crimes, and the vicious cycle repeats again. If we could stop them at the petty crime stage, by using IoT technologies, as illustrated in the bullying example, they probably would never grow into serious criminals. So finally, Utopia—a dream that so many people from so many countries have fought so hard for so long—might be achievable through a technology revolution! Imagine that!

But just like many things in the life, IoT is not, of course, without controversy. There are many concerns about IoT. On the top of the list are security and privacy. If you can access your home appliances remotely, someone else could also access them remotely. There are already many reports on hacking into cameras, meters, household appliances, phones, and cars. So security has to be the top priority of any IoT developments.

Privacy is another concern. There will be tons of our information available, such the name, the date of birth, gender, the address, the telephone number, the credit cards, the things we do, the products we buy etc. Who owns this information, and who can access this information? Do you really want everyone to know where you are and what you do? Do schoolchildren want to wear a wire for us to hear all their conversations? (Not likely!) Do you really want everyone to know who you are calling, and what you are talking about? Do you want your entire lives digitally recorded, on the off-chance a criminal might be caught? Privacy is one of the most important human rights. No one wants to live in a Big Brother state. So the whole community must have input, to make sure we have the balance right—that innocent, law-abiding citizens can enjoy their right of privacy, while the police can have enough information to fight the crime and prevent terrorist attacks.

2.4 Potential IoT Applications

2.4.1 Home

Smart homes will be probably the most popular IoT applications. Smart home, or home automation, is an extension of building automation, with which we can monitor and control heating, ventilation and air conditioning (HVAC), lighting, appliances, and security systems. By connecting all the home appliances, we can automate many daily routines, such as automatically turning on and off the lights and heating, starting or stoping cooking and washing, and so on. With the smart grid and smart meters, we can reduce the energy usages and utility bill, and with security systems, we can make the home more secure by automatically detecting, and hopefully deterring, intrusion using various of infrared, motions, sound, vibration sensors as well as alarm systems.

Smart home can also make elderly and disabled people more comfortable and safer at home. With the IoT, we can collect and analyze data from elderly and disabled people to diagnose diseases, predict potential risks, identify or prevent accidents such as falls, open or lock the door (or windows) remotely, and let family members monitor them remotely. With the IoT, it is also possible to get elderly and disabled people more connected to the outside world and reduce their sense of loneliness.

The smart home market was predicted to have a market value over US$137 billion by the year 2023, according to Markets and Markets (see marketsandmarkets.com, July 2017, Report Code: SE 3172).

2.4.2 Healthcare

The IoT make it possible for remote health monitoring and emergency notification systems. A very popular approach is through wearable technologies. These wearable devices can collect a range of health data, such as heart rate, body temperature, and blood pressure, which can then be wirelessly transmitted to a remote site for storage and further analysis. This also enables telehealth/telemedicine, i.e., to diagnose or treat patients remotely.

2.4.3 Transport

The IoT can significantly improve transport systems. With all the cars connected, it is much easier to plan your journey, avoid traffic jams, find a parking space, and reduce traffic accidents. The driverless cars will no doubt have the biggest impact. Many companies—such as Tesla, Google, Uber, Volvo, Volkswagen, Audi, and General Motors—are actively developing and promoting them. The driverless cars can make our journey more enjoyable, and possibly much safer. Getting a driving license could soon be a thing of past!

The IoT can also benefit public transport. By connecting all the information boards and advertising billboards at train stations and airports, it helps passengers to get regular updates, and in the event of an accident, to detect problems quickly and cutting maintenance costs. By improving end-to-end visibility, warehouse management, and fleet management, the IoT will also benefit the logistics industry.

2.4.4 Energy

By integrating sensors and actuators, it is likely to reduce energy consumption of all the energy-consuming devices. The IoT will also modernize the power industry infrastructure, to improve efficiency and productivity.

2.4.5 Manufacture

The application of IoT in industry often referred as Industry 4.0, or the fourth Industrial Revolution (Figure 2.3). The first Industrial Revolution took place in eighteenth century, when steam engine mobilized the industrial production. The second Industrial Revolution took place in earlier nineteenth century, when electric power powered mass production. The third Industrial Revolution, or the Digital Revolution, took place at the end of nineteenth century, when electronics and IT further automated production. Industry 4.0 builds on cyber-physical systems that tightly integrate machines, software, sensors, Internet, and users together. It will create smart factories, in which machines can use self-optimization, self-configuration, and even artificial intelligence to complete complex tasks in order to deliver vastly superior cost-efficiencies and better-quality goods or services.

2.4.6 Environment

By deploying environmental sensors, we can measure and monitor air quality, water quality, soil conditions, radiation, and hazardous chemicals more efficiently. We can

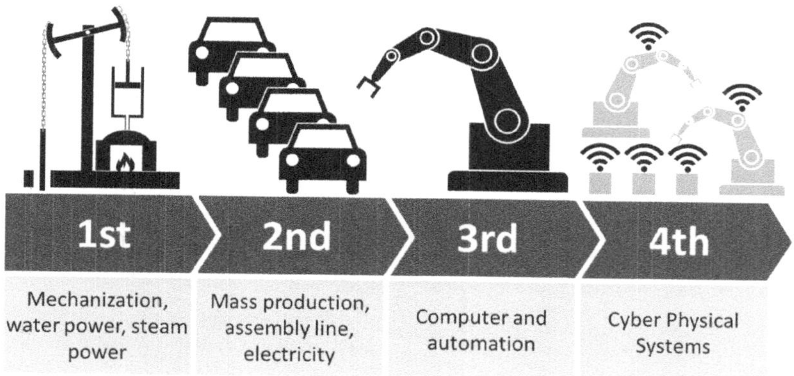

Figure 2.3 Four Industrial Revolutions. (*Source:* https://commons.wikimedia.org/wiki/Category:Industry_4.0#/media/File:Industry_4.0.png)

also predict earthquakes and tsunamis better, and detect forest fire, snow avalanches, landslides quicker. All these will help us to protect our environment better. By tagging wild animals, especially endangered species, we can study and understand better the behavior of the animals, and therefore provide better protections and safer habitats. The IoT will also enable smart farming, which will provide 24/7 visibility into soil and crop health, and help farmers to optimize the usage of fertilizers and plant protection products. This will again inevitably have a positive impact on environment.

2.5 Summary

This chapter introduces the concept of the Internet of Things (IoT), explains how IoT works, and how IoT will change the way we live. It also introduces some potential IoT applications.

2.6 Chapter Review Questions

Q2.1 What is the Internet of Things (IoT)?

Q2.2 How does IoT works?

Q2.3 What are potential IoT applications?

Q2.4 What is Industry 4.0?

3

IoT Enabling Technologies

Tell me and I forget, teach me and I may remember, involve me and I learn.
- Benjamin Franklin

3.1 Sensors and Actuators

A sensor is a device that converts a physical parameter to an electrical output. A sensor is a type of transducer. Sensors can be divided into analog sensors and digital sensors. Analog sensors give output in the format of voltages and currents. Microntrollers will need ADC (analog-to-digital converter) to read in the data from analog sensors. Many newer sensors are digital sensors, i.e., they give output in digital format, using protocols such as I2C (Inter-Integrated Circuit), SPI (Serial Peripheral Interface), and UART (universal asynchronous receiver/transmitter) etc. Digital sensors are excellent for embedded systems, as they bypass the need for ADC, and make the circuit much simpler. Examples include temperature sensors, humidity sensors, pressure sensors, smoke sensors, sound and light sensors, etc.

An actuator is a device that converts an electrical signal to a physical output, i.e., motion. An actuator can be controlled by electric voltage or current, pneumatic or hydraulic pressure, or even human power. In embedded systems, actuators are mainly controlled by electricity. When the control signal is received, the actuator converts the electric energy into mechanical motion. Actuators can create a linear motion, rotary motion or oscillatory motion. Examples of actuators include electric motors, piezoelectric actuators, pneumatic actuators, step motors, and door lock actuators etc.

3.2 Communications

Apart from conventional communication technologies such as Ethernet, WiFi, and Bluetooth, there are many other technologies that can be used for communications in the Internet of Things.

Designing Embedded Systems and the Internet of Things (IoT) with the ARM® Mbed™, First Edition. Perry Xiao

Companion website: www.wiley.com\go\xiao\designingembeddedsystemandIoTwitharmmbed

3.2.1 RFID and NFC (Near-Field Communication)

Radio-frequency identification (RFID) is a technology that can uniquely identify and track tags attached to objects using radio frequency electromagnetic waves. A RFID system typically includes a tag, a reader, and an antenna. The reader sends an interrogating signal to the tag via the antenna, and the tag responds with its unique information. RFID tags can be either active or passive. Active RFID tags have their own power source and therefore can be read over a long range (up to 100 meters). Passive RFID tags do not have their own power source. They are powered by the electromagnetic energy transmitted from the RFID reader. Therefore, they can only be read over a short distance (<25 m). RFID primarily operate at the following frequency ranges, as shown in Table 3.1.

Near-field communication (NFC) is a communication technology that operates at the same frequency (13.56 MHz) as HF RFID. Different from RFID, NFC is based on peer-to-peer communication, which means that a NFC device can be either a reader or a tag. This unique ability has made NFC a popular choice for contactless payment, ID cards, and travel card etc. NFC devices typically communicate within 4 cm (2 in.) of each other. NFC is now available on most new smart phones. NFC smart phones can be used for contactless payment, as well as for passing along information (contact info or photographs) from one smart phone to the other by tapping the two devices together.

https://en.wikipedia.org/wiki/Near_field_communication
https://en.wikipedia.org/wiki/Radio-frequency_identification

3.2.2 Bluetooth Low Energy (BLE)

Bluetooth low energy (BLE) is a newer member of the Bluetooth family, based on Bluetooth 4.0 standards. Similar to classic Bluetooth, BLE also operates in the 2.4 GHz ISM band, but uses a simpler modulation system. BLE remains in sleep mode constantly

Table 3.1 RFID Frequency Bands.

Band	Range	Data Speed	Tags
Low frequency (LF): 125–134.2 kHz	10 m	low	passive
High frequency (HF): 13.56 MHz	10 cm–1 m	low to moderate	passive
Ultra high frequency (UHF): 433 MHz	1–100 m	moderate	passive or active
Ultra high frequency (UHF): 856 MHz–960 MHz	1–12 m	moderate to high	passive or active
Microwave: 2.45–5.8 GHz	1–2 m	high	active
Microwave: 3.1–10 GHz	<200 m	high	active

except for when a connection is initiated, and it therefore consumes much less power. BLE hit the market in 2011, and is marketed as Bluetooth Smart. BLE is designed to provide much reduced power consumption and cost while maintaining a similar communication range. BLE typically operates at a range about 100 m, with a data rate about 1 Mbits/s.

Following are the typical BLE applications:

- Heart rate monitors
- Blood pressure monitors
- Blood glucose monitors
- Fibit-like devices
- Industrial monitoring sensors
- Geography-based, targeted promotions (iBeacon)
- Proximity sensing

https://www.bluetooth.com/

3.2.3 LiFi

Light Fidelity (LiFi) is a novel, wireless, bidirectional, high-speed communication technology based on rapidly modulated visible light. It is a type of Visible Light Communications (VLC) system. Similar to WiFi, LiFi transmits data using electromagnetic waves. But instead of using radio waves (MHz – GHz), it uses visible light (~THz). LiFi uses household LED (light emitting diodes) light bulbs as transmitters. By varying the electric current supplied to a LED light bulb at extremely high speeds, data can be encoded as the rapid brightness changes, which can then be picked up by a photodetector (photodiode). These rapid changes are too quick to be noticed to human eyes; therefore, LiFi does not affect the main function of LED lights—lighting. LiFi has a huge advantage in term of infrastructure, as LED light bulbs are increasingly used in buildings, streets, and vehicles. It can operate at an impressive speed of up to 224 gigabits per second, and it is insensitive to electromagnetic interference. LiFi cannot penetrate walls, which means it can only operate at a short range, but at the same time, this makes it less likely to be hacked. There are already products on the market that can provide light and connectivity at the same time.

http://purelifi.com/

3.2.4 6LowPAN

6LoWPAN stands for IPv6 (Internet Protocol Version 6) over Low power Wireless Personal Area Networks (WPAN). It is a basically a low-power, low data rate, wireless mesh network based on IEEE 802.15.4 standards, using IPv6 as the communication protocol. Comparing with other local area networks, 6LoWPAN has a distinct advantage, i.e., it is based TCP/IP open standards, including TCP, UDP, HTTP, COAP, MQTT, and websockets etc. It has end-to-end IPv6 addressable nodes, and can be easily connected to the Internet directly. It is also self-healing because of mesh routing. 6LoWPAN has been used in wireless sensor networks, lights, and meters.

https://datatracker.ietf.org/wg/6lowpan/charter/

3.2.5 ZigBee

ZigBee is a high-level communication technology for low-power, low-data-rate personal area networks, such as sensor networks, home automations, and medical devices. ZigBee is based on IEEE 802.15.4 standard. It has a transmission distance of 10–100 meters and needs to be line of sight. It operates in the industrial, scientific, and medical (ISM) radio bands, i.e., 868 MHz in Europe, 915 MHz in the United States and Australia, 784 MHz in China, and 2.4 GHz in the rest of the world. ZigBee has a data rate ranging from 20 kbit/s (868 MHz band) to 250 kbit/s (2.4 GHz band). ZigBee networks are normally cheaper than other wireless networks such as Bluetooth or WiFi. ZigBee has been used for wireless light switches, electrical meters (smart grid, demand response, etc.), and industrial equipment monitoring etc.

So what is the difference between ZigBee and 6LoWPAN? Well, ZigBee has been around longer, and therefore has been adopted more widely than 6LoWPAN. ZigBee is no doubt still the most popular low-cost, low-power wireless mesh networking standard available today. However, 6LoWPAN is catching up and becoming more attractive, since it is IP-based, particularly with IPv6 support. This makes it easier to integrate with the rest of the Internet. To date, many semiconductor companies (e.g., Texas Instruments, Freescale, and Atmel etc.) are making 802.15.4 chips that support both ZigBee and 6LoWPAN.

http://www.zigbee.org/

3.2.6 Z-Wave

Z-Wave is a wireless communication technology that is primarily used for home automation, such as controlling and automating lights and appliances. It can be used as a security system or to monitor and control your property remotely. Z-Wave operates at unlicensed industrial, scientific, and medical (ISM) band, i.e., 868.42 MHz in Europe, 908.42 MHz in the United States and Canada, and other frequencies in other regions. Z-Wave is designed to provide reliable, low-latency transmission at a range of about 100 meters, with data rates up to 100 kbit/s. A Z-Wave network normally includes a primary controller and a collection of devices (up to 232).

http://www.z-wave.com/

3.2.7 LoRa

LoRa is a long-range communication technology that is intended for low-power, long-distance communications of battery powered IoT devices—that is, low-power wide area network (LPWAN). It supports secure bidirectional communications of networks with millions and millions of devices.

https://www.lora-alliance.org/

Table 3.2 gives a quick comparison of different wireless communication technologies. LiFi and WiFi potentially offer the highest data rates, while cellular and LoRa offer the longest distances.

Table 3.2 Comparison of Different Technologies.

	Standard	Frequency	Range	Data Rate
LiFi	Similar to 802.11	400–800 THz	<10 m	<224 Gbps
WiFi	802.11a/b/g/n/ac	2.4 GHz and 5 GHz	~50 m	<1 Gbps
Cellular	GSM/GPRS/EDGE (2G), UMTS/HSPA (3G), LTE (4G), 5G	900, 1800, 1900, and 2100 MHz 2.3, 2.6, 5.25, 26.4, and 58.68 GHz	<200 km	<500 kps (2G), <2 Mbps (3G), <10 Mbps (4G) <100 Mbps (5G)
Bluetooth	Bluetooth 4.2	2.4 GHz	50–150 m	1 Mbps
RFID/NFC	ISO/IEC 18000-3	13.56 MHz	10 cm	100–420 kbps
6LowPAN	RFC6282	2.4 GHz and ~1 GHz	<20 m	20–250 kbps
ZigBee	ZigBee 3.0 based on IEEE802.15.4	2.4 GHz	10–100 m	250 kbps
Z-Wave	Z-Wave Alliance ZAD12837 / ITU-T G.9959	868.42 MHz and 908.42 MHz	<100 m	<100 kbps
LoRa	LoRaWAN	868 MHz and 915 MHz	<15 km	0.3–50 kbps

3.3 Protocols

Protocols, or communication protocols, are a set of rules that allow devices to communicate with each other. Protocols define the syntax, semantics, and synchronization of communication. A close analogy to protocols is human languages. There are many communication protocols available for IoT applications. Following are commonly used protocols: HTTP, Websocket, and MQTT.

3.3.1 HTTP

The Hypertext Transfer Protocol (HTTP) is the communication protocol behind the World Wide Web (WWW). It is based on client–server architecture, and operates in a request and response fashion (Figure 3.1). HTTP uses TCP (transmission control protocol) to provide reliable connections. HTTP is stateless, as the client and server do not

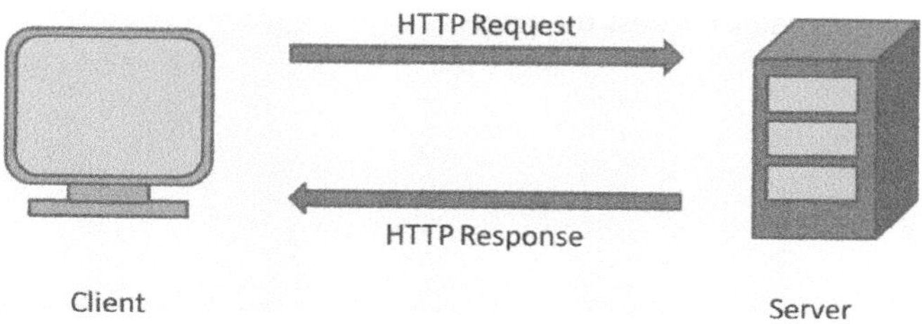

Figure 3.1 The HTTP protocol.

maintain a connection during the communication. The current version is HTTP/1.1 and the previous version is HTTP/1.0. The newer version HTTP/2 is coming soon, which will have many new features, such as server push, to minimize the number of clients' requests and increase speed.

Following is an example of HTTP request message:

```
GET /index.html HTTP/1.1
Host: www.mbed.com
Connection: keep-alive
User-agent: Mozilla/4.0
Accept-language: en
```

Following is an example of HTTP response message:

```
HTTP/1.1 200 OK
Server: nginx/1.7.10
Date: Sun, 12 Feb 2017 12:21:57 GMT
Content-Type: text/html
Content-Length: 185
Connection: close
Location: https://www.mbed.com/

<html>
<head><title>... ...</title></head>
<body>
... ...
</body>
</html>
```

3.3.2 WebSocket

WebSocket is a communication protocol designed for web browsers and web servers, but unlike HTTP, WebSocket provides full-duplex communication over a single TCP connection. WebSocket is stateful, as the client and server do maintain a connection during the communication. The WebSocket makes more interaction between a browser and a web server possible, enables real-time data transfer and streams of messages. To date, WebSocket is implemented in all major web browsers, e.g., Firefox 6, Safari 6, Google Chrome 14, Opera 12.10 and Internet Explorer 10.

Following is an example of WebSocket request message:

```
GET ws://websocket.test.com/ HTTP/1.1
Host: websocket.test.com
Upgrade: websocket
Connection: Upgrade
Origin: http://test.com
```

Following is an example of WebSocket response message:

```
HTTP/1.1 101 WebSocket Protocol Handshake
Date: Mon, 16 Jan 2017 16:54:12 GMT
Connection: Upgrade
Upgrade: WebSocket
```

Further Information about WebSocket

https://www.websocket.org/

3.3.3 MQTT

MQ Telemetry Transport (MQTT) is a lightweight, machine-to-machine communication protocol designed for IoT devices by IBM. MQTT is based on a publisher–subscriber model, where the publisher publishes data to a server (also called broker), and the subscriber subscribes to the server and receives data from the server. The MQTT broker is responsible for distributing messages and can be somewhere in the Clouds. See Figure 3.2.

For IoT devices, MQTT offers many advantages over HTTP and WebSocket, which requires a server constantly running, which requires more computing power, more bandwidth and more energy consumption. They are not purposely designed for IoT devices, where response times, throughput, lower battery use, and lower bandwidth are key design criteria. MQTT is purposely design as a "lightweight" messaging protocol, features faster response and throughput, and lower battery and bandwidth usage.

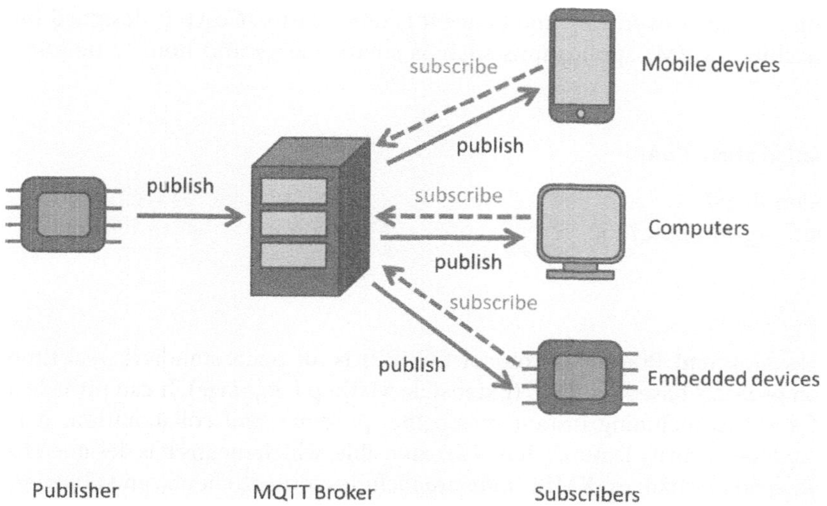

Figure 3.2 The MQTT protocol.

Although MQTT broker also need to be running all the time, but IoT devices (publishers and subscribers) are lightweight.

MQTT also allows to prioritize messages, such as QoS (Quality of Service):

0: The client/server will deliver the message once, with no confirmation required.
1: The client/server will deliver the message at least once, confirmation required.
2: The client/server will deliver the message exactly once by using a handshake process.

Appendix B has more details on how to download and set up a MQTT broker using the popular HiveMQ software.

Further Information about MQTT

http://mqtt.org/
http://www.hivemq.com/resources/getting-started/
http://www.hivemq.com/plugin/mqtt-message-log/
http://www.hivemq.com/blog/hivemq-mqtt-websockets-support-message-log-plugin-2-min
http://www.hivemq.com/plugin/file-authentication/
http://www.hivemq.com/demos/websocket-client/

3.3.4 CoAP

The Constrained Application Protocol (CoAP) is a specialized application layer protocol for constrained IoT devices, i.e., devices with limited computing power, power consumption, and network connectivity, etc. It is based on request and response messages, similar to HTTP, but it uses UDP (user datagram protocol) rather TCP (transmission control protocol). Although UDP does not provide reliable transmissions, it is much simpler, has much smaller overhead, and hence it is much faster. CoAP is designed for machine-to-machine (M2M) applications such as smart energy and home / building automation.

Further Information about CoAP

http://coap.technology/
https://tools.ietf.org/html/rfc7252

3.3.5 XMPP

Extensible Messaging and Presence Protocol (XMPP) is an open standard, real-time communication protocol based on XML (Extensible Markup Language). It can provide a wide range of services including instant messaging, presence and collaboration. It is decentralized and has security features. It is also extensible, which means it is designed to grow and accommodate changes. XMPP software includes servers, clients, and libraries.

Further Information about XMPP

https://xmpp.org/

3.4 Node-RED

Node-RED is a web-based open source software tool developed by IBM, which can be used to connect hardware devices over the Internet. Figure 3.3 shows the Node-RED homepage. For example, with Node-RED, you can connect your mbed development board to the Internet, read the sensor values, display it in a chart, in a web page, in an email, or a Twitter message. You can also send commands back to the development board to perform some control. It is a graphic-based programming tool, which uses functional blocks called *nodes* to build the program. All you need to do is to wire up the nodes and configure them. This makes many programming tasks remarkably simple and easy to implement. Figure 3.4 shows a simple WebSocket-based chat program implemented in Node-RED.

Node-RED is a great tool for IoT projects. It uses JavaScript to create functions, and allows user to import and export programs using JSON (JavaScript Object Notation), which is a lightweight open standard for data exchange.

There are many ways of using Node-RED, the most straight forward way is to use Node-RED from the IBM Watson IoT Platform—Bluemix, as shown in Figure 3.5. More details are available in Chapter 12.

Alternatively, you can also use Node-RED on Raspberry Pi; see Appendix C for more details.

Further Information about Node-RED

https://nodered.org/
https://flows.nodered.org/
https://nodered.org/docs/getting-started/first-flow

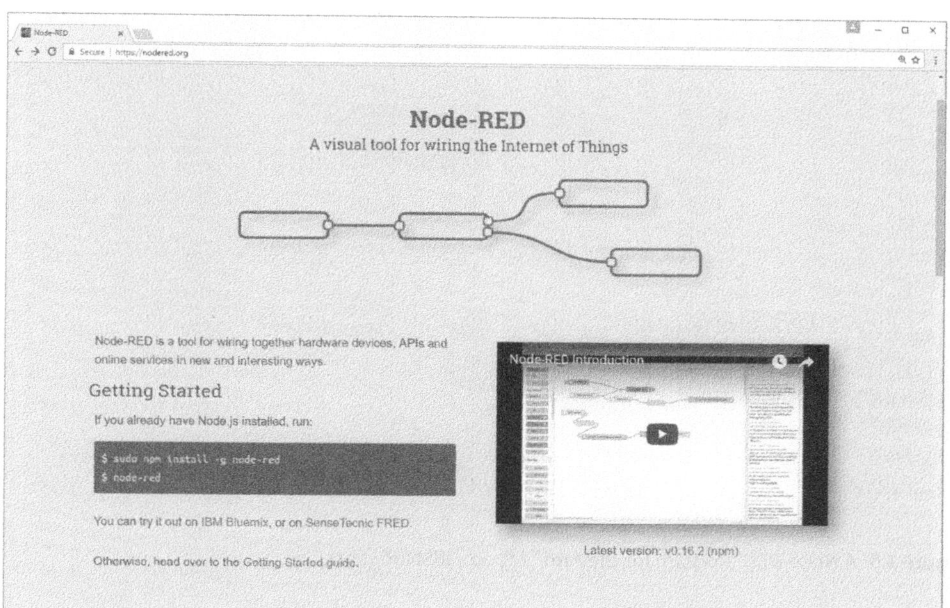

Figure 3.3 The Node-RED website.

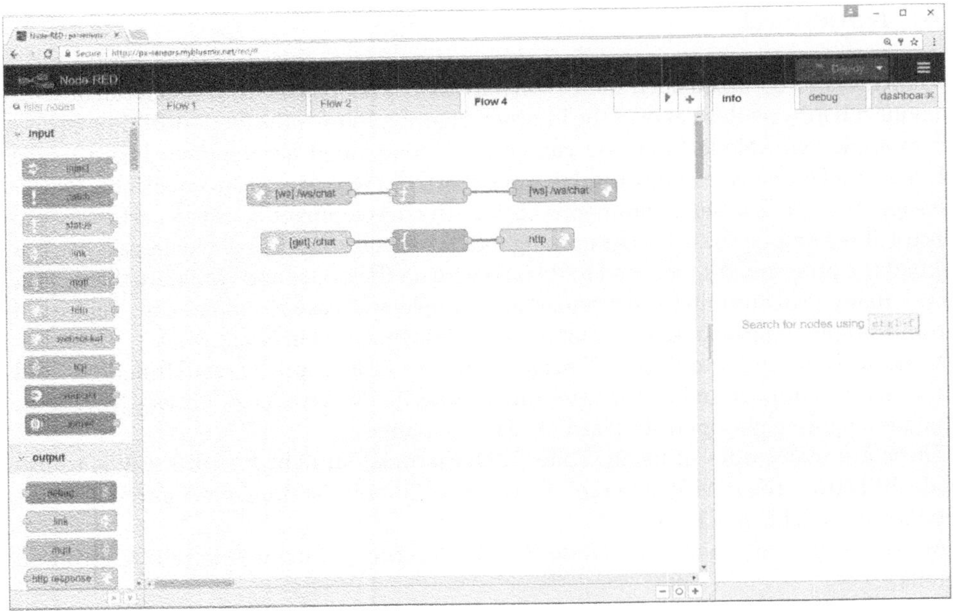

Figure 3.4 A simple WebSocket chat program written in Node-RED.

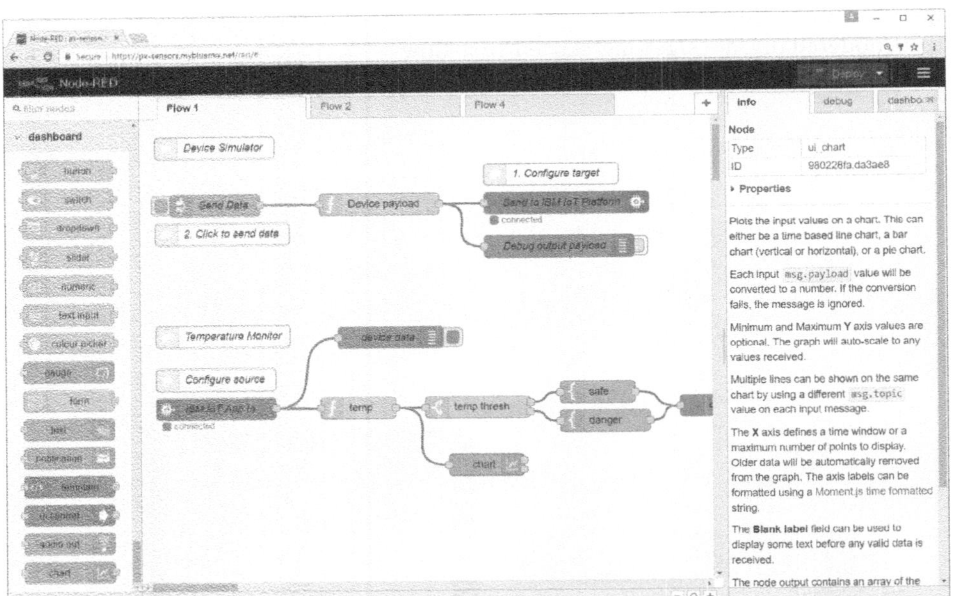

Figure 3.5 A Node-RED program for the Arm® Mbed™ IBM IoT starter kit.

3.5 Platforms

IoT platforms connect the sensors and data network to one another, integrating with backend applications to provide insights using backend applications to make sense of plethora of data generated by hundreds of sensors. With Iot platforms, you can connect and monitor your devices and sensors, display and analyze sensor data, control your devices, and develop software applications for your devices. Following is a list of commonly used IoT platforms.

3.5.1 IBM Watson IoT—Bluemix (http://www.ibm.com/internet-of-things/)

IBM Watson is a cloud-based computer system that combines artificial intelligence (AI) and sophisticated analytical software for optimal performance as a "question answering" machine. IBM Watson also supports IoT applications. The IBM Watson IoT platform Bluemix allows users to build IoP applications quickly, and connect IoT devices easily and securely. Figure 3.6 shows the schematic diagram of IBM Watson IoT platform from the IBM Watson website. Device-specific SDKs are available for Embedded C, JavaScript, Python, iOS, Android and Arduino Yún. The Arm® Mbed™ FRDM-K64F development kit used in this book can be easily connected to the IBM Watson IoT platform. The IBM Water IoT platform also provides real-time insights that contextualize and analyze real-time IoT data.

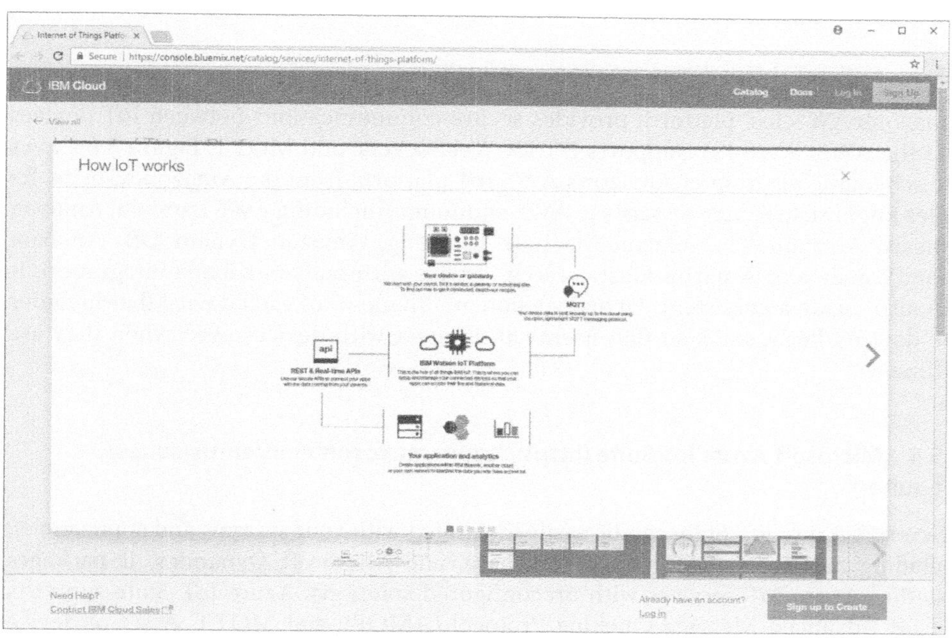

Figure 3.6 The schematic diagram of IBM Watson IoT platform from IBM website. (*Source:* https://console.ng.bluemix.net/catalog/services/internet-of-things-platform/)

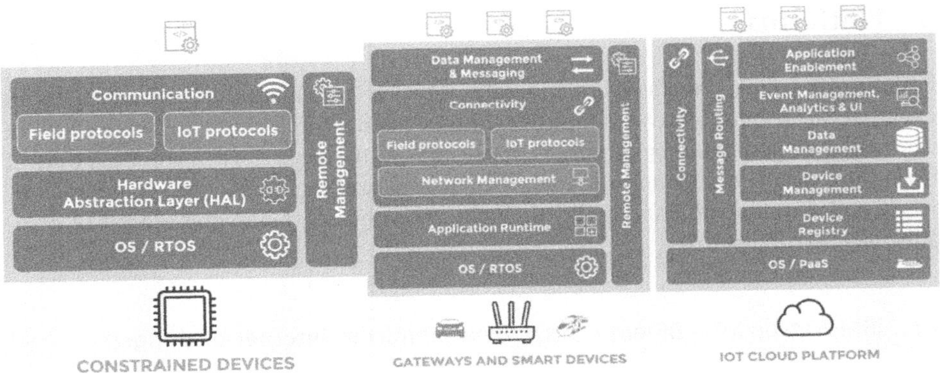

Figure 3.7 The software stacks of Eclipse IoT platform for constrained devices, gateways, and cloud platforms. (*Source:* https://iot.eclipse.org/devices/)

3.5.2 Eclipse IoT (https://iot.eclipse.org/)

This is an open source platform developed by The Eclipse Foundation. Eclipse IoT provides the technology needed to build IoT devices, gateways, and cloud platforms. See Figure 3.7 for the corresponding three software stacks. Eclipse IoT provides open source implementations of IoT standards and protocols, open-source frameworks and services that will be used by IoT solutions, and tools for IoT developers.

3.5.3 AWS IoT (https://aws.amazon.com/iot/)

Amazon's AWS IoT platform provides secure communications between IoT devices and the AWS. AWS IoT supports HTTP, WebSockets, and MQTT. Figure 3.8 shows the schematic diagram of Amazon's AWS IoT platform from the Amazon website. Its Rules Engine can route messages to AWS endpoints, including AWS Lambda, Amazon Kinesis, Amazon S3, Amazon Machine Learning, Amazon DynamoDB, Amazon CloudWatch, and Amazon Elasticsearch Service with built-in Kibana integration. It can also create a persistent, virtual version, or "shadow," of each device that includes the device's latest state, so that users can interact with devices even when they are offline.

3.5.4 Microsoft Azure IoT Suite (https://azure.microsoft.com/en-us/suites/iot-suite/)

Microsoft Azure IoT Suite can be easily integrated with your systems and applications, including Salesforce, SAP, Oracle Database, and Microsoft Dynamics. It packages together Azure IoT services with preconfigured solutions. Azure IoT Suite supports HTTP, Advanced Message Queuing Protocol (AMQP), and MQTT. A set of device SDKs for .NET, JavaScript, Java, C and Python are available. Figure 3.9 shows the schematic diagram of Microsoft Azure IoT solution architecture.

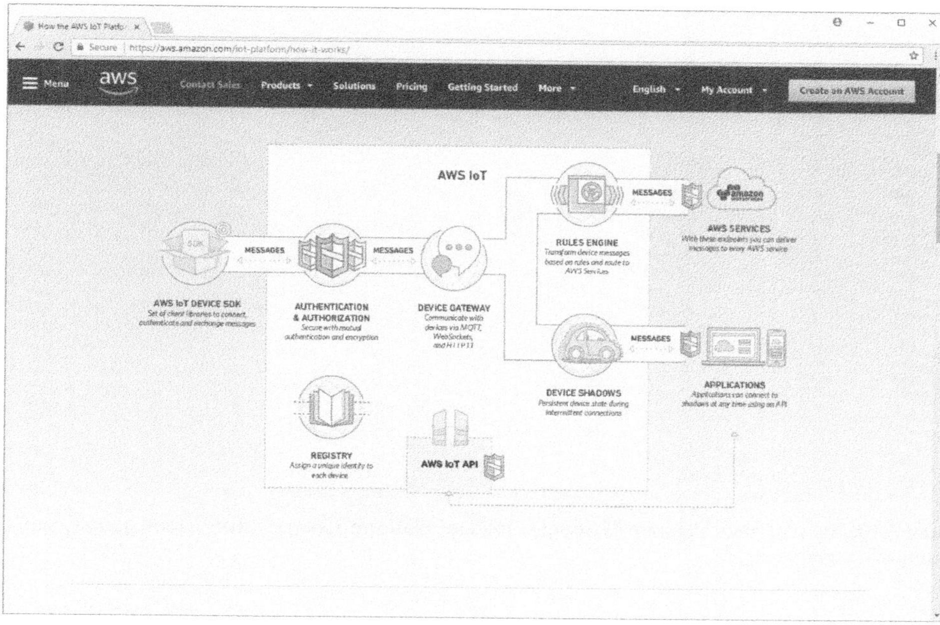

Figure 3.8 The schematic diagram of Amazon AWS IoT platform from the Amazon website. (*Source:* https://aws.amazon.com/iot-platform/how-it-works/)

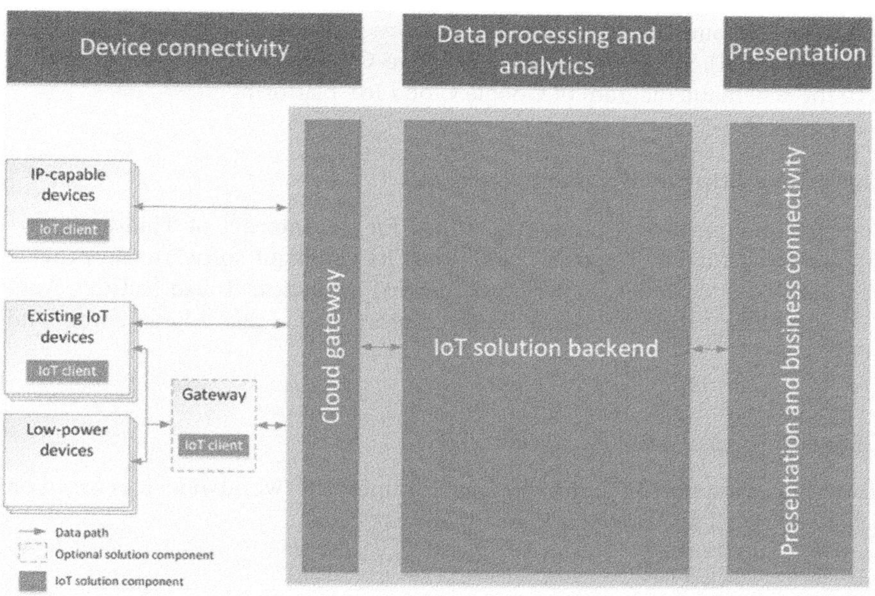

Figure 3.9 The schematic diagram of Microsoft Azure IoT solution architecture. (*Source:* https://docs.microsoft.com/en-us/azure/iot-hub/iot-hub-what-is-azure-iot)

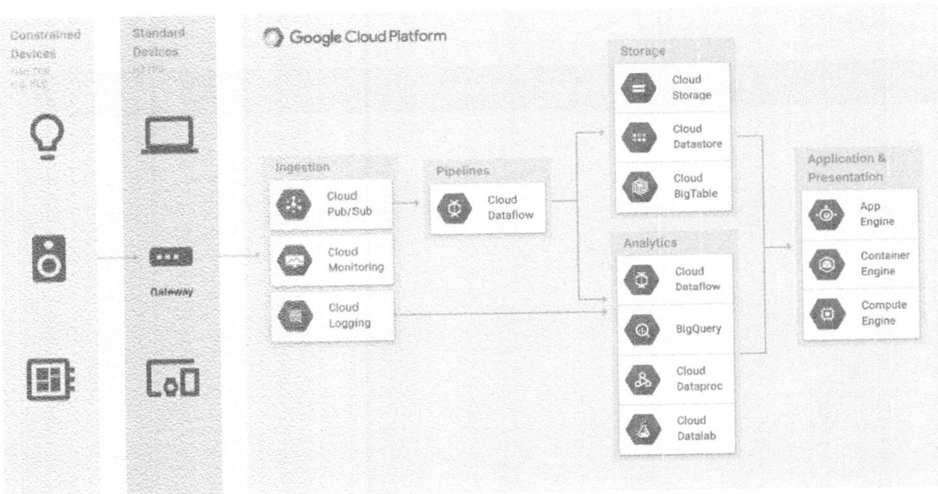

Figure 3.10 The schematic diagram of Google Cloud IoT platform. (*Source:* https://cloud.google.com/solutions/iot-overview)

3.5.5 Google Cloud IoT (https://cloud.google.com/solutions/iot/)

Google Cloud IoT takes advantage of Google's heritage of web-scale processing, analytics, and machine intelligence. It utilizes Google's global fiber network (70 points of presence across 33 countries) for ultra-low latency. Software libraries are available for Go, Java (Android), .NET, JavaScript, Objective-C (iOS), PHP, Python and Ruby. Figure 3.10 the schematic diagram of Google Cloud IoT platform.

3.5.6 ThingWorx (https://www.thingworx.com/)

ThingWorx is a complete development platform for the Internet of Things. ThingWorx enables powerful, enterprise IoT solutions. Its Coldlight software can provide automated predictive analytics, as well as other IoT analytics. It also features Augmented Reality Integration (Vuforia Studio Enterprise), Edge Microserver, and "Always On" SDK.

3.5.7 GE Predix (https://www.predix.com/)

GE Predix platform supports over 60 regulatory frameworks worldwide. It is based on Pivotal Cloud Foundry, and provides many Predix Services.

3.5.8 Xively (https://www.xively.com/)

Xively supports connections using native MQTT and WebSockets MQTT. It provides a C client library for use on devices. Xively provides an application for integrating connected products into the Salesforce Service Cloud.

3.5.9 macchina.io (https://macchina.io/)

This is an open-source-based IoT platform that implements a web-enabled, modular, and extensible JavaScript and C++ runtime environment. It is based on the POCO C++ Libraries and the V8 JavaScript engine. It is also based on a powerful plug-in and services model. It includes HTTP(S) and MQTT clients and SQLite as its embedded database.

3.5.10 Carriots (https://www.carriots.com/)

This platform provides integrations with Arduino, Raspberry Pi, and other DIY hardware platforms

It uses its HTTP RESTful API to push and pull XML or JSON encoded data. It deploys and scales from tiny prototypes to thousands of devices.

3.6 Summary

This chapter introduces the IoT enabling technologies, including sensors and actuators, communications, protocols, and various of IoT platforms.

3.7 Chapter Review Questions

Q3.1 What are sensors and actuators?

Q3.2 What is BLE?

Q3.3 How does LiFi work?

Q3.4 What is 6LowPAN?

Q3.5 What is Arm® Mbed™?

Q3.6 What is WebSocket?

Q3.7 What is WebSocket?

Q3.8 What is MQTT?

Q3.9 What is Node-RED?

Q3.10 What are IoT platforms?

Part II

Arm® Mbed™ Development

In this Part:

4

Getting Started with Arm® Mbed™

Success consists of going from failure to failure without loss of enthusiasm.
- Winston Churchill

4.1 Introduction

The current version of Arm® Mbed™ OS (operating system) is version 5.7. As shown in the Arm® Mbed™ documentation website (https://os.mbed.com/docs) (Figure 4.1). You can have three ways to get started with Arm® Mbed™ development.

- Online compiler
- Command line interface (mbed CLI)
- Third-party development environment

The easiest and quickest way is Arm® Mbed™ online compiler, e.g., the web-based compiler (https://os.mbed.com/docs/v5.6/tools/arm-mbed-online-compiler.html). This is what this book is focused on.

For the Arm® Mbed™ CLI, you will need to download and install the Arm® Mbed™ CLI software (https://os.mbed.com/docs/v5.6/tools/mbed-cli.html). It takes some effort, but the advantage is that it can work offline—that is, without Internet connection!

There are also a lot of third-party development environments available, including Keil uVision, DS-5, LPCXpresso, GCC, IAR Systems, and Kinetic Design Studio. More details can be found at https://os.mbed.com/docs/v5.6/tools/exporting.html

For more details:

https://os.mbed.com/
https://os.mbed.com/docs/v5.6/tools/index.html

4.2 Hardware and Software Required

4.2.1 Hardware

To get started, you will need:

- The Arm® Mbed™ Ethernet IoT Starter Kit, which includes an mbed Freedom FRDM-K64F development board and mbed application shield.

Designing Embedded Systems and the Internet of Things (IoT) with the ARM® Mbed™, First Edition. Perry Xiao

Companion website: www.wiley.com\go\xiao\designingembeddedsystemandIoTwitharmmbed

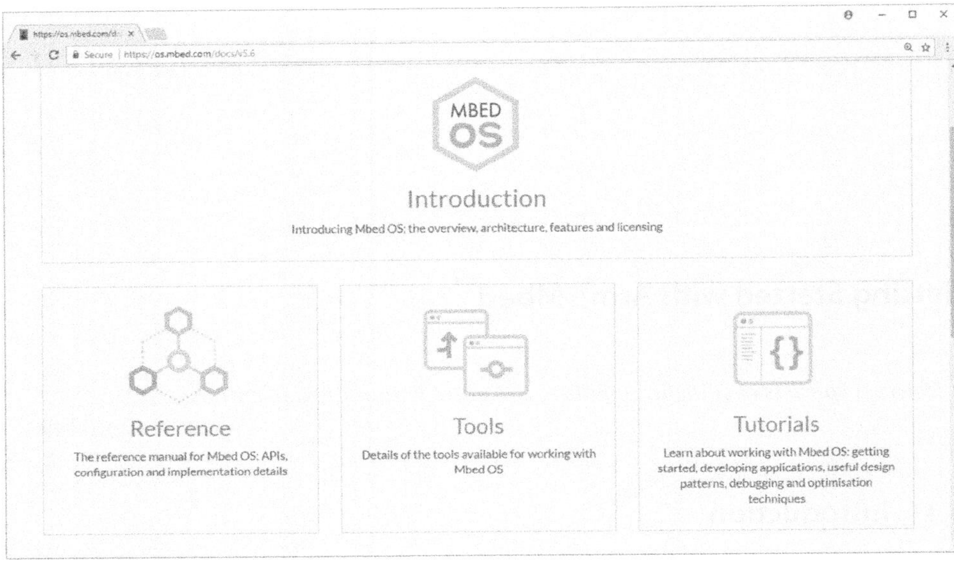

Figure 4.1 The Arm® Mbed™ documentation website.

- Micro USB cable
- Breadboard with jumper wires

4.2.2 Software

Although you don't need any software to compile and run your application on the Arm® Mbed™ devices when you are using an online compiler, you do need some software to communicate with the devices. Depending on your computer, you might need to install *serial port driver* and *Terminal software.*

Serial Port Driver

When you connect your mbed device to your computer, it can appear as a serial port, also called a virtual COM port. On Mac and Linux, this will happen automatically. For Windows, you need to install a serial port driver.

Just go to the following *Windows Serial Configuration* web page (see Figure 4.2) and follow the instructions to download and install the serial port driver.

https://os.mbed.com/handbook/Windows-serial-configuration

Terminal Software

You will also need to install terminal software, through which you can receive and send data to your mbed device. Just go to the following Arm® Mbed™ "*Terminals*" website (see Figure 4.3), and follow the instructions to download and install terminal software.

https://os.mbed.com/handbook/Terminals

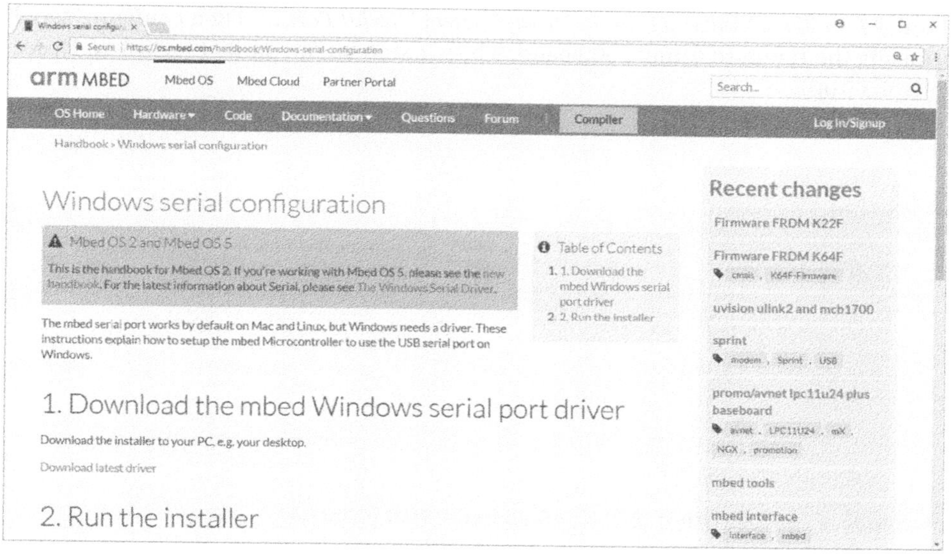

Figure 4.2 The Windows Serial Configuration website.

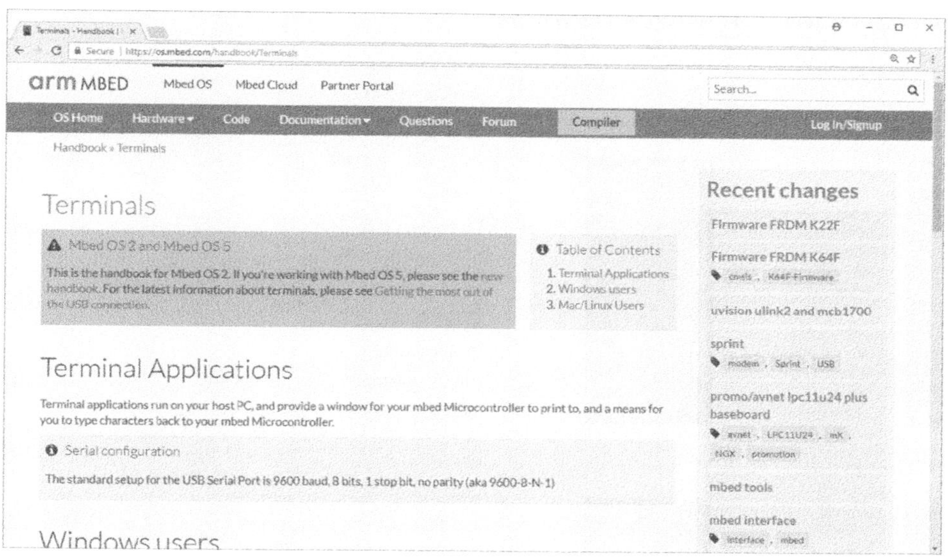

Figure 4.3 The Arm® Mbed™ Terminals website.

There are several popular terminal software available. In this book, the majority of the examples are based on *Tera Term* terminal software (http://sourceforge.jp/projects/ttssh2/files) in the Microsoft Windows environments, as Tera Term terminal software can automatically recognize which serial port that your mbed FRDM-K64F development board is connected to (Figure 4.4).

From "*Tera Term*" software menu "*Setup*" select "*Serial Port...*" Then configure serial port using the standard setup: 9600 baud, 8 bits, 1 stop bit, no parity (9600-8-N-1); see Figure 4.5 (top).

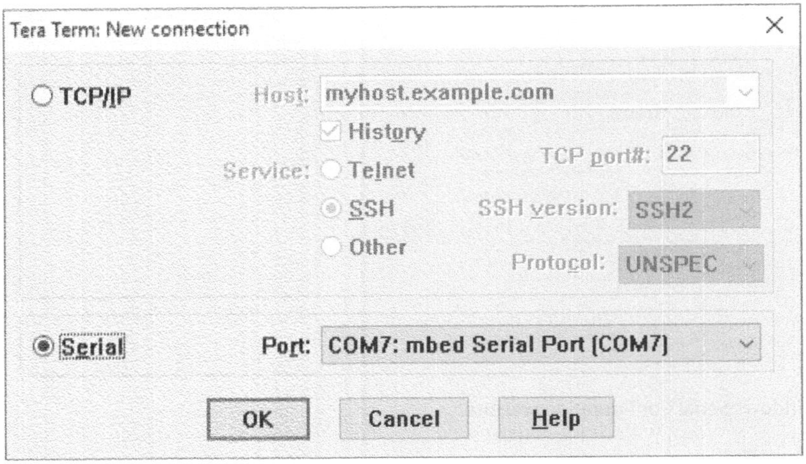

Figure 4.4 The Tera Term new connection window.

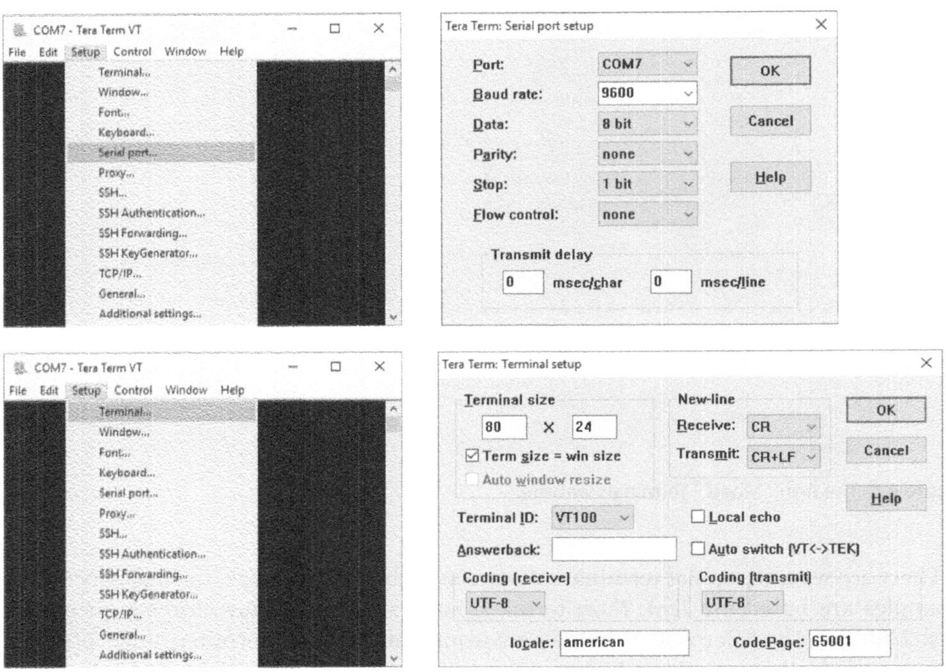

Figure 4.5 The Tera Term Serial port configuration (top) and Terminal configuration (bottom).

By default, Tera Term only transmits "*\r*" (CR, carriage return) when you press the "*Enter*" key. It is better to configure it to transmit a "*\n*" (NL, new line) as well, then the Arm® Mbed™ Serial read function "*gets()*" should terminate once it receives the "*\n*". See Chapter 7, section 7.1 for more about serial communications.

To configure the transmission, from "*Tera Term*" software menu "*Setup*" select "*Terminal...*". Then configure "Transmit:" as "CR+NL"; see Figure 4.5 (bottom).

Other popular terminal software

Putty.exe: https://the.earth.li/~sgtatham/putty/latest/w32/putty.exe
Arduino Serial Monitor: https://www.arduino.cc/en/main/software

4.3 Your First Program: Blinky LED

4.3.1 Connect the Mbed to a PC

Connect the Arm® Mbed™ FRDM-K64F development board to a computer using micro USB cable, there are two micro USB port on the board, make you are using the one on the right side, next to the "*Reset*" button (Figure 4.6). It will then appear as a standard USB memory drive, in this case, it is in drive G.

4.3.2 Click "*mbed.htm*" to Log In

Double-click the file "*mbed.htm*"—your web browser will then open a Login / Signup page (Figure 4.7). If you have an account, just log in; if you don't have an account, just sign up by following the instructions.

Figure 4.6 The FRDM-K64F board and the Arm® Mbed™ USB drive (G:) window.

Figure 4.7 The Arm® Mbed™ Login / Signup window.

Alternatively, you can also go the mbed developer website, https://os.mbed.com/, and click the "*Compiler*" menu on the top.

4.3.3 Add the FRDM-K64F Platform to Your Compiler

After login, you will be redirected to FRDM-K64F development board home page, which has all the details of the device (Figure 4.8). Click "*Add to your mbed Compiler*" button on the right-hand side. This will add the FRDM-K64F development board platform into your compiler, so that you can start writing code for the device. Each Arm® Mbed™ development board is a platform, so you will need to add different platforms for different mbed development board.

4.3.4 Import an Existing Program

Further down the page, there is an "*Open existing Project*" section (Figure 4.9). Click "*Import Program*" button to import the existing "*mbed_blinky*" project into your compiler.

The default project name is "*mbed_blinky*" (Figure 4.10), but you can change it to any name you prefer. Click the "*Import*" button, which will bring you to the online compiler web page.

Figure 4.11 shows the program online compiler web page. The "*main.cpp*" is the main C++ file that defines what your program is going to do. In this example, in the "*main.cpp*" file, it first includes the "*mbed.h*" header file, then defines LED1 as the digital output. In the "*main()*" function, it uses a "*while*" loop to switch the LED1 on, wait for 0.2 second, then switch the LED1 off, and wait another 0.2 seconds.

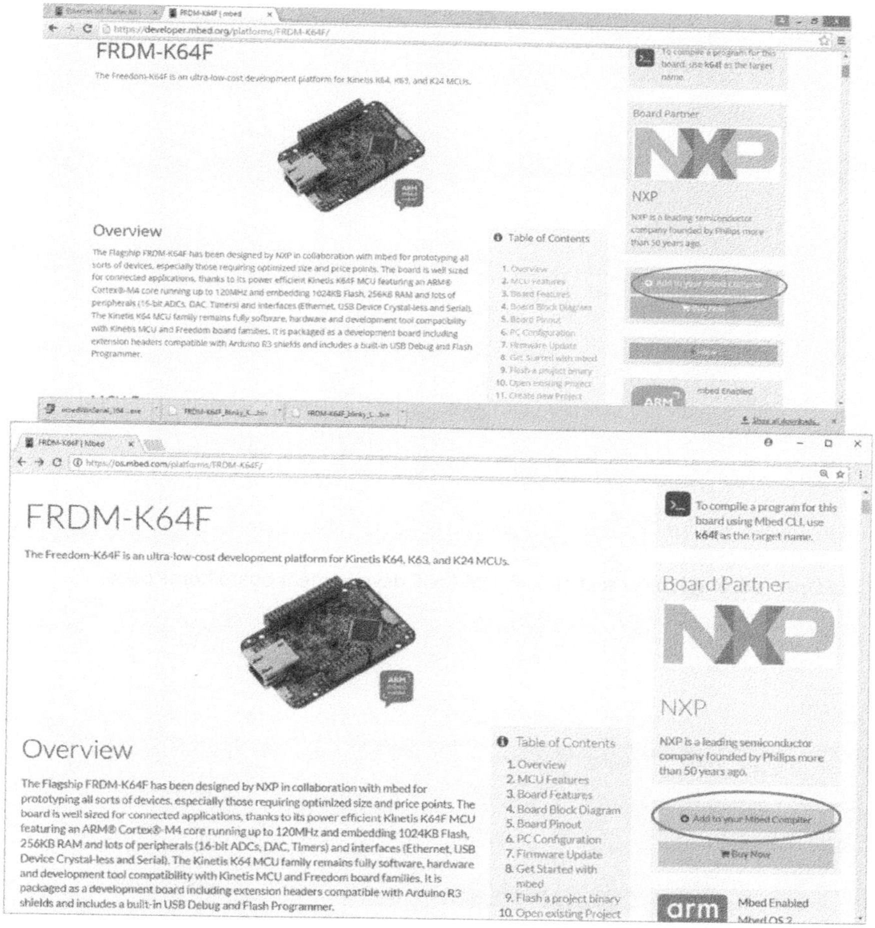

Figure 4.8 The FRDM-K64F development board home page.

```
*************************************************************
//Example 4.1
#include "mbed.h"                //include mbed.h header

DigitalOut myled(LED1);          //define LED1 as digital output

int main() {                     //main function
    while(1) {                   // while loop
        myled = 1;               //switch LED1 off
        wait(0.2);               //wait 0.2 seconds
        myled = 0;               //switch LED1 on
        wait(0.2);               // wait 0.2 seconds
    }
}
*************************************************************
```

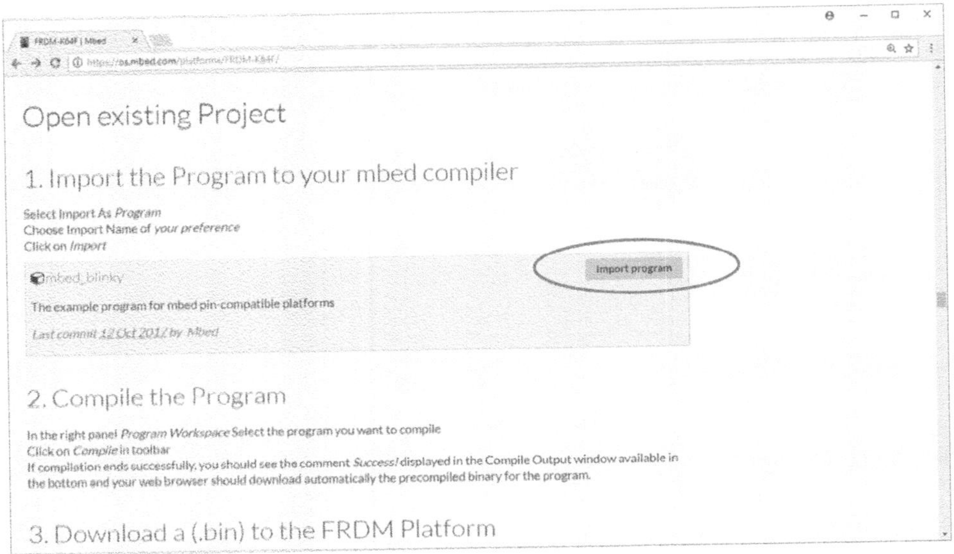

Figure 4.9 The Open existing Project section in FRDM-K64F development board home page.

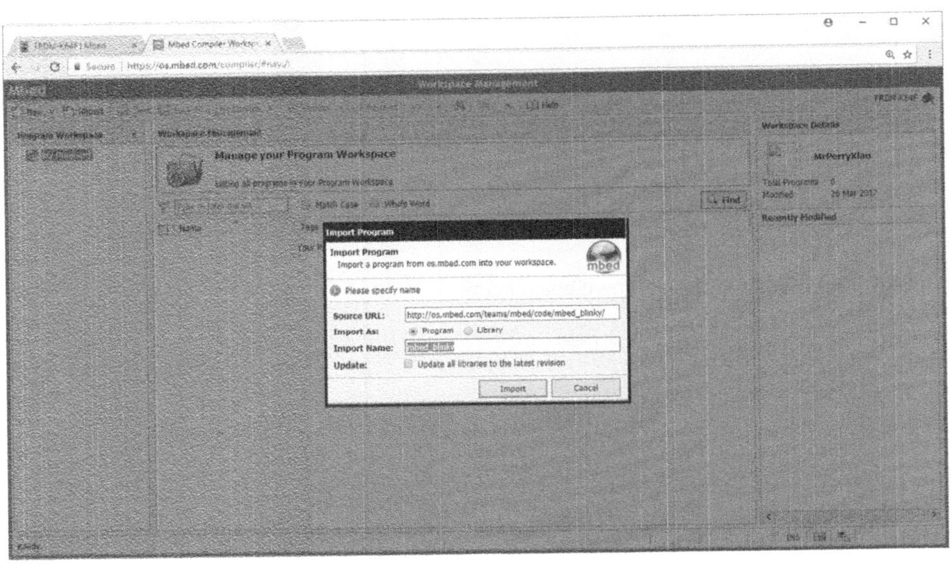

Figure 4.10 The Import Program pop-up window.

FRDM-K64D has only one RGB LED, LED1 (also called LED_RED) here refers to the red color of RGB LED. Similarly, LED2 (or LED_GREEN) and LED3 (or LED_BLUE) refer to green and blue colors.

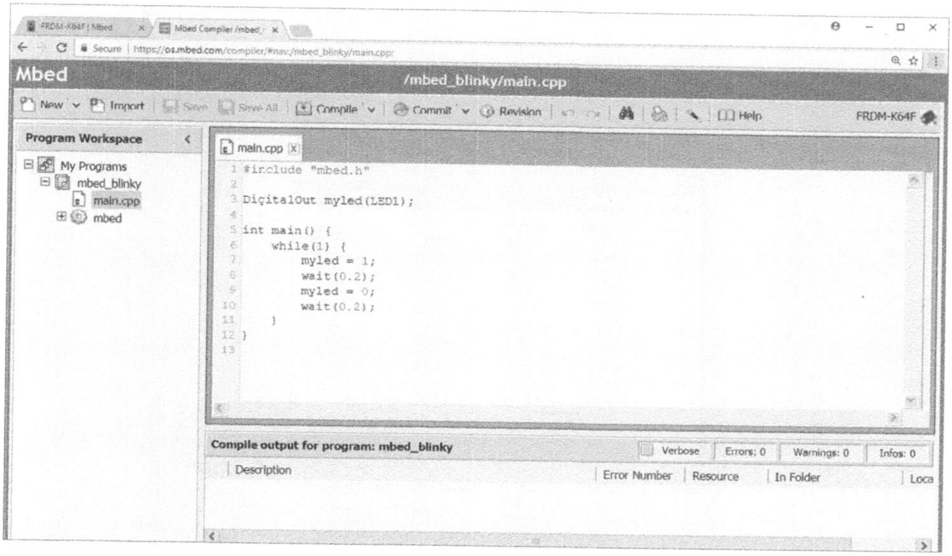

Figure 4.11 The *"mbed_blinky"* program online compiler web page.

4.3.5 Compile, Download, and Run Your Program

Click the *"Compile"* button to compile the program. If successful, a file called *"mbed_blinky_K64F.bin"* will be created and downloaded to the default download folder. Copy the file to FRDM-F64K USB drive and press the reset button to run your program! Now you should see the red LED blinking!

4.3.6 What Next?

Congratulations! You have just successfully run your first program. Next, you can try to download and run other existing programs from:

https://os.mbed.com/teams/FRDM-K64F-Code-Share/code/

You can also create your own programs.

4.4 Create Your Own Program

From your online compiler, you can create a new program by clicking the *"New"* button. A *"Create new program"* pop-up window will appear (Figure 4.12). Make sure you select the right platform (FRDM-K64F) and right template. I have found both *"gpio example for the Freescale freedom platform"* and *"mbed OS Blinky LED Helloworld!"* are good templates to start with. You can then easily modify the code to do what you want to do.

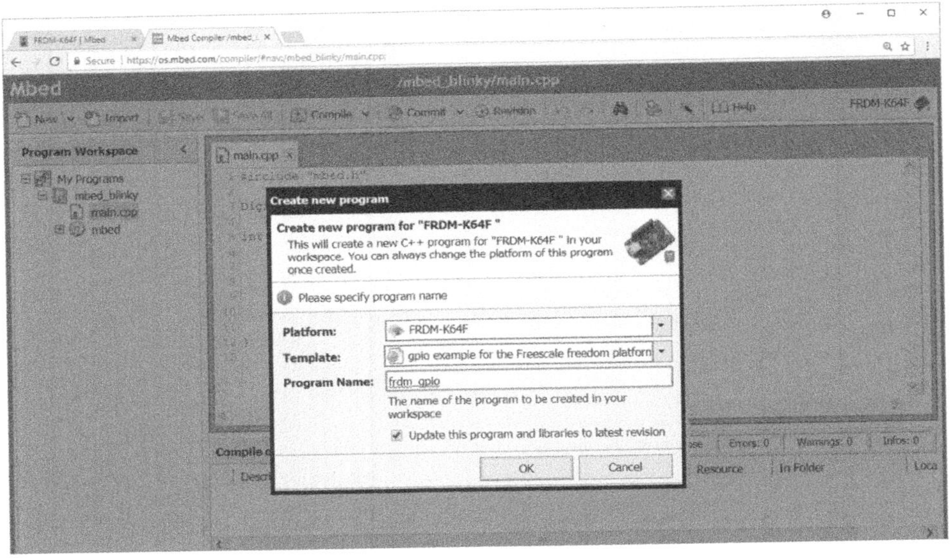

Figure 4.12 The Create new program pop-up window.

4.5 C/C++ Programming Language

The Arm® Mbed™ program uses C++ programming language. This is different from C programming language, which was originally developed by Dennis Ritchie for UNIX operating systems at "AT&T's Bell Laboratory" of USA back in the 1970s. C is a low-level powerful programming language, but it lacks many modern features. C++ is a newer language based on C, developed by Bjarne Stroustrup, also at Bell Laboratory in the 1980s. C++ has many features, such as easier memory management and object-oriented programming. All the functions in C are also available in C++.

4.6 Functions and Modular Programming

When you write simple programs, you can just put all the code inside the "*int main()*" function, as shown in Example 4.1. However, when your program is getting longer and complex, it is better to separate some of the reusable code into functions. Functions are also called subroutines, procedures or methods. With functions, you can easily reuse the code, and make the "*int main()*" function much simpler—hence, reducing the programming complexity.

Following is a simple function example. It does exactly the same as Example 4.1, but uses a functions called "*void flashled(double t)*" to flash the LED every *t* seconds.

```
**************************************************************
// Example 4.2
#include "mbed.h"

DigitalOut myled(LED1);

void flashled(double t) {
        myled = 1;
        wait(t);
        myled = 0;
        wait(t);
}
int main() {
    while(1) {
        flashled(0.2);
    }
}
**************************************************************
```

You can also put the "*void flashled(double t)*" function after the "*int main()*" function, as shown in Example 4.3. In that case, you will need to declare the functions in the beginning, before the "*int main()*" function. The declaration statements for functions are called prototypes.

```
**************************************************************
// Example 4.3
#include "mbed.h"

DigitalOut myled(LED1);

void flashled(double t);

int main() {
    while(1) {
        flashled(0.2);
    }
}
void flashled(double t) {
        myled = 1;
        wait(t);
        myled = 0;
        wait(t);
}
**************************************************************
```

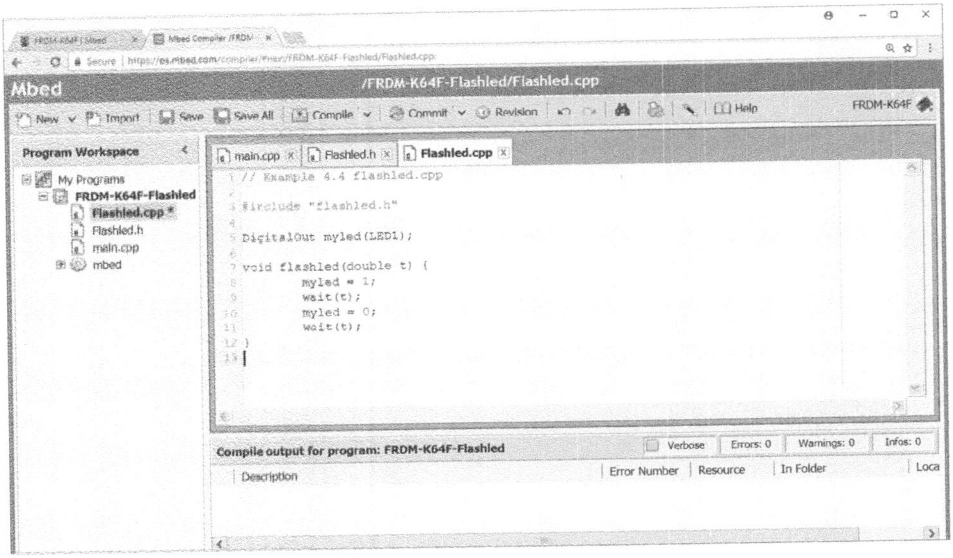

Figure 4.13 The program with "main.cpp," "flashled.cpp," and "flashled.h" files.

Exercise 4.1

Add an extra input variable to the "*void flashled(double t)*" function, so that it becomes "*void flashled(int n, double t)*" and it blinks different LED depending on the input value *n*.

For large projects, you can also separate code into different files. This is called modular programming. The following example separates the flash led functions into "*flashled.cpp*" and "*flashled.h*" files, as shown in Figure 4.13. You can add a new file from the online compiler by right-clicking your program and select "*New File....*" The header file, i.e., "*.h" file, is mainly for declarations, such as compiler directives, variable declarations and function prototypes. The "*cpp*" file is for implementing the functions. In this case, the header file "*flashled.h*" is used to join multiple files together.

```
*****************************************************************
// Example 4.4  main.cpp
#include "flashled.h"

int main() {
    while(1) {
        flashled(0.2);
    }
}
*****************************************************************
```

```
*************************************************************
//Example 4.4 flashled.h

#ifndef FLASHLED_H
#define FLASHLED_H

#include "mbed.h"
void flashled(double t);

#endif
*************************************************************
```

```
*************************************************************
// Example 4.4 flashled.cpp

#include "flashled.h"

DigitalOut myled(LED1);

void flashled(double t) {
        myled = 1;
        wait(t);
        myled = 0;
        wait(t);
}
*************************************************************
```

Exercise 4.2

Add an extra input variable to the *"void flashled(double t)"* function so that it becomes *"void flashled(int n, double t)"* and it blinks different LED, depending on the input value *n*.

Further Information about Functions and Modular Programming

https://os.mbed.com/media/uploads/robt/mbed_course_notes_-_modular_design.pdf

4.7 Manage Platforms

From your online compiler, you can select your platform by clicking the platform icon on the top-right corner. From the pop-up window (Figure 4.14), you can get the full technical details of the FRDM-K64F development board and its pin layout. You can also select a different platform or add more platforms.

But to remove a platform, you will need to go back to the Arm® Mbed™ development board web page and click the "*Remove*" button on the left-hand side of the page (Figure 4.15).

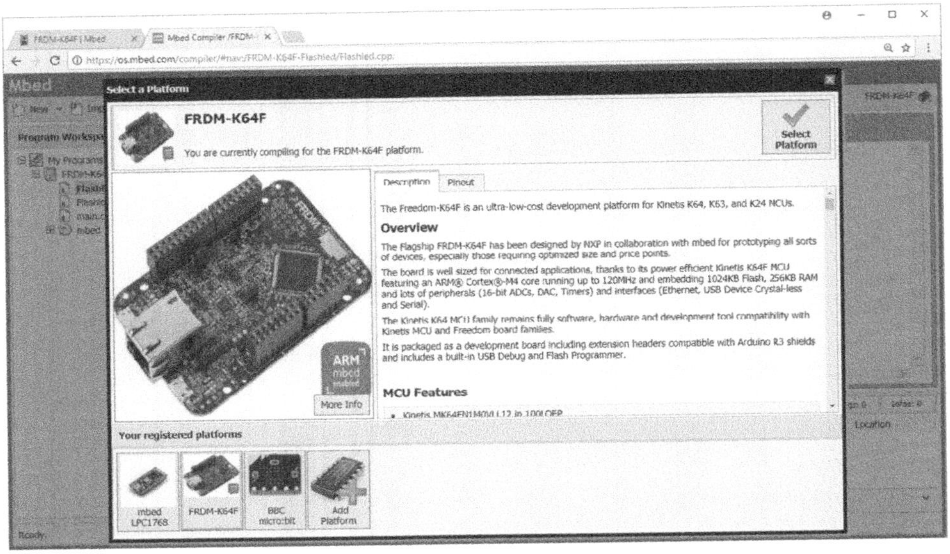

Figure 4.14 The Arm® Mbed™ manage platforms pop-up window.

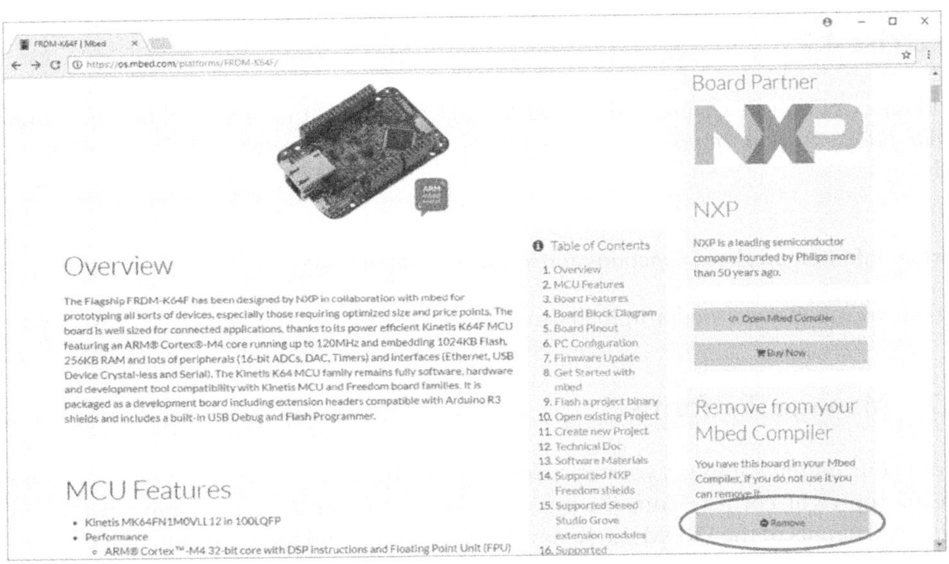

Figure 4.15 The Remove platform section in FRDM-K64F development board home page.

4.8 Clone Your Program

If you want to create a new program based on the existing one, you can clone your program, i.e., make a copy of the existing program. Just select the program you would like to clone, right-click to display the pop-up menu, and select "*Clone...*" (Figure 4.16). Then select the new name that you would like to save the cloned program as Figure 4.17.

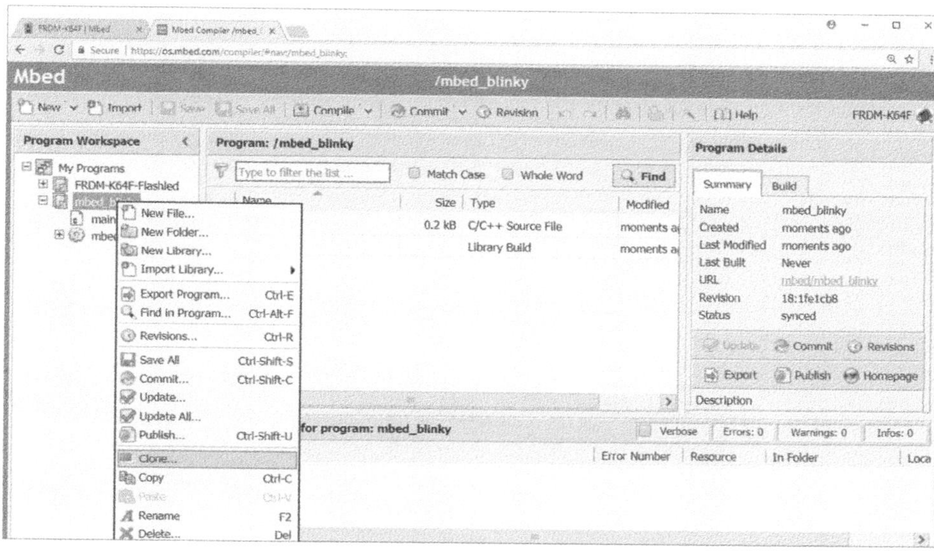

Figure 4.16 The Clone program menu in the program web page.

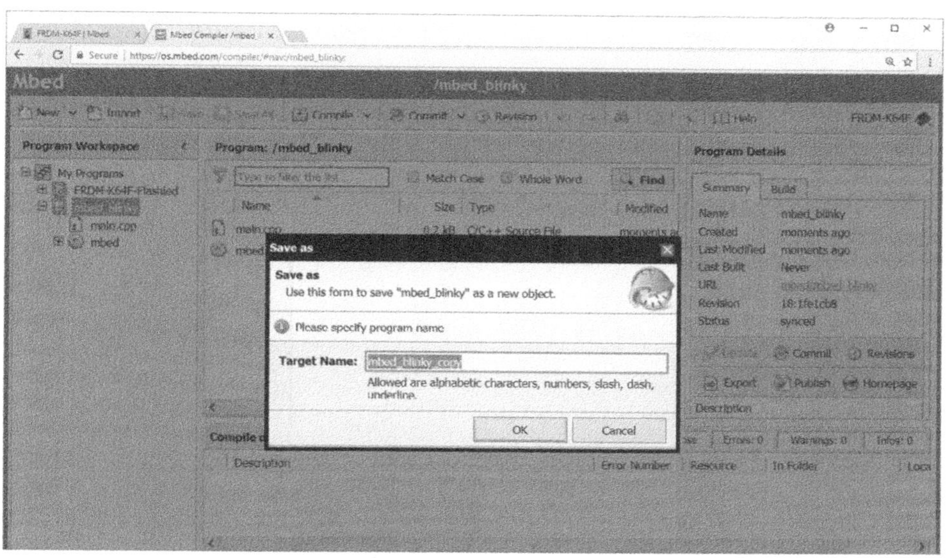

Figure 4.17 The Save as pop-up window during cloning.

4.9 Search and Replace

You can search in your program by clicking the "*Find*" button on the top, or pressing "*CTRL+F*" keys. A search and replace tool bar will then appear (Figure 4.18). You can use it to search and replace within your current file. The "*Advanced*" button on the right side allows you to search all the files within your project folder (Figure 4.19).

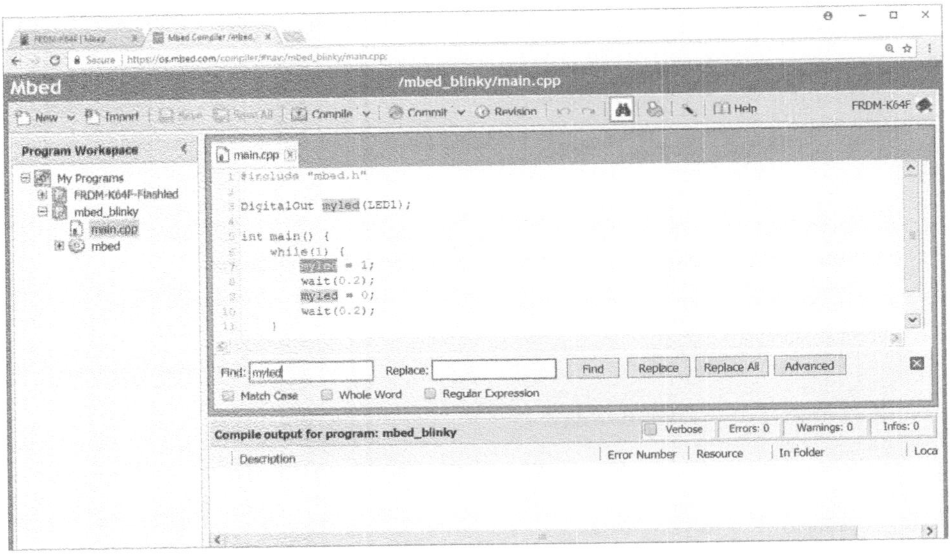

Figure 4.18 The search and replace in your program.

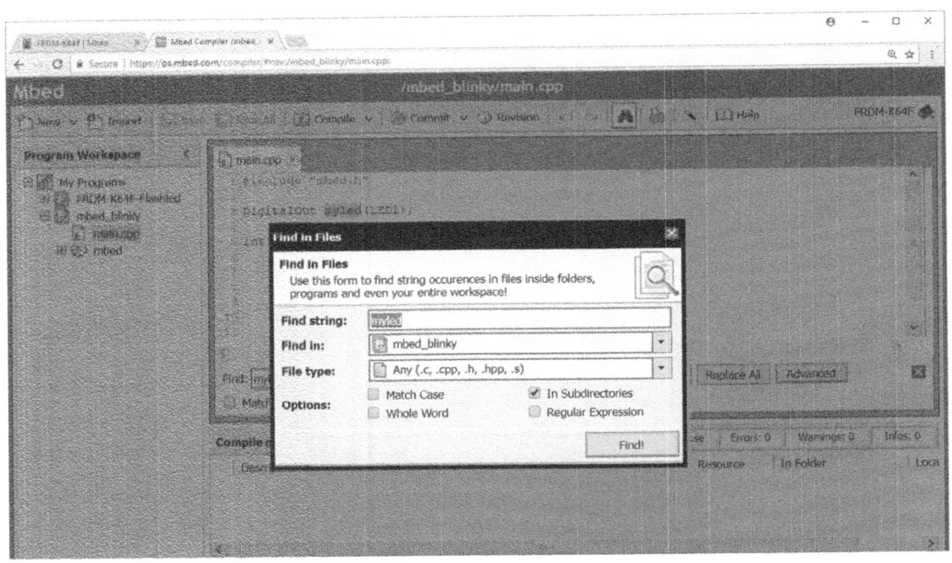

Figure 4.19 The advanced search in all files in your program.

4.10 Compile Your Program for Multiple Platforms

Although this book is focused on the FRDM-K64F development board, most codes are compatible with other platforms, such as NXP LPC1768. All you need to do is to put the platform-specific code in the "**#if defined() #elif defined()**" structure; see the following code.

```
**************************************************************
// Example 4.5

#include "mbed.h"

#if defined(TARGET_K64F)
   //FRDM-K64F code here

#elif   defined(TARGET_LPC1768)
   //LPC1768 code here

#elif defined(TARGET_LPC4330_M4)
   //LPC4330 code here

#endif


int main() {
    while(1) {
        //common code here
    }
}
**************************************************************
```

The platform-specific codes are mostly related to pin settings. Table 4.1 shows a comparison of the pinout between FRDM-K64F and LPC 1768 boards. When you compile the program, just make sure to select the correct platform. Mode details will be available in the next few chapters.

4.11 Delete Your Program

From your online compiler, to delete your program, just select the program you want to delete, and right-click the mouse. From the right-click drop-down menu, select "*Delete...*" (Figure 4.20). Simple!

Table 4.1 The pin comparison of FRDM-K64F and LPC 1768.

	FRDM-K64F	LPC 1768
LED	LED1 (LED_RED), LED2(LED_GREEN), LED3 (LED_BLUE) LED4 (LED_RED)	LED1, LED2, LED3, LED4
Digital inputs/outputs	D0, D1, D2, ..., D15	P5, P6, ..., P14
Analog inputs	A0, A1, A2, A3, A4, A5	P15, P16, ..., P20
Analog outputs	DAC0_OUT	P18
PWM (pulse width modulation)	A4, A5, D3, D5, D6, ..., D13	P21, P22, ..., P26
I2C	D14, D15 (SCA, SCL)	P9, P10 (SCA, SCL)
SPI	D11, D12, D13 (MOSI, MISO, SCLK) PTD4 (CS)	P5, P6, P7 (MOSI, MISO, SCLK) P8 (CS)
Serial	D1, D0 (Tx, Rx)	P9, P10 (Tx, Rx)

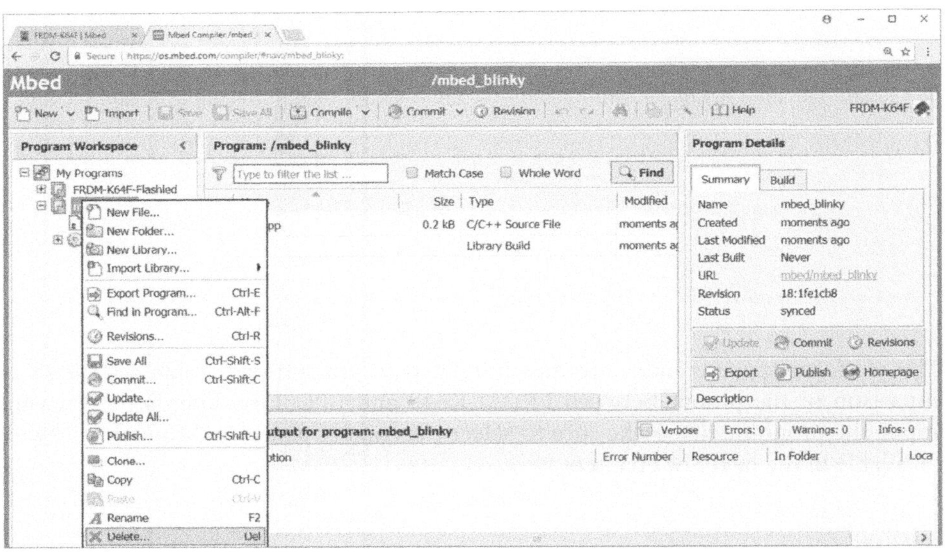

Figure 4.20 The Delete menu in the program page.

4.12 Disaster Recovery Procedure

In the event of a disaster, i.e., a faulty program etc., where you cannot see your mbed USB drive anymore, you can use the following procedure to recover:

- Unplug the FRDM-K64F board.
- Hold the reset button down.
- While holding the reset button, replug in the FRDM-K64F board.

The mbed USB drive should reappear. Keep holding the reset button until the new program is saved onto the USB drive.

In the worst-case scenario, when even the new program cannot solve the problem, you will probably need to reload the Firmware; see next section for details.

The following page has more details on how to deal with "dead" mbed devices.

https://os.mbed.com/cookbook/deadmbed

4.13 Upgrade Firmware

As of this writing, the latest firmware version for the FRDM-K64F is 0226. You can check the firmware version by either opening the DETAILS.TXT file if present on your mbed board or opening the MBED.HTM file with a text editor.

If you need to upgrade your firmware, or simply recover from a disaster, as described in the previous section, the following web page has all the details. Figure 4.21 is the screenshot of the web page.

https://os.mbed.com/handbook/Firmware-FRDM-K64F

You will basically need two steps:

1) **Enter Bootloader Mode**
   You can enter the Bootloader mode by unplugging the FRDM-K64F board, press and hold the reset button, replug the board, and release the reset button. You board should be mounted on your computer as "*Bootloader*" drive (Figure 4.22).
2) **Download and Upgrade the Firmwire**
   Download the latest firmwire from the website; copy and paste it into the "*Bootloader*" drive.

The latest firmware for LPC1768 and LPC11U24 can be found in:

https://os.mbed.com/handbook/Firmware-LPC1768-LPC11U24

4.14 Help

From your online compiler, you can get help by clicking the "*Help*" menu. It has all the details on how to get started, how to import programs and libraries, as well as collaborations, API documents, publishing your code, exporting your code and shortcuts (Figure 4.23).

Further Information on Help

https://os.mbed.com/docs

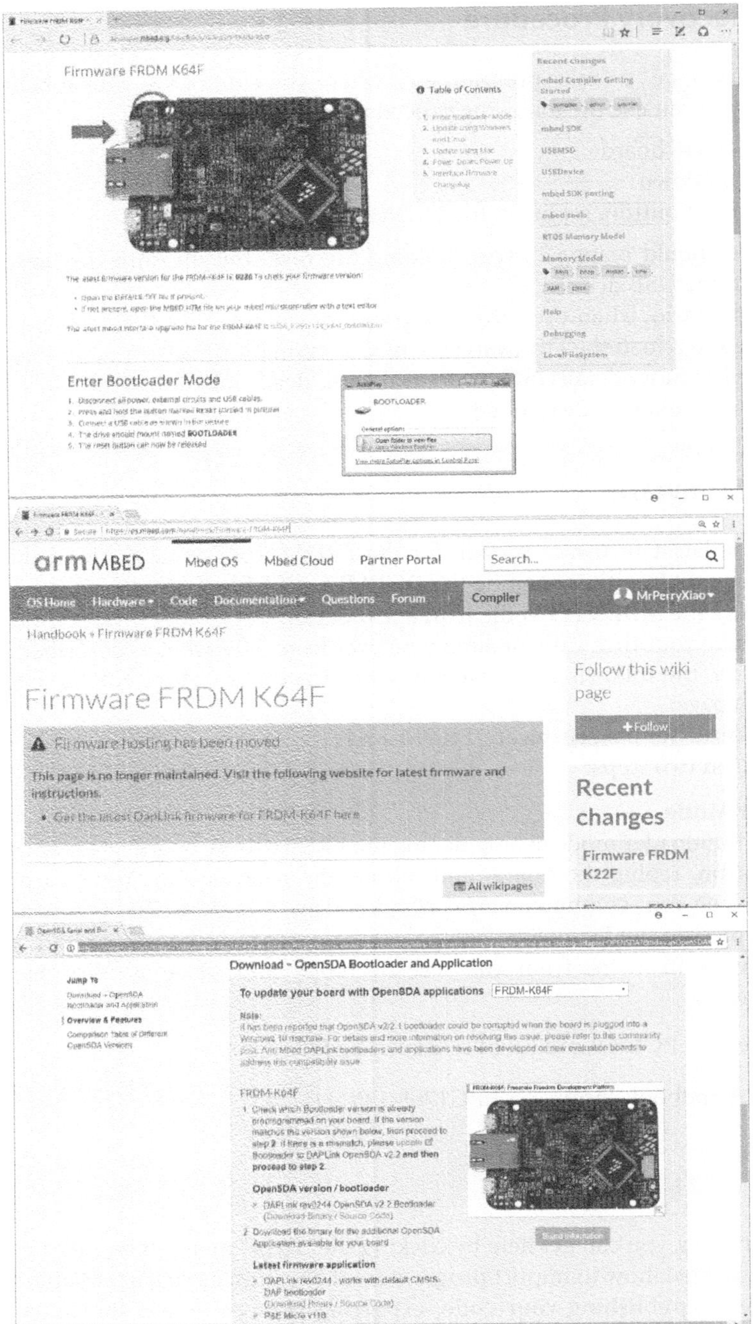

Figure 4.21 The mbed web page for upgrading FRDM-K64F firmwire (top) and following the link to NXP firmware page (bottom).

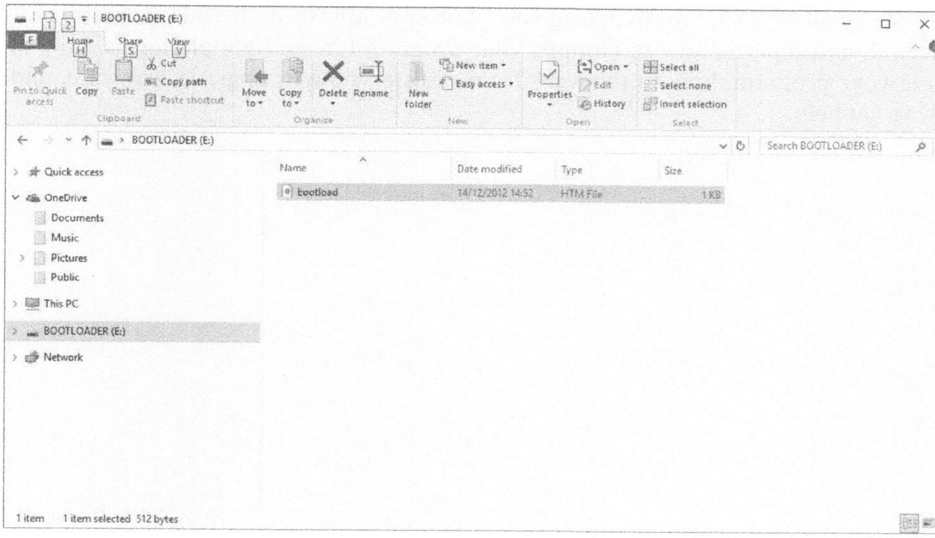

Figure 4.22 The mbed board in Bootloader mode.

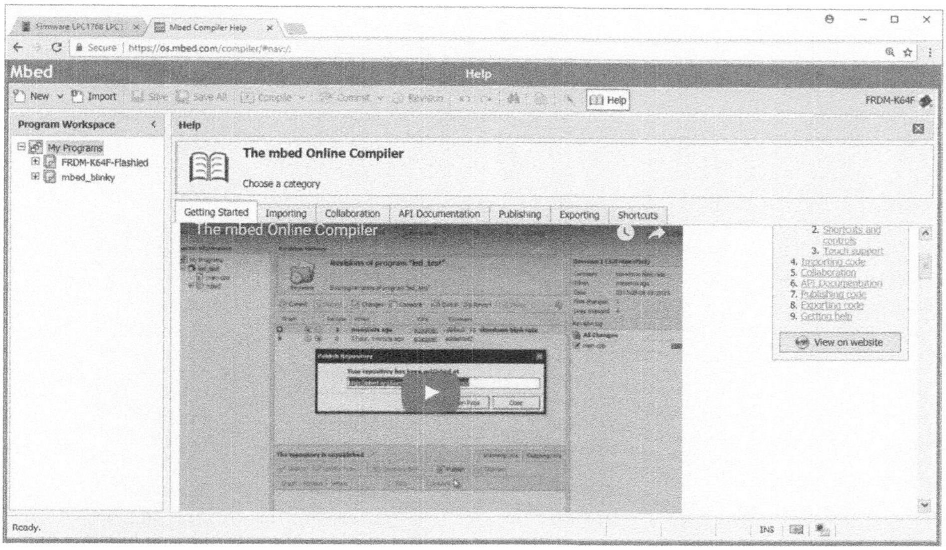

Figure 4.23 The online Help window.

4.15 Summary

This chapter describes the hardware and software required for the Arm® Mbed™ development. It also illustrates the steps to get started with the Arm® Mbed™ development, how to import and run your first LED blinking Hello World program, and how to create a new project, how to manage the platforms, how to clone your program, how to search and replace, how to compile your program for multiple platforms, how to delete your program, how to recovery from disaster, how to upgrade firmware, and how to get help.

5

Inputs and Outputs

Whether you think you can, or you think you can't—you're right.

- Henry Ford.

5.1 Digital Inputs and Outputs

Digital inputs and outputs are used to read in and write out digital values (i.e., 0 or 1). The mbed uses a power rail of 3.3 volts, with 0 volts representing 0 (or off), and 3.3 volts representing 1 (or on).

5.1.1 Digital Inputs

Connect the mbed FRDM-K64F development board to your computer. From the online compiler, create a new project, call it "*FRDM-K64F_DigitalIn*," and change the "*main.cpp*" content as shown in Figure 5.1.

The "*#include "mbed.h"*" line includes the mbed header file into the program, which provides all the functions of mbed. The "*DigitalIn din(D7)*" line creates a digital input from pin D7 and associates it with a variable called *din*. There are 16 digital pins in FRDM-K64F, ranging from D0, D1..., to D15. In the "*main()*" function, the "*while(1)*" represents an indefinite loop. This is typical for microcontrollers, as they need to work continuously all the time. Inside the loop, "*din.read()*" read the value from the digital input. As it is digital, the value is either 0 or 1. The "*printf()*" prints the results out; "*%d*" means print a integer type variable value here. "*\n\r*" means go to a new line after printing. "*wait(0.25)*" means wait for 0.25 second. By default, "*printf()*" prints to computer serial port, this is very useful, as you can view the results using a Terminal software, such as "*Tera Term*" (Figure 5.2). There will more about serial communications in the next chapter.

Designing Embedded Systems and the Internet of Things (IoT) with the ARM® Mbed™, First Edition. Perry Xiao

Companion website: www.wiley.com\go\xiao\designingembeddedsystemandIoTwitharmmbed

```
**************************************************************
// Example 5.1

#include "mbed.h"

DigitalIn   din(D7);

int main(void)
{
    while (1) {
        printf("%d\n\r",din.read());
        wait(0.25);
    }
}
**************************************************************
```

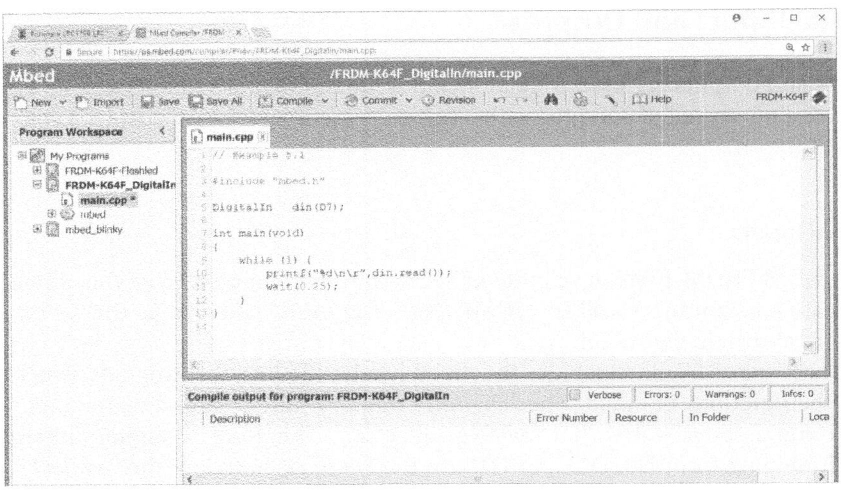

Figure 5.1 The "*FRDM-K64F_DigitalIn*" program page.

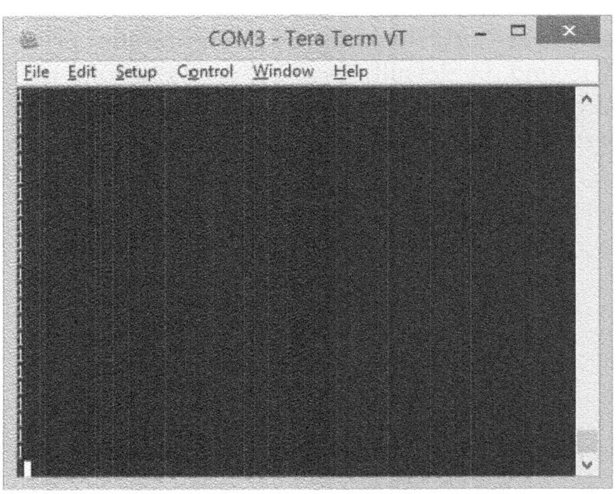

Figure 5.2 The Tera Term outputs.

Digital inputs are very useful to reading digital values, such as the outputs from a push button and PIR (Passive Infrared) sensor, as illustrated in Figure 5.3.

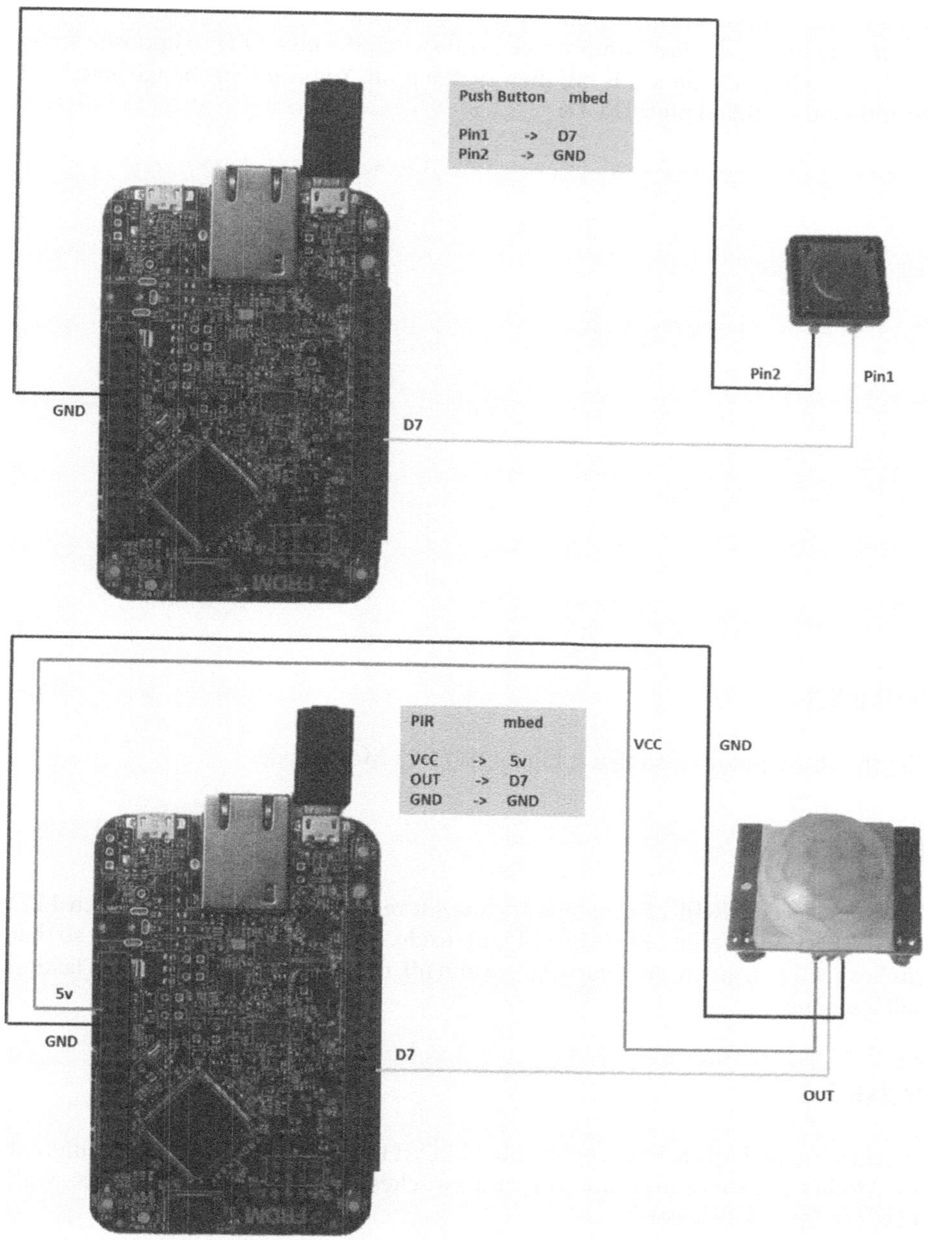

Figure 5.3 The schematic circuit diagram of *FRDM-K64F* board with a push button (top) and PIR sensor (bottom).

5.1.2 Digital Outputs

For digital outputs, create another new project, call it "*FRDM-K64F_DigitalOut,*" then modify the "*main.cpp*" file as follows. In this case, the line "*DigitalOut led(LED_BLUE)*" creates a digital output for blue color of RGB LED, and associates it with a variable called *led*. The line "*led = !led*" simply mean you switch the blue LED to opposite state, if it is one, switch it off, and if it is off, then switch it on. You can also change blue LED to any of the other digital pins: D0, D1, ... D15.

```
****************************************************************
// Example 5.2

#include "mbed.h"

DigitalOut   led(LED_BLUE);

int main(void)
{
    while (1) {
        led = !led;
        wait(0.5f);
    }
}
****************************************************************
```

Exercise 5.1

Modify the above program so that it blinks "SOS" in Morse code.

Exercise 5.2

FRDM-K64F has an RGB LED, which includes a red LED (LED_RED), a green LED (LED_GREEN), and a blue LED (LED_BLUE) inside. Modify the above program so that it switches each red, green, and blue LED on and off, one after another, with each lasting for half a second.

Exercise 5.3

By switching on and off each red, green, blue LED, it is possible to create $2^3 = 8$ different colors. Modify the above program so that it switches the 8 colors in a sequence, with each lasts for quarter a second.

Alternatively, you can also connect an external LED, as shown in Figure 5.4, where LED's long leg (+) is connected to **D7**, and short leg (-) is connected to the GND. Following is the example code to flash the LED.

```
*************************************************************
// Example 5.3

#include "mbed.h"

DigitalOut   led(D7);

int main(void)
{
    while (1) {
        led = !led;
        wait(0.5f);
    }
}
*************************************************************
```

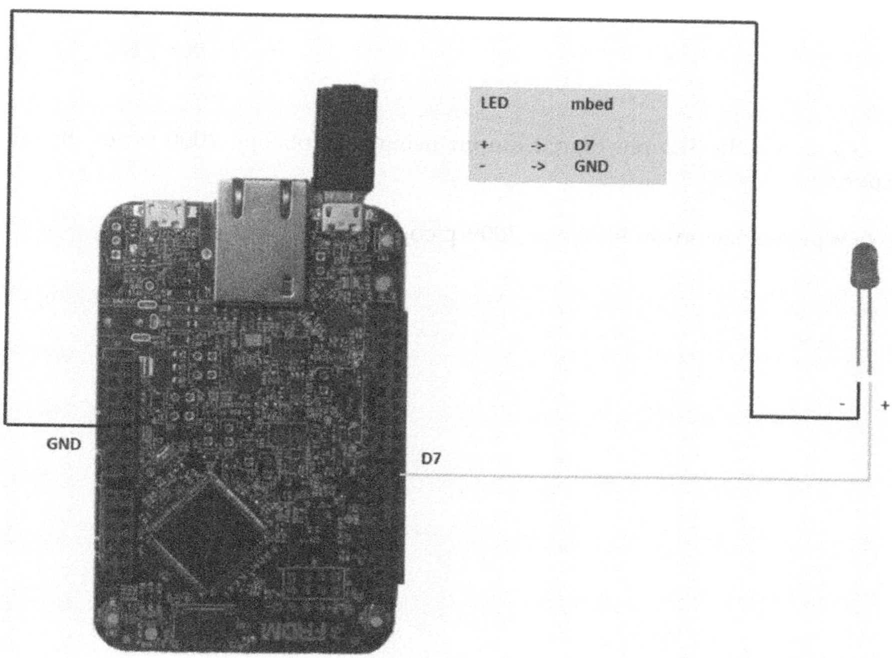

Figure 5.4 The schematic circuit diagram of the *FRDM-K64F* board with a LED.

Exercise 5.4

Based on above example, use three LEDs, red, green, and yellow, and light them up in a traffic light pattern.

You can also check the maximum digital output frequency by switching on and off a digital pin without any delays, as shown in the following example, which sets **D2** pin as a digital output. You can observe the changes of the output by using an oscilloscope.

```
************************************************************
// Example 5.4

#include "mbed.h"

DigitalOut   dout(D2);

int main(void)
{
    while (true) {
        dout = !dout;
    }
}
************************************************************
```

Figure 5.5 shows the **D2** pin digital output using a PicoScope 2000 series digital oscilliscope.

https://www.picotech.com/oscilloscope/2000/picoscope-2000-overview

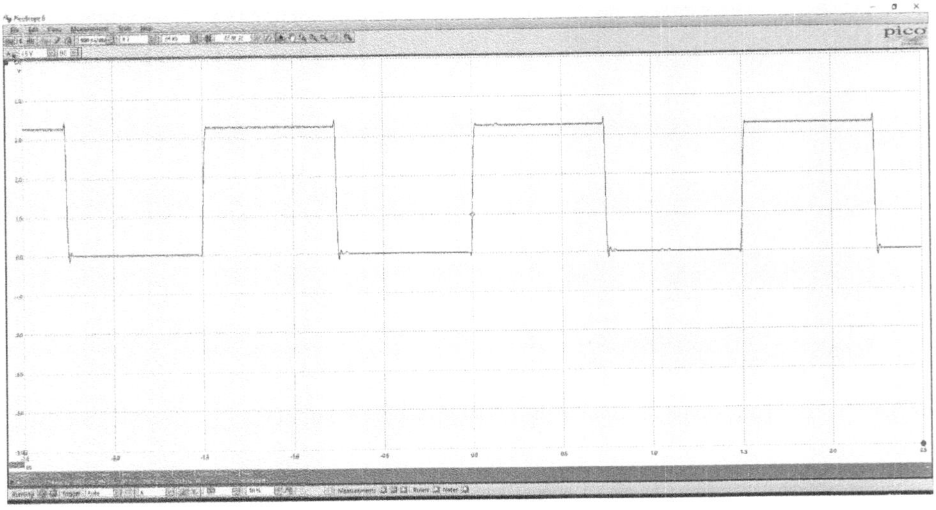

Figure 5.5 The *FRDM-K64F* digital output using PicoScope.

The results show that it is possible to set digital outputs as fast as 666.7 Hz.

Now you can combine the digital inputs and outputs to do something interesting. The following example reads the digital pin D7, reverses its value (*dout = !din;*), and sets it for D8 pin for output. The *"printf()"* prints out the two pin values, separated by tab "\t", to the computer serial port, as shown in the "*Tera Term*" screenshot (Figure 5.6). Again, "*%d*" means to print the numbers as integers, and "\n\r" or "\r\n" means to insert a new line after printing.

```
***************************************************************
// Example 5.5

#include "mbed.h"

DigitalIn     din(D7);
DigitalOut    dout(D8);

int main(void)
{
    while (1) {
        dout = !din;
        printf("%d \t %d \n\r", din.read(), dout.read());
        wait(0.5f);
    }
}
***************************************************************
```

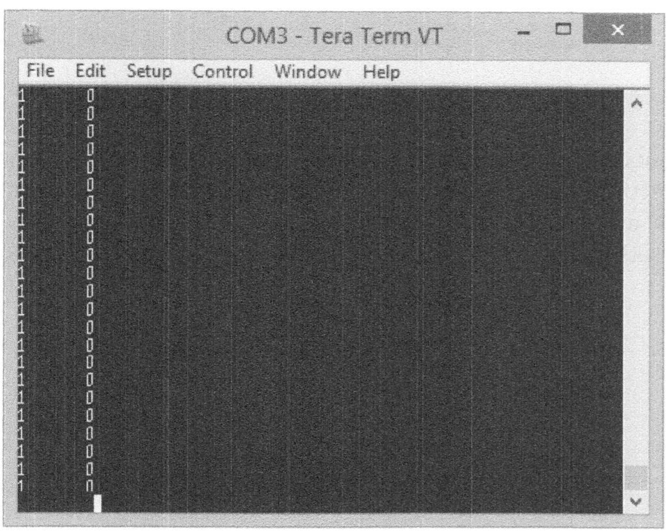

Figure 5.6 The Tera Term outputs.

Exercise 5.5

Modify the above program so that it reads two digital inputs from pins D6 and D7, performs the logical AND, and sets it to D9 pin for output.

The above code can be modified for both FRDM-K64F and LPC 1768 boards. So on FRDM-K64F it uses D7 and D8 digital pins, while on LPC 1768 it uses P11 and P12 pins.

```
***********************************************************
// Example 5.6

#include "mbed.h"

#if defined(TARGET_K64F)
   DigitalIn    din(D7);
   DigitalOut   dout(D8);
#elif defined(TARGET_LPC1768)
   DigitalIn    din(P11);
   DigitalOut   dout(P12);
#endif

int main(void)
{
    while (1) {
        dout = !din;
        printf("%d \t %d \n\r", din.read(), dout.read());
        wait(0.5f);
    }
}
***********************************************************
```

Apart from the digital inputs and digital outputs, you can also set a digital pin as both input and output, i.e., bidirectional, as illustrated in the following example. It first sets pin D7 (or P11 in LPC1768) as input, waits for 0.5 second, reads and prints its value, then sets the pin as output, sets its value to 1 (i.e., 3.3 V), prints out the value, and waits for another 0.5 seconds.

```
***********************************************************
// Example 5.7

#include "mbed.h"

#if defined(TARGET_K64F)
   DigitalIn    din(D7);
#elif defined(TARGET_LPC1768)
   DigitalIn    din(p11);
#endif
```

```
int main(void)
{
    while (1) {
          pin.input();
          wait(0.5f);
          printf("Input: %d \n\r", pin.read());
          pin.output();
          pin = 1;
          printf("Output: %d \n\r", pin.read());
          wait(0.5f);
    }
}
*************************************************************
```

5.1.3 BusIn, BusOut, and BusInOut

In mbed, "*BusIn,*" "*BusOut,*" and "*BusInOut*" interfaces allow you to create a number of DigitalIn pins that can be read and/or written as one value. In the following "*BusIn*" example, it reads pins D3, D4, D5, D6 (or for LPC1768 are P12, P13, P14, P15) as one value. D3 is the least significant bit (LSB), and D6 is the most significant bit (MSB). Any of the numbered mbed pins can be used as a DigitalIn in the "*BusIn,*" "*BusOut,*" and "*BusInOut.*"

```
*************************************************************
// Example 5.8

#include "mbed.h"

#if defined(TARGET_K64F)
   BusIn nibble(D3, D4, D5, D6 );
#elif defined(TARGET_LPC1768)
   BusIn nibble(p12, p13, p14, P15);
#endif

int main() {
    while(1) {
        // read the bus and mask out unused bits
        int v = (nibble & nibble.mask());
        printf("%d\r\n",v);
        wait(1);
    }
}
*************************************************************
```

The "*BusOut*" can create a number of DigitalIn pins that can be written as one value. In the following example, RGB LED (for LPC1768 are LED1, LED2, LED3) will light up as binary values from 0 to 7.

```
************************************************************
// Example 5.9

#include "mbed.h"

#if defined(TARGET_K64F)
   BusOut nibble(LED_RED, LED_GREEN, LED_BLUE);
#elif defined(TARGET_LPC1768)
   BusOut nibble(LED1, LED2, LED3);
#endif

int main() {
    while(1) {
        for(int i=0; i<8; i++) {
            nibble = i;
            wait(0.5);
        }
    }
}
************************************************************
```

The "*BusInOut*" can create a number of DigitalIn pins that can be read and written as one value. In the following example, it creates a bus of four pins: D3, D4, D5, D6 (for LPC1768 are P12, P13, P14, P15). It first sets the bus as output mode, and writes value 0xF to it, i.e., all pins are set high. It waits for 0.25 second, then sets the bus as input mode, waits for another 0.25 second, and reads the value from the bus and prints it to computer serial port. The "*%X*" means to print the value in hexadecimal format.

```
************************************************************
// Example 5.10

#include "mbed.h"

#if defined(TARGET_K64F)
   BusInOut bio(D3, D4, D5, D6);
#elif defined(TARGET_LPC1768)
   BusInOut bio(p12, p13, p14, p15);
#endif

int main() {
    while(1) {
        bio.output();
        bio = 0xF;
        wait(0.25);
        bio.input();
        wait(0.25);
        // read the bus and mask out unused bits
        int v = (bio & bio.mask());
```

```
        printf("%X\n\r",v);
    }
}
*************************************************************
```

Further Information on "*BusIn*," "*BusOut*," and "*BusInOut*":

https://docs.mbed.com/docs/mbed-os-api-reference/en/latest/APIs/io/DigitalIn/
https://docs.mbed.com/docs/mbed-os-api-reference/en/latest/APIs/io/DigitalOut/
https://docs.mbed.com/docs/mbed-os-api-reference/en/latest/APIs/io/DigitalInOut/
https://docs.mbed.com/docs/mbed-os-api-reference/en/latest/APIs/io/BusIn/
https://docs.mbed.com/docs/mbed-os-api-reference/en/latest/APIs/io/BusOut/
https://docs.mbed.com/docs/mbed-os-api-reference/en/latest/APIs/io/BusInOut/

5.2 Analog Inputs and Outputs

5.2.1 Analog Inputs

Analog inputs are for reading in voltage values (0–3.3 V) from the pins, with 12-bit and 16-bit resolution, and sampling rate can go up to 800 ksps. The analog pins are A0, A1, .., A5. The analog outputs are for setting voltage values (0–3.3 V) to pins for outputs. The analog pin is DAC0_OUT.

Connect the mbed FRDM-K64F development board to your computer. From the online compiler, create a new project, call it "*FRDM-K64F_AnalogIn,*" and change the "*main.cpp*" content as shown. The "*AnalogIn ain(A1)*" line creates an analog input from pin A1 and associates it with a variable called *ain*. There are six analog pins in FRDM-K64F, ranging from A0, A1..., to A5. Inside the loop, "*ain.read()*" read a floating-point value from the analog input, as a fractional percentage. The "*printf()*" prints the results out, "*%10.3f*" means print a float-type variable value here, using minimum 10 spaces and 3 decimal points. *f* means floating-point number. "*wait_ms(500)*" is another wait function, which means waiting for 500 millisecond.

```
*************************************************************
// Example 5.11

#include "mbed.h"

AnalogIn   ain(A1);

int main(void)
{
    while (1) {
        printf("%10.3f\n\r", ain.read());
        wait_ms(500);
    }
}
*************************************************************
```

You can also read in 16-bit values, as shown in the following example. In this case, "*ain.read_u16()*" reads a 16-bit normalized value from pin A0 and assigns it to variable *v*, which is an "*uint16_t*" type of integer. "*%04X*" means print the value as a four-digit hexadecimal value. The "*0x*" in the front simply adds 0x in front of the hex number, e.g., 0×56.

```
***************************************************************
// Example 5.12

#include "mbed.h"

AnalogIn    ain(A0);

int main(void)
{
    uint16_t v;
    while (1) {
        v = ain.read_u16();
        printf("0x%04X\n\r",v);
        wait_ms(500);
    }
}
***************************************************************
```

Analog input is very useful for reading in the voltage values from sensors, such as analog temperature sensor (LM35) and light-dependent resistor (LDR). Figure 5.7 shows typical setups for FRDM-K64F board with a temperature sensor and with a LDR sensor.

5.2.2 Analog Outputs

For analog output, from the online compiler, create a new project, call it "*FRDM-K64F_AnalogOut,*" and change the "*main.cpp*" content as shown. The "*aout.write()*" write to the analog output pin with a floating-point value, represents as a fractional percentage. In this case, it writes 0.5 × 3.3 = 1.65 volts to the pin. The same commend can also expressed as "*aout = 0.5f.*" Figure 5.8 shows the Tera Term outputs of the program.

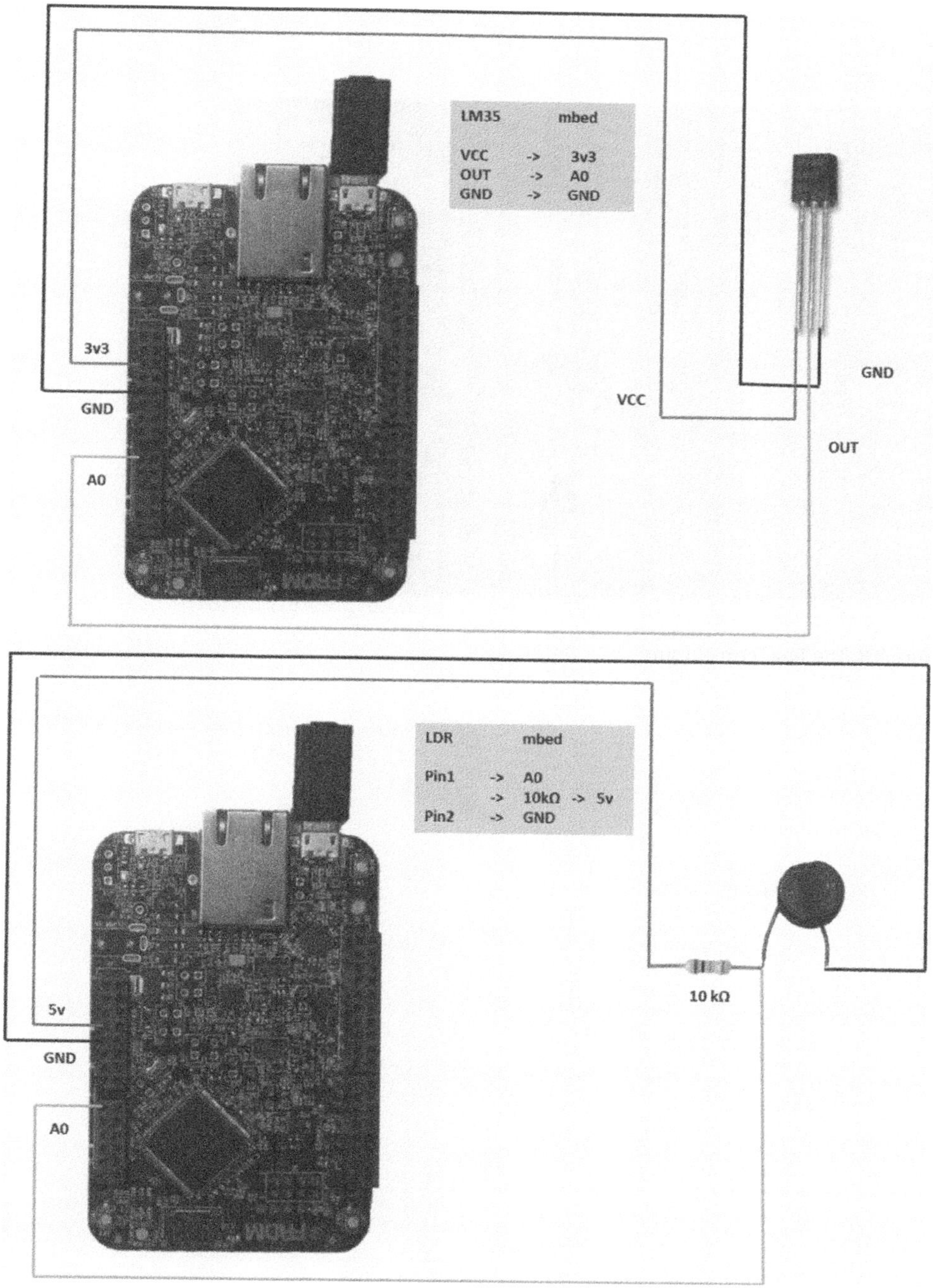

Figure 5.7 The schematic circuit diagram *FRDM-K64F* board with a temperature sensor (top) and LDR sensor (bottom).

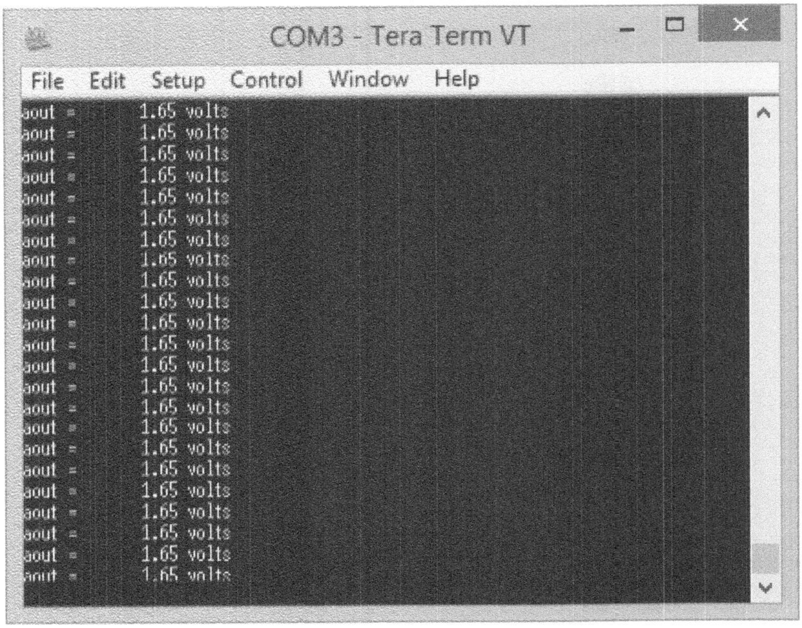

Figure 5.8 The Tera Term outputs.

```
*************************************************************
// Example 5.13

#include "mbed.h"

AnalogOut  aout(DAC0_OUT);

int main(void)
{
    while (1) {
        aout.write(0.5f);          // or  aout = 0.5f;
        printf("aout = %10.2f volts\n\r", aout.read() * 3.3f);
        wait(1.0f);
    }
}
*************************************************************
```

Exercise 5.6

Modify the above program so that it reads the analog input from pins A0, multiply it by 10, and set it to the analog output pin DAC0_OUT for output.

The following example uses a for loop to set the analog output pin DAC0_OUT in a seesaw format. It starts with 0.0 × 3.3 volts, increased by 0.1 × 3.3 volts each time, all the way up to 1.0 × 3.3 volts, then starts all over again.

```
***********************************************************
// Example 5.14

#include "mbed.h"

AnalogOut  aout(DAC0_OUT);

int main(void)
{
    while (1) {
        for (float i = 0.0f; i < 1.0f; i += 0.1f) {
            aout = i;
            printf("aout = %10.2f volts\n", aout.read() * 3.3f);
            wait(0.2f);
        }
    }
}
***********************************************************
```

Exercise 5.7

Modify the above program, so that it can create a sine wave on the analog output pin DAC0_OUT pin.

The following is a multiple platform example for both FRDM-K64F and LPC 1768 boards. It simply reads the analog input, A0 on FRDM-K64F, and P15 on LPC 1768, then assigns the value to analog output pin, DAC0_OUT on FRDM-K64F, and P18 on LPC 1768.

```
****************************************************************
// Example 5.15

#include "mbed.h"

#if defined(TARGET_K64F)
   AnalogIn    ain(A0);
   AnalogOut   aout(DAC0_OUT);
#elif defined(TARGET_LPC1768)
   AnalogIn    ain(p15);
   AnalogOut   aout(p18);
#endif

int main(void)
{
    while (1) {
        aout = ain.read();
        printf("%10.2f \n\r", aout.read());
        wait(0.5f);
    }
}
****************************************************************
```

For more details about Analog Inputs and Outputs

https://docs.mbed.com/docs/mbed-os-api-reference/en/latest/APIs/io/AnalogIn/
https://docs.mbed.com/docs/mbed-os-api-reference/en/latest/APIs/io/AnalogOut/

5.3 Pulse Width Modulation (PWM)

Pulse width modulation, or PWM, is popular a technique for microcontrollers getting analog results with digital means. It first creates a square wave signal with a fixed frequency. Then, by varying the width of the pulses, you can change the output power. For example, PWM can be used to control the LED light intensity, or to control the motor speed.

To use PWM, from the online compiler, create a new project, call it "*FRDM-K64F_PWM,*" and change the "*main.cpp*" content as shown. The "*PWMOut pout(D9)*" define D9 pin as PWM output, the "*pout.period(2.0f)*" specify the period as 2 seconds, and the "*pout.write(0.5f)*" specify the duty cycle is 0.5 or 50%, i.e., 1 second. This code generates a fixed PWM output of 2-second pulses with 50% duty cycle. Modify the code, to change the duty cycle to 10% and 90%, and use an oscilloscope to observe the changes in pulse width.

```
***************************************************************
// Example 5.16

#include "mbed.h"

PwmOut pout(D9);

int main() {
    pout.period(2.0f);
    pout.write(0.50f);
    while(1);
}
***************************************************************
```

If you want to change the PWM output while it's running, you will need to modify the while loop. As shown in the following example, it sets the period as 1 second, and within the loop, it first sets the duty cycle to 0.5 second, waits for 5 seconds, then sets the duty cycle to 0.1 second. You can either use *"pout.pulsewidth(0.5f)"* or *"pout.write(1.0f)"* to specify the duty cycle. The difference is that *"pout.pulsewidth(0.5f)"* specifies in seconds, while *"pout.write(1.0f)"* specifies in percentage. If you put an LED across the D9 pin and the ground, you will see the LED on brightly for 5 seconds, then dim for 5 seconds.

```
***************************************************************
// Example 5.17

#include "mbed.h"

#if defined(TARGET_K64F)
   PwmOut pout(D9);
#elif defined(TARGET_LPC1768)
   PwmOut pout(p26);
#endif


int main() {
    pout.period(1.0f);
    while(1){
       pout.pulsewidth(0.5f);    //or    pout.write(1.0f);
       wait(5.0f);
       pout.pulsewidth(0.1f);    //or    pout.write(0.2f);
       wait(5.0f);
    }
}
***************************************************************
```

Exercise 5.8

Imagine that you connected a potentiometer to analog input A0. Modify the above program so that it reads the value from A0, and change the PWM pulse width accordingly.

A very useful application of PWM is to drive a servo motor. In this example, you will need to import Servo library:

https://os.mbed.com/users/simon/code/Servo/docs/36b69a7ced07//classServo.html

```
*************************************************************
// Example 5.18

#include "mbed.h"
#include "Servo.h"

#if defined(TARGET_K64F)
   PwmOut pout(D9);
#elif defined(TARGET_LPC1768)
   PwmOut pout(p21);
#endif

int main() {
    for(float p=0; p<1.0; p += 0.1) {
        myservo = p;
        wait(0.2);
    }
}
*************************************************************
```

Further Information about PWM

https://docs.mbed.com/docs/mbed-os-api-reference/en/latest/APIs/io/PwmOut/

5.4 Accelerometer and Magnetometer

FRDM-K64F has an onboard six-axis combo accelerometer and magnetometer sensor (FXOS8700Q). To use a sensor, from the online compiler, create a new project, call it "*FRDM-K64F_FXOS8700Q*," and change the "*main.cpp*" content as shown. Figure 5.9 shows the program page from the online compiler.

```
*************************************************************
// Example 5.19

#include "mbed.h"
#include "FXOS8700Q.h"

I2C i2c(PTE25, PTE24);
FXOS8700QAccelerometer acc(i2c, FXOS8700CQ_SLAVE_ADDR1);

int main(void)
{
    motion_data_units_t acc_data;
    acc.enable();
    printf("FXOS8700QAccelerometer Who Am I= %X\r\n", acc.whoAmI());
    while (true) {
        acc.getAxis(acc_data);
        printf("%1.4ff %1.4ff %1.4ff \r\n", acc_data.x, acc_
data.y, acc_data.z);

        wait(1.0f);
    }
}
*************************************************************
```

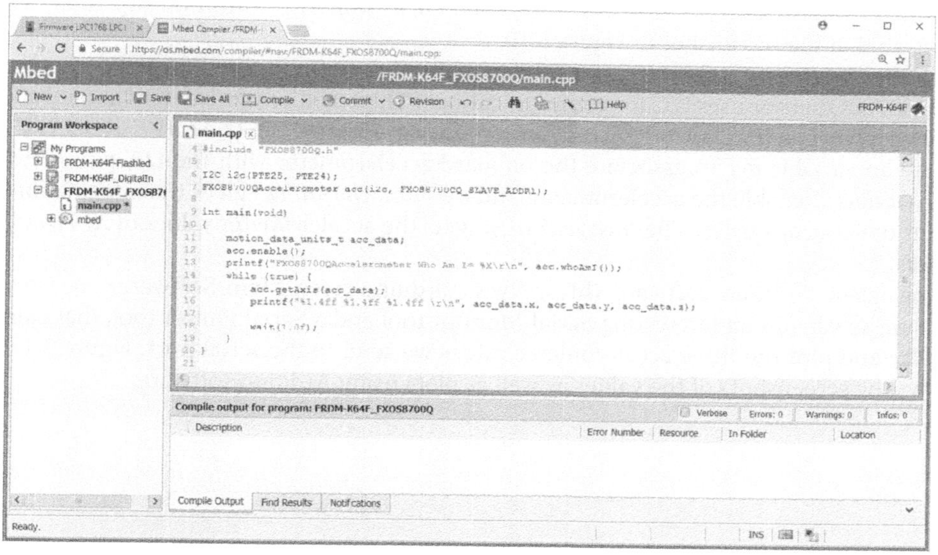

Figure 5.9 The "*FRDM-K64F_FXOS8700Q*" program page.

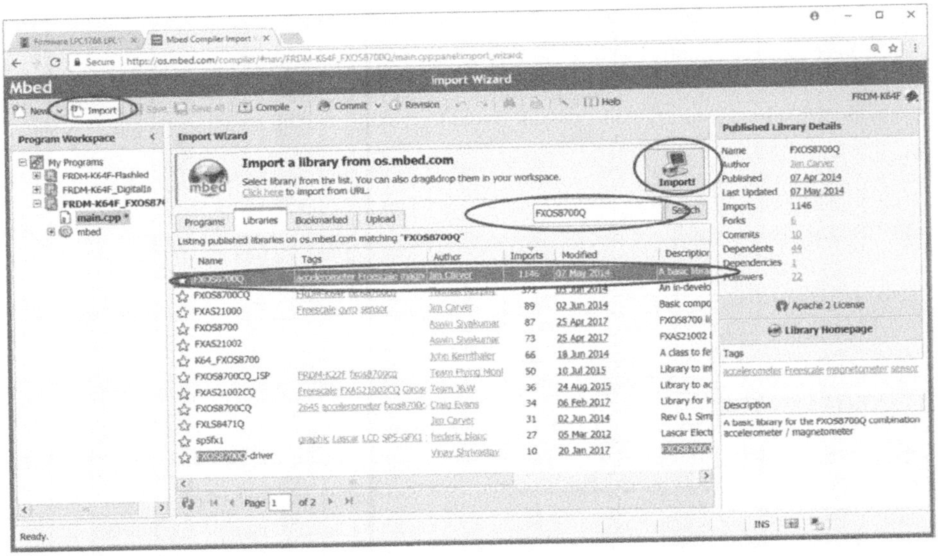

Figure 5.10 The Import library wizard in *"FRDM-K64F_FXOS8700Q"* program page.

You will need to import a library called *"FXOS8700Q"*:

https://os.mbed.com/teams/NXP/code/FXOS8700Q/

You can import it into your project by clicking the *"Import!"* button, and from the *"Libraries"* tab, search for *"FXOS8700Q"*, then click *"Import!"* button on top right, as illustrated in the screenshot in Figure 5.10.

In the *"main.cpp"*, the *"I2C i2c(PTE25, PTE24);"* defines the I2C pins, as the FXOS8700Q sensor uses I2C for communications. There will be more details about I2C next chapter. The *"FXOS8700QAccelerometer acc(i2c, FXOS8700CQ_SLAVE_ADDR1);"* create an variable *acc* to associate the onboard accelerometer with the I2C pins. The *"acc.enable();"* enable the accelerometer, and the *"acc.whoAmI();"* gives the information about the accelerometer. The *"acc.getAxis();"* gets the accelerometer values of *X*, *Y*, and *Z* axes.

The latest Arduino software (https://www.arduino.cc/en/Main/Software), version 1.6.8 or newer, has an interesting Serial Monitor tool and a Serial Plotter tool, that can display and plot the three accelerometer values we send to the serial port. Figure 5.11 shows the screenshots of the values as well as plots using Arduino software.

Apart from "*acc.getAxis(),*" you can also use "*acc.getX(),*" "*acc.getY(),*" and "*acc.getZ()*" to get the accelerometer values of the *X*, *Y*, and *Z* axes, as shown in the next example.

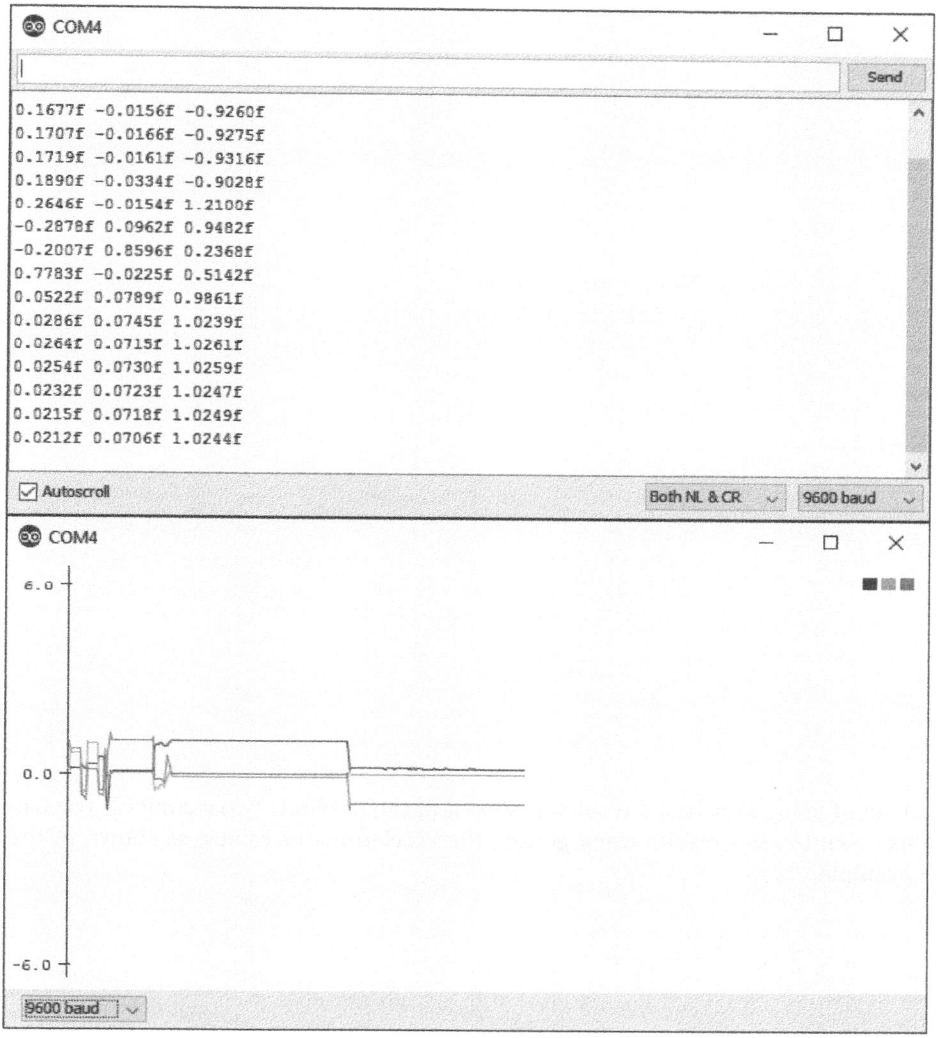

Figure 5.11 The Serial Monitor tool (top) and Serial Plotter tool (bottom) outputs in Arduino software.

```
*************************************************************
// Example 5.20

#include "mbed.h"

#include "FXOS8700Q.h"

I2C i2c(PTE25, PTE24);
FXOS8700QAccelerometer acc(i2c, FXOS8700CQ_SLAVE_ADDR1);

int main(void)
{
    motion_data_units_t acc_data;
    float faX, faY, faZ, tmp_float;

    acc.enable();
    printf("FXOS8700QAccelerometer Who Am I= %X\r\n", acc.whoAmI());
    while (true) {
        acc.getX(faX);
        acc.getY(faY);
        acc.getZ(faZ);
        printf("%1.4ff %1.4ff %1.4ff\r\n", faX, faY, faZ);
        printf("%1.4ff %1.4ff %1.4ff\r\n", acc.getX(tmp_float),
acc.getY(tmp_float), acc.getZ(tmp_float));
        wait(1.0f);
    }
}
*************************************************************
```

Instead of using unit-based results as shown in the previous two examples, you can also use count-based results when getting the accelerometer values, as shown in the next example.

```
*************************************************************
// Example 5.21

#include "mbed.h"
#include "FXOS8700Q.h"

I2C i2c(PTE25, PTE24);
FXOS8700QAccelerometer acc(i2c, FXOS8700CQ_SLAVE_ADDR1);

int main(void)
{
    motion_data_counts_t acc_raw;
    int16_t raX, raY, raZ, tmp_int;
```

```
    acc.enable();
    printf("FXOS8700QAccelerometer Who Am I= %X\r\n", acc.whoAmI());
    while (true) {
        acc.getAxis(acc_raw);
        printf("ACC: X=%06dd Y=%06dd Z=%06dd \r\n", acc_raw.x,
acc_raw.y, acc_raw.z);
        acc.getX(raX);
        acc.getY(raY);
        acc.getZ(raZ);
        printf("ACC: X=%06dd Y=%06dd Z=%06dd \r\n", raX, raY, raZ);
        printf("ACC: X=%06dd Y=%06dd Z=%06dd \r\n", acc.
getX(tmp_int), acc.getY(tmp_int), acc.getZ(tmp_int));
        wait(5.0f);
    }
}
****************************************************************
```

Similarly, the following example illustrates how to get readings from a magnetometer by using unit-based results.

```
****************************************************************
// Example 5.22

#include "mbed.h"
#include "FXOS8700Q.h"

Serial pc(USBTX, USBRX);
I2C i2c(PTE25, PTE24);
FXOS8700QMagnetometer mag(i2c, FXOS8700CQ_SLAVE_ADDR1);

int main(void)
{
    motion_data_units_t mag_data;

    mag.enable();
    printf("FXOS8700QMagnetometer Who Am I= %X\r\n", mag.whoAmI());
    while (true) {
        // unit-based results
        mag.getAxis(mag_data);
        printf("%4.1ff %4.1ff %4.1ff\r\n", mag_data.x, mag_
data.y, mag_data.z);
        wait(0.5f);
    }
}
****************************************************************
```

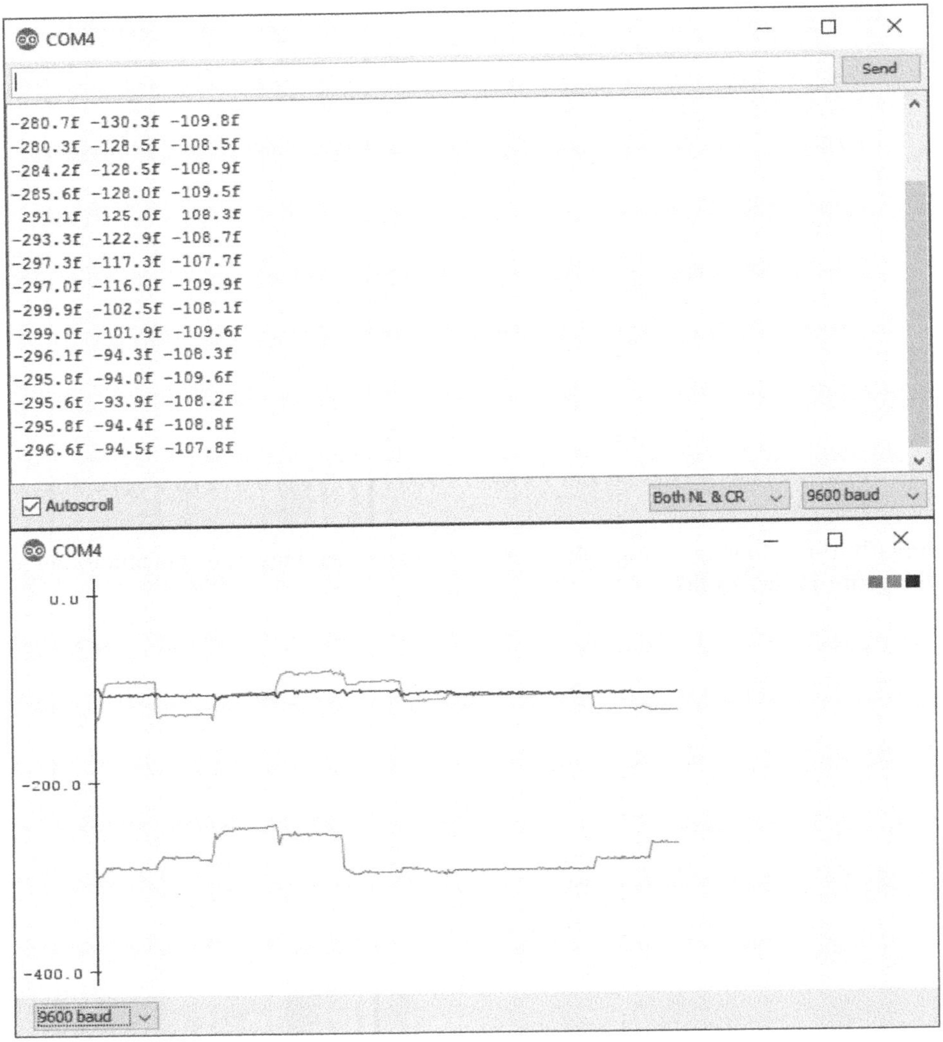

Figure 5.12 The Serial Monitor tool (top) and Serial Plotter tool (bottom) outputs in Arduino software.

Again, you can use Arduino software to show the values sent to serial port, and plot them using Serial Plotter, as illustrated Figure 5.12.

Again, apart from "***acc.getAxis(),***" you can also use "***acc.getX(),***" "***acc.getY(),***" and "***acc.getZ()***" to get the accelerometer values of *X*, *Y*, and *Z* axes, as shown in the next example.

```
**************************************************************
// Example 5.23

#include "mbed.h"
#include "FXOS8700Q.h"

Serial pc(USBTX, USBRX);
I2C i2c(PTE25, PTE24);
FXOS8700QMagnetometer mag(i2c, FXOS8700CQ_SLAVE_ADDR1);

int main(void)
{
    motion_data_units_t mag_data;
    float fmX, fmY, fmZ, tmp_float;

    mag.enable();
    printf("FXOS8700QMagnetometer Who Am I= %X\r\n", mag.whoAmI());
    while (true) {
        // unit-based results
        mag.getAxis(mag_data);
        printf("MAG: X=%4.1ff Y=%4.1ff Z=%4.1ff\r\n",
mag_data.x, mag_data.y, mag_data.z);
        mag.getX(fmX);
        mag.getY(fmY);
        mag.getZ(fmZ);
        printf("MAG: X=%4.1ff Y=%4.1ff Z=%4.1ff\r\n", fmX, fmY, fmZ);
        printf("MAG: X=%4.1ff Y=%4.1ff Z=%4.1ff\r\n", mag.
getX(tmp_float), mag.getY(tmp_float), mag.getZ(tmp_float));
        wait(5.0f);
    }
}
**************************************************************
```

Following is the same example, but using count-based results.

```
**************************************************************
// Example 5.24

#include "mbed.h"
#include "FXOS8700Q.h"

Serial pc(USBTX, USBRX);
I2C i2c(PTE25, PTE24);
FXOS8700QMagnetometer mag(i2c, FXOS8700CQ_SLAVE_ADDR1);
```

```
int main(void)
{
   motion_data_counts_t mag_raw;
   int16_t rmX, rmY, rmZ, tmp_int;

   mag.enable();
   printf("FXOS8700QMagnetometer Who Am I= %X\r\n", mag.whoAmI());
   while (true) {
      // count-based results
      mag.getAxis(mag_raw);
      printf("MAG: X=%06dd Y=%06dd Z=%06dd\r\n", mag_raw.x, mag_
raw.y, mag_raw.z);
      mag.getX(rmX);
      mag.getY(rmY);
      mag.getZ(rmZ);
      printf("MAG: X=%06dd Y=%06dd Z=%06dd\r\n", rmX, rmY, rmZ);
      printf("MAG: X=%06dd Y=%06dd Z=%06dd\r\n", mag.getX(tmp_int),
mag.getY(tmp_int), mag.getZ(tmp_int));
      wait(5.0f);
    }
}
*****************************************************************
```

Exercise 5.9

Modify the above program so that it can read both the accelerometer and the magnetometer values.

NXP LPC1768 development does not include an onboard accelerometer or magnetometer, so this section's code is not applicable to NXP LPC1768.

Further Information about Accelerometer and Magnetometer Sensors

https://os.mbed.com/teams/NXP/code/FXOS8700Q/

5.5 SD Card

FRDM-K64F board has an onboard SD card socket. To use an SD card, from the online compiler, create a new project, call it *"FRDM-K64F_SDCard,"* and change the *"main.cpp"* content as shown. You will need to import a library called *"SDFileSystem"*:
https://os.mbed.com/users/mbed_official/code/SDFileSystem/

You can import it into your project, by clicking the *"Import!"* button, and from the *"Libraries"* tab, search for "SD card," as illustrated in Figure 5.13.

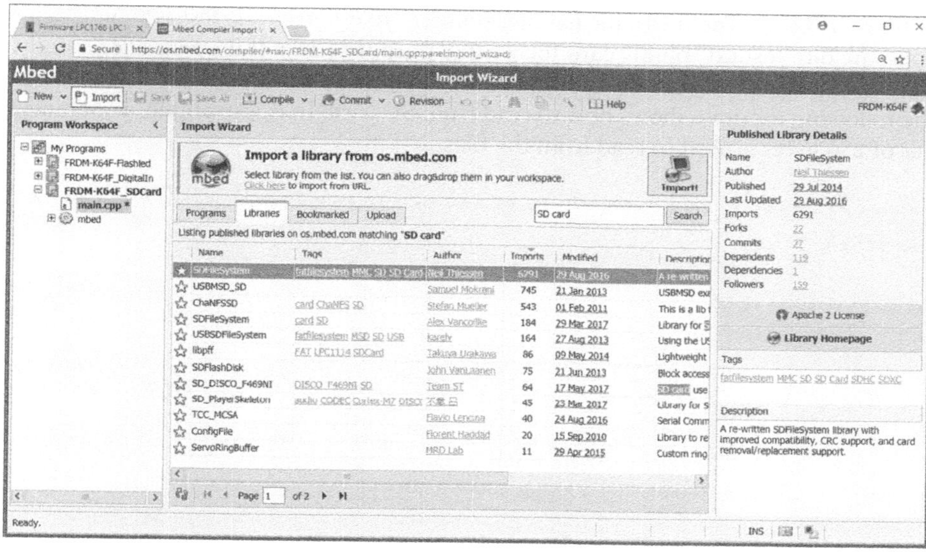

Figure 5.13 The Import Wizard for "SDFileSystem" library.

The following example shows how to write to an SD card, as shown in Figure 5.14. If you have ever written file read and write code in C Language, you will find the syntax is almost identical. The *"SDFileSystem sd(PTE3, PTE1, PTE2, PTE4, "sd");"* line specifies microcontroller pins that are connected to SD card module; in this case, they are SPI (Serial Peripheral Interface Bus) pins. We will cover SPI interface in greater detail in

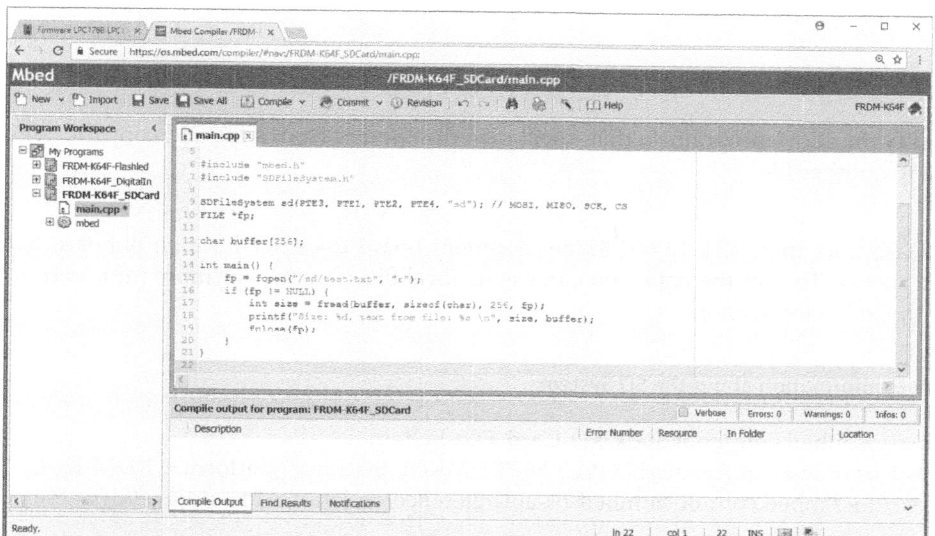

Figure 5.14 The *"FRDM-K64F_SDCard"* program page.

Chapter 6. "*FILE *fp;*" defines a file handler pointer. "*fp = fopen("/sd/test.txt", "w");*" tries to open the file "test.txt" on SD card for writing purpose ("w"). If it opens successfully, it will write "Hello World" to the file, and if not, it won't do anything.

The following example shows how to read from SD card. In this example, a character buffer of 256 bytes is used to read from the file.

```
****************************************************************
// Example 5.25

#include "mbed.h"
#include "SDFileSystem.h"

SDFileSystem sd(PTE3, PTE1, PTE2, PTE4, "sd"); // MOSI, MISO, SCK, CS
FILE *fp;

char buffer[256];

int main() {
    fp = fopen("/sd/test.txt", "r");
    if (fp != NULL) {
        int size = fread(buffer, sizeof(char), 256, fp);
        printf("Size: %d, text from file: %s \n", size, buffer);
        fclose(fp);
    }
}
****************************************************************
```

Exercise 5.10

Modify the above program so that it will read the content from a file and copy the content to another file.

Please note that NXP LPC1768 development board does not have an onboard SD card socket. To save the data, you can use its local file systems. For more information, see the following section.

Further Information about the SD System

https://os.mbed.com/cookbook/SD-Card-File-System
https://os.mbed.com/teams/NXP/code/FRDMK64_SDCard/?platform=FRDM-K64F
https://docs.mbed.com/docs/mbed-os-api-reference/en/latest/APIs/storage/filesystem/

5.6 Local File System (LPC1768)

With mbed NXP LPC1768 and LPC11U24, you can save data files to a specific area of flash memory installed on the mbed. This is the area you can see from mbed USB drive on your computer. Following is a simple example to write to a text file in the local file system and to read from it. The "LocalFileSystem local("local")" declaration defines the local file system. The "FILE* fp1 = fopen("/local/log.txt","w")" opens the "log.txt" file for writing purpose ("w"), and similarly, the "FILE* fp2 = fopen("/local/log.txt","r")" opens the file for reading purpose ("r"). The "fclose()" closes the file. The "fputs()" writes text to the file, and the "fgets()" reads from the file.

```
****************************************************************
// Example 5.26

#include "mbed.h"
Serial pc(USBTX,USBRX);
LocalFileSystem local("local");
char rs[256];
int main ()
{
    FILE* fp1 = fopen("/local/log.txt","w");
    fputs("Hello World", fp1);
    fclose(fp1);

    FILE* fp2 = fopen ("/local/log.txt","r");
    fgets(rs,256,fp2);
    fclose(fp2);
    pc.printf("text data: %s \n\r",rs);
}
****************************************************************
```

Following is an example program that reads analog input A0 (or P19 for LPC1768) for 10 times and saves it to a log file. It also uses a Timer to record the time lapsed. In this example, we use *"FILE* fp = fopen("/local/log.txt","a")"* to append to the file (*"a"*), as writing to file (*"w"*) will overwrite the existing file content.

Please be aware that when the microcontroller is reading or writing to a file, LPC1768 will be marked as "removed" on the host computer. This is normal, and it will reappear when all file handles are closed or the microcontroller program exits.

```
*************************************************************
// Example 5.27

#include "mbed.h"

#if defined(TARGET_K64F)
  AnalogIn   ain(A0);
#elif defined(TARGET_LPC1768)
  AnalogIn   ain(p19);
#endif

Timer t;
LocalFileSystem local("local");   // define local file system
int main() {
   t.start();                     // start the timer
   for(int i=0;i<10;i++)
   {
      FILE* fp = fopen ("/local/log.txt","a");
      fprintf(fp,"time=%.3fs: Ain =%.3f \n\r",t.read(),ain.read());
      fclose(fp);                 // close file
      wait(1);
   }
}
*************************************************************
```

Further Information about the Local File System

https://os.mbed.com/handbook/LocalFileSystem
https://os.mbed.com/media/uploads/robt/mbed_course_notes_-_memory_and_data.pdf

5.7 Interrupts

Interrupt is a very useful way to trigger an event according to the change of an input. Computer mouse and keyboard typically use interrupt to work. To use interrupts, from the online compiler, create a new project, call it "*FRDM-K64F_Interrupts*" and change the "*main.cpp*" content as shown. The "*InterruptIn button(sw2);*" define Switch 2 as interrupt input. The "*flip()*" function flips the LED1 on and off. The "*button.rise(&flip);*" attached the "*flip()*" function address to the rising edge of Switch 2 button. So the program will do nothing most of the time, until you press the Switch 2 button, which will switch the LED1 (red color) one and off. Because the interrupt is handled by the microcontroller automatically, so you don't need to put "*button.rise(&flip);*" into the while loop.

```
************************************************************
// Example 5.28

#include "mbed.h"

InterruptIn button(sw2);
DigitalOut led(LED1);

void flip() {
    led = !led;
}

int main() {
    button.rise(&flip);
    while(1);
}
************************************************************
```

Exercise 5.11

Modify the above program so that it uses two switches and two LEDs such that when you press Switch 2, it flips the green LED and when you press Switch 3 it flips blue LED.

Further Information about Interrupts

https://docs.mbed.com/docs/mbed-os-api-reference/en/5.1/APIs/io/InterruptIn/

5.8 Summary

This chapter introduces the inputs and outputs of FRDM-K64F development board, which includes digital inputs and outputs, bus inputs and outputs, analog inputs and outputs, PWM, six-axis combo accelerometer and magnetometer sensor, SD card, local file system (LPC1768), and interrupts.

6

Digital Interfaces

The best preparation for tomorrow is doing your best today.

- H. Jackson Brown Jr.

Digital interfaces are used by microcontrollers to communicate directly with other devices.

6.1 Serial

Serial interface is one of the most commonly used and most popular communication interfaces, due to its simplicity and easiness to use. Serial interface uses two pins to communicate, Rx and Tx, for receiving data and for transmitting data, respectively. The default setting for serial interface is, baud rate: 9600; data bits: 8; stop bits: 1; and parity: no (9600-8-N-1).

To use serial interface, from the mbed online compiler, create a new project, call it *"FRDM-K64F_Serial"*, and change the *"main.cpp"* content as shown. The *"Serial pc(USBTX, USBRX);"* specifies the microcontroller to PC serial communication using standard USBTX and USBRX pins. The *"pc.printf("Hello World\n");"* will send "Hello World" to computer using serial port.

```
*****************************************************************
// Example 6.1

#include "mbed.h"

Serial pc(USBTX, USBRX);

int main() {
    pc.printf("Hello World\n");
    while(1);
}
*****************************************************************
```

Designing Embedded Systems and the Internet of Things (IoT) with the ARM® Mbed™, First Edition. Perry Xiao

Companion website: www.wiley.com\go\xiao\designingembeddedsystemandIoTwitharmmbed

As we saw in the last chapter, the *"printf()"* also sends the information to computer serial port; therefore, the following example will work exactly same as the above code.

```
*************************************************************
// Example 6.2

#include "mbed.h"

int main() {
    printf("Hello World\n");
    while(1);
}
*************************************************************
```

You can use *"pc.getc()"* to read a character from a computer serial port, and use *"pc.putc()"* to write a character-to-computer serial port. The following example will read a character from computer and echo it back, as long as there is something readable (*pc.readable()*).

```
*************************************************************
// Example 6.3

#include "mbed.h"

Serial pc(USBTX, USBRX);

int main() {
    while(1) {
        if(pc.readable()) {
            pc.putc(pc.getc());
        }
    }
}
*************************************************************
```

Exercise 6.1

Modify the above program so that it can read lowercase characters from a computer serial port, convert them to uppercase, and echo them back.

You can also use *"gets()"* function to read a number of characters from a computer serial port. The following example reads maximally 256 characters each time. To make *"gets()"* work efficiently, it is important to configure your terminal software (Tera Term, putty.exe, Arduino serial monitor etc.) to transmit the data ending with a "*\n*" (NL, new line) character. In this way, *"gets()"* will be able to read variable length of data, as it will stop whenever it reads a "*\n*". See Chapter 5, section 5.2.2, for Tera Term terminal configuration.

```
************************************************************
// Example 6.4

#include "mbed.h"

Serial pc(USBTX, USBRX);

int main() {
    while(1) {
        if(pc.readable()) {
            char buff [256];
            pc.gets(buff, 256);
            pc.printf("%s\n\r", buff);
        }
    }
}
************************************************************
```

If the incoming data contains integer, float, or double numbers, you can also use the *"sscanf()"* function to extract the numbers out. The following example shows how to read three float numbers from a computer serial port; the numbers are separated by |.

```
************************************************************
// Example 6.5

#include "mbed.h"

Serial pc(USBTX, USBRX);

int main() {
    float re[3]={1.0,0.0,0.0};
    while(1) {
        if(pc.readable()) {
            char buff [256]="";
            pc.gets(buff, 256);
            pc.printf("%s\n\r", buff);
            sscanf (buff,"%f|%f|%f",&re[0],&re[1],&re[2]);
        }
        pc.printf("%10.3f\t%10.3f\t%10.3f\n\r", re[0],re[1],re[2]);
    }
}
************************************************************
```

Exercise 6.2

Modify the above program so that it can read three integer numbers from computer serial port; numbers are separated by " , ".

Apart from communicating with the computer, you can also communicate with other devices using a serial port. The following example uses D4 and D5 pins as a serial interface, and sends "Hello World" to the interface. In this case, the baud rate used is 115,200.

```
*************************************************************
// Example 6.6

#include "mbed.h"

Serial dev(D4, D5);          // Tx and Rx

int main() {
    dev.baud(115200);
    dev.printf("Hello World\n");
}
*************************************************************
```

Exercise 6.3

Modify the above program so that it can read characters from a computer serial port, send them to the device serial interface (D4, D5), and vice versa.

Exercise 6.4

Modify the above program so that it can communicate with another FRDM-K64F board (or LPC1768 board) through a serial interface (D4, D5).

Further Information about Serial Interface

https://docs.mbed.com/docs/mbed-os-api-reference/en/latest/APIs/interfaces/digital/Serial/

6.2 SPI

The Serial Peripheral Interface (SPI) bus is a synchronous serial communication interface specification used for short-distance communication, primarily in embedded systems. The interface was developed by Motorola and has become a de facto standard. Typical applications include Secure Digital (SD) cards and liquid crystal displays (LCD).

SPI devices communicate in full duplex mode using a master–slave architecture with a single master. The master device originates the frame for reading and writing. Multiple

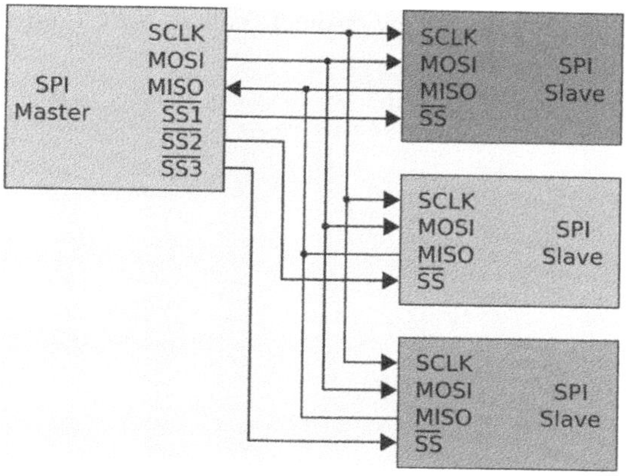

Figure 6.1 The SPI communication protocol. (*Source:* https://en.wikipedia.org/wiki/Serial_Peripheral_Interface_Bus#/media/File:SPI_three_slaves.svg)

slave devices are supported through selection with individual slave select (SS) lines, as shown in Figure 6.1.

To use an SPI interface, you will need two devices, one as the master and one as the slave. In this example, we will use a FRDM-K64F board as the master, and a LPC1768 board as the slave.

For the master, from the online compiler, create a new project; call it "*FRDM-K64F_SPI*" and change the "***main.cpp***" content as shown.

```
**************************************************************
// Example 6.7

#include "mbed.h"

SPI spi(PTD2, PTD3, PTD1); // mosi, miso, sclk
DigitalOut cs(PTD0);

int main() {
    cs = 1;

    spi.format(8,3);
    spi.frequency(1000000);

    cs = 0;

    spi.write(0x8F);

    int whoami = spi.write(0x00);
    printf("WHOAMI register = 0x%X\n", whoami);

    cs = 1;
}
**************************************************************
```

For the slave, from the online compiler, create a new project; call it *"LPC1768_SPISlave"* and change the *"main.cpp"* content as shown.

```
*************************************************************
// Example 6.8

#include "mbed.h"

SPISlave device(p5, p6, p7, p8);  // mosi, miso, sclk, ssel

int main() {
   device.reply(0x00);                  // Prime SPI with first reply
   while(1) {
        if(device.receive()) {
            int v = device.read();   // Read byte from master
            v = (v + 1) % 0x100;     // Add one to it, modulo 256
            device.reply(v);         // Make this the next reply
        }
   }
}
*************************************************************
```

Exercise 6.5

Modify the above two programs so that the SPI server can read the digital input pin D0 and send a value to SPI clients.

Further Information about SPI Interface:

https://docs.mbed.com/docs/mbed-os-api-reference/en/latest/APIs/interfaces/digital/SPI/
https://docs.mbed.com/docs/mbed-os-api-reference/en/latest/APIs/interfaces/digital/SPISlave/

6.3 I2C

The I2C (Inter-Integrated Circuit) is a multi-master, multi-slave, single-ended, serial computer bus invented by Philips Semiconductor (now NXP Semiconductors). It is typically used for attaching lower-speed peripheral ICs to processors and microcontrollers in short-distance, intra-board communication. I2C uses only two bidirectional open-drain lines, serial data acquisition (SDA) and serial clock line (SCL), pulled up with resistors. Typical voltages used are +5 V or +3.3 V, although systems with other voltages are permitted, as shown in Figure 6.2.

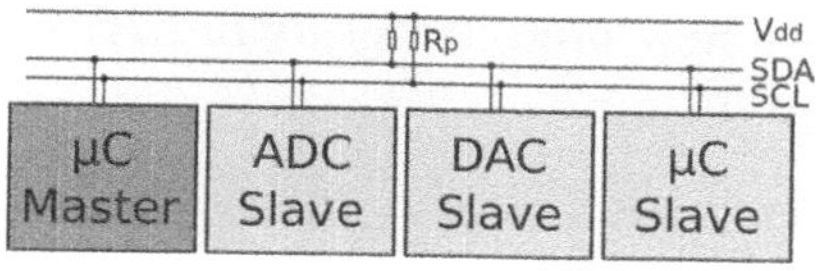

Figure 6.2 The I2C communication protocol. (*Source:* https://en.wikipedia.org/wiki/I%C2%B2C#/media/File:I2C.svg)

To use the I2C interface, you will need two devices, one as the master and one as the slave. In this example, we will use a FRDM-K64F board as the master and an LPC1768 board as the slave.

For the master, from the online compiler, create a new project; call it "*FRDM-K64F_I2C*" and change the "*main.cpp*" content as shown.

```
*************************************************************
// Example 6.9

#include "mbed.h"

// Read temperature from LM75BD

I2C i2c(PTD25, PTD24);          //SDA and SCL

const int addr = 0x90;

int main() {
    char cmd[2];
    while (1) {
        cmd[0] = 0x01;
        cmd[1] = 0x00;
        i2c.write(addr, cmd, 2);

        wait(0.5);

        cmd[0] = 0x00;
        i2c.write(addr, cmd, 1);
        i2c.read(addr, cmd, 2);

        float tmp = (float((cmd[0]<<8)|cmd[1]) / 256.0);
        printf("Temp = %.2f\n", tmp);
    }
}
*************************************************************
```

For the slave, from the online compiler, create a new project; call it "*LPC1768_SPISlave*" and change the "*main.cpp*" content as shown.

```
*************************************************************
// Example 6.10

#include <mbed.h>

I2CSlave slave(p9, p10);          //SDA and SCL

int main() {
   char buf[10];
   char msg[] = "Slave!";

   slave.address(0xA0);
   while (1) {
       int i = slave.receive();
       switch (i) {
           case I2CSlave::ReadAddressed:
             slave.write(msg, strlen(msg) + 1);  // Includes null char
             break;
           case I2CSlave::WriteGeneral:
             slave.read(buf, 10);
             printf("Read G: %s\n", buf);
             break;
           case I2CSlave::WriteAddressed:
             slave.read(buf, 10);
             printf("Read A: %s\n", buf);
             break;
       }
       for(int i = 0; i < 10; i++) buf[i] = 0;   // Clear buffer
   }
}
*************************************************************
```

Exercise 6.6

Modify the above two programs so that the I2C server can send 10 data to I2C clients.

Further Information about I2C Interface:

https://docs.mbed.com/docs/mbed-os-api-reference/en/5.1/APIs/interfaces/digital/I2C/

https://docs.mbed.com/docs/mbed-os-api-reference/en/5.1/APIs/interfaces/digital/I2CSlave/
https://docs.mbed.com/docs/mbed-os-api-reference/en/latest/APIs/interfaces/digital/I2C/
https://docs.mbed.com/docs/mbed-os-api-reference/en/latest/APIs/interfaces/digital/I2CSlave/

6.4 CAN

CAN or controller area network is a bus standard that allows microcontrollers and devices to communicate with each other without going through a host computer. It is a message-based protocol, designed originally for multiplex electrical wiring within automobiles, but is also used in many other contexts. Development of the CAN bus started in 1983 at Robert Bosch GmbH.

CAN is a multi-master serial bus standard for connecting electronic control units [ECUs] also known as nodes. Two or more nodes are required on the CAN network to communicate. The complexity of the node can range from a simple I/O device up to an embedded computer with a CAN interface and sophisticated software. The node may also be a gateway allowing a standard computer to communicate over a USB or Ethernet port to the devices on a CAN network. All nodes are connected to each other through a two-wire bus. The wires are 120 Ω nominal twisted pair, as shown in Figure 6.3.

The following example sends a counter from one CAN bus (can1) and listens for a packet on the other CAN bus (can2). Each bus controller should be connected to a CAN bus transceiver. These should be connected together at a CAN bus. In this example, the *Ticker* interface is used to set up a recurring interrupt to repeatedly call the "*send()*" function at a specified rate. More details about the Ticker interface can be found in Chapter 9, section 9.2.

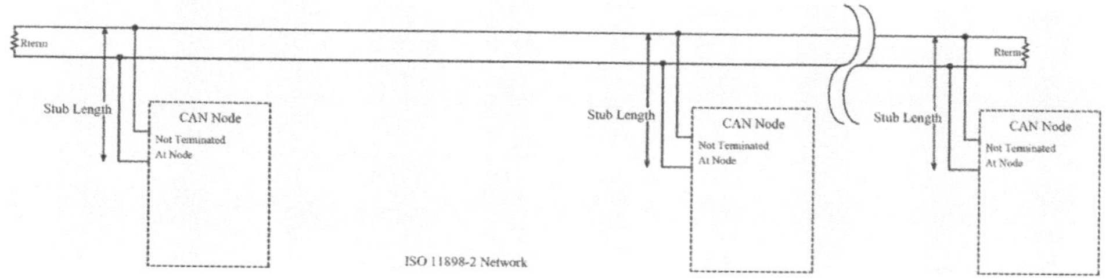

Figure 6.3 The CAN communication protocol. (*Source:* https://en.wikipedia.org/wiki/CAN_bus#/media/File:CAN_ISO11898-2_Network.png)

```
*************************************************************
// Example 6.11

#include "mbed.h"

Ticker ticker;
DigitalOut led1(LED1);
DigitalOut led2(LED2);
CAN can1(p9, p10);
CAN can2(p30, p29);
char counter = 0;

void send() {
    printf("send()\n");
    if(can1.write(CANMessage(1337, &counter;, 1))) {
        printf("wloop()\n");
        counter++;
        printf("Message sent: %d\n", counter);
    }
    led1 = !led1;
}

int main() {
    printf("main()\n");
    ticker.attach(&send;, 1);
    CANMessage msg;

    while(1) {
        printf("loop()\n");
        if(can2.read(msg)) {
            printf("Message received: %d\n", msg.data[0]);
            led2 = !led2;
        }
        wait(0.2);
    }
}
*************************************************************
```

Exercise 6.7

Modify the above program so that it reads the analog pin A0, and sends the value to CAN1.

Further Information about the CAN Interface:

https://docs.mbed.com/docs/mbed-os-api-reference/en/latest/APIs/interfaces/digital/CAN/

6.5 Summary

This chapter introduces digital interfaces, such as serial, SPI, I2C, and CAN, which are used by microcontrollers to communicate directly with other devices.

7

Networking and Communications

I never did a day's work in my life. It was all fun.

- Thomas A. Edison

7.1 Ethernet

The FRDM-K64F development board comes with an onboard Ethernet socket. So the simplest way to connect to the Internet is through the Ethernet.

From the Arm® Mbed™ online compiler, create a new program *"FRDM-F64F_NetworkInfo"* (Figure 7.1). Copy the following code into *"main.cpp."*

```
*****************************************************************
// Example 7.1

#include "mbed.h"
#include "EthernetInterface.h"
#include "rtos.h"

EthernetInterface eth;

int main()
{
    eth.init();
    eth.connect();
    printf(" IP address: %s \r\n",eth.getIPAddress());
    printf(" Network Mask: %s \r\n",eth.getNetworkMask());
    printf(" MAC address: %s \r\n",eth.getMACAddress());
    printf(" Gateway address: %s \r\n",eth.getGateway());

    while(1){};
}
*****************************************************************
```

Designing Embedded Systems and the Internet of Things (IoT) with the ARM® Mbed™, First Edition. Perry Xiao

Companion website: www.wiley.com\go\xiao\designingembeddedsystemandIoTwitharmmbed

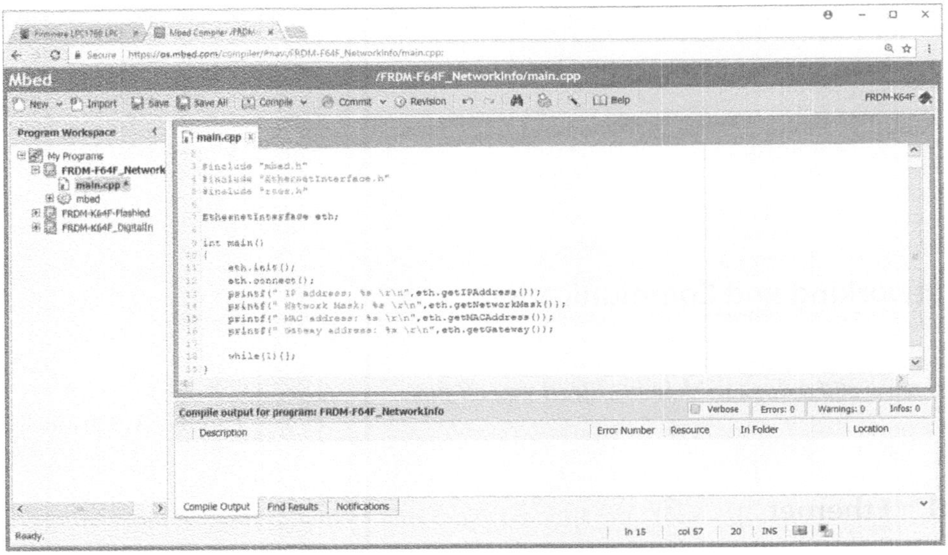

Figure 7.1 The EthernetInterface program.

This example illustrates how to initialize the Ethernet, connect the Ethernet, and get network information such as IP address, subnet mask, MAC address, and gateway address.

In this program, you will need to import two libraries:

1) "*EthernetInterface*" library (https://os.mbed.com/users/mbed_official/code/EthernetInterface/)
2) "*mbed-rtos*" library (https://os.mbed.com/users/mbed_official/code/mbed-rtos/)

To import a library into a program, just click the "*Import!*" button on top of the online compiler. Search for "*EthernetInterface*" library and click the "*Import!*" button (Figure 7.2). A pop-up confirmation window will appear. Make sure all the information is correct, then click the "*Import!*" button (Figure 7.3). Figure 7.4 shows how to search and import "*mbed-rtos*" library. More details about import and export libraries and programs will be available in Chapter 10.

Please note that as of this writing, there are two compilation errors in the latest "*EthernetInterface*" library (revision 54:183490eb1b4a, 14 Jan 2017), as shown in Figure 7.5.

Just comment out the two lines to solve the errors (Figure 7.6).

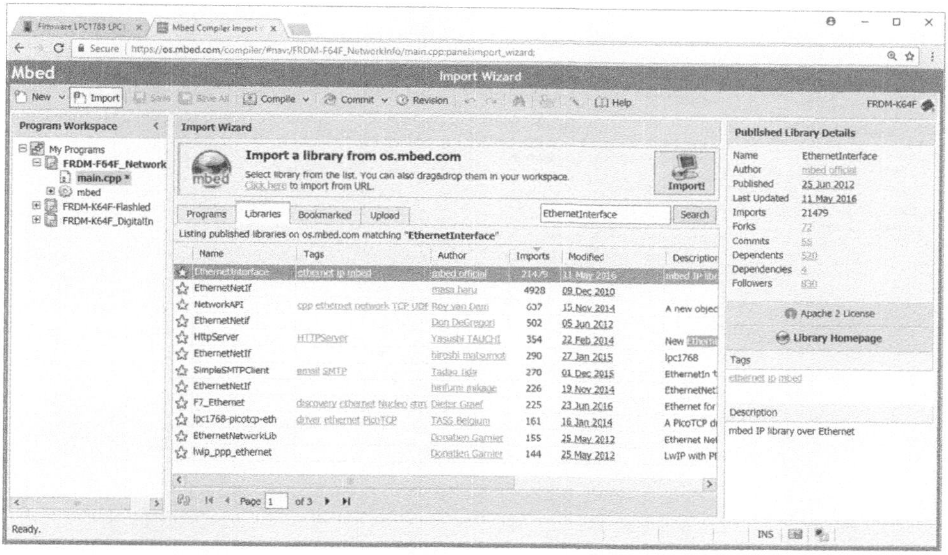

Figure 7.2 Import "*EthernetInterface*" library into the program.

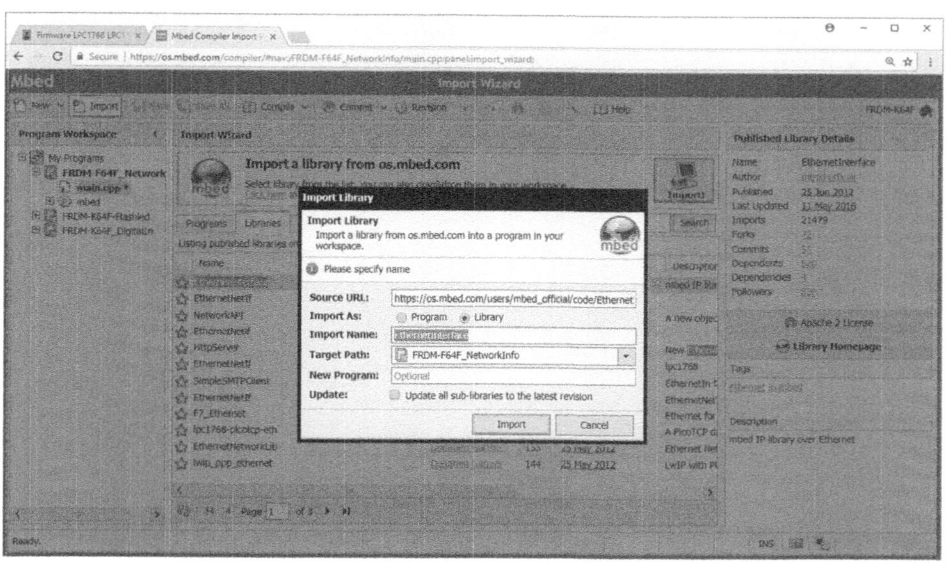

Figure 7.3 Import Library pop-up window.

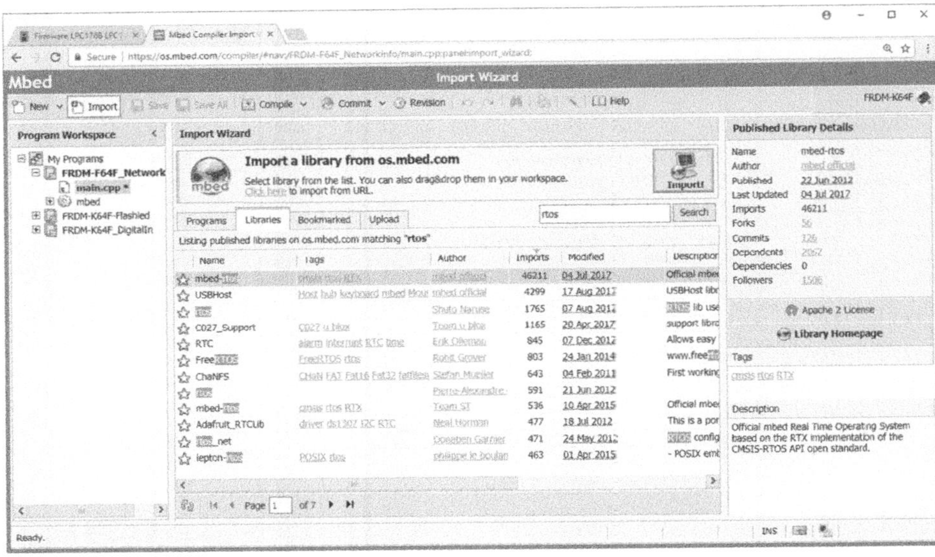

Figure 7.4 Import "*mbed-rtos*" library into the program.

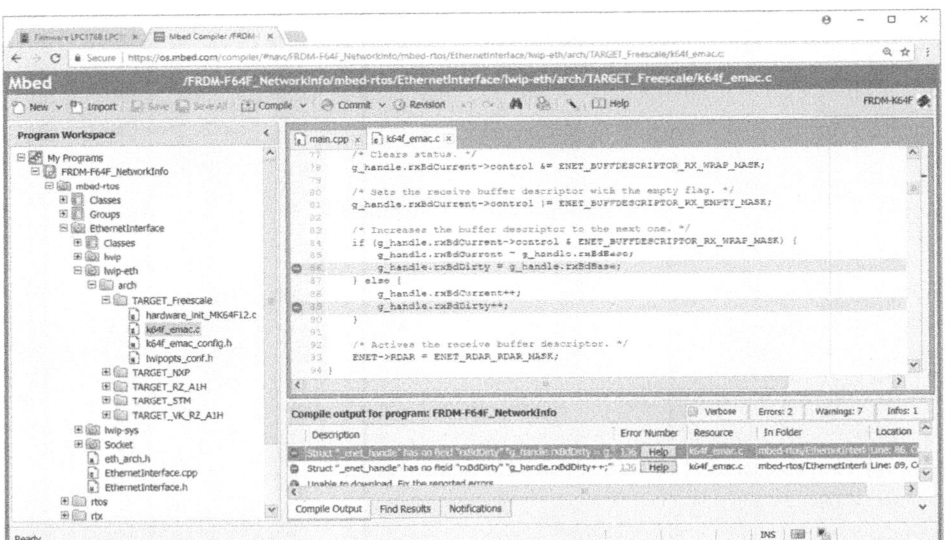

Figure 7.5 The compilation errors in the "*EthernetInterface*" library.

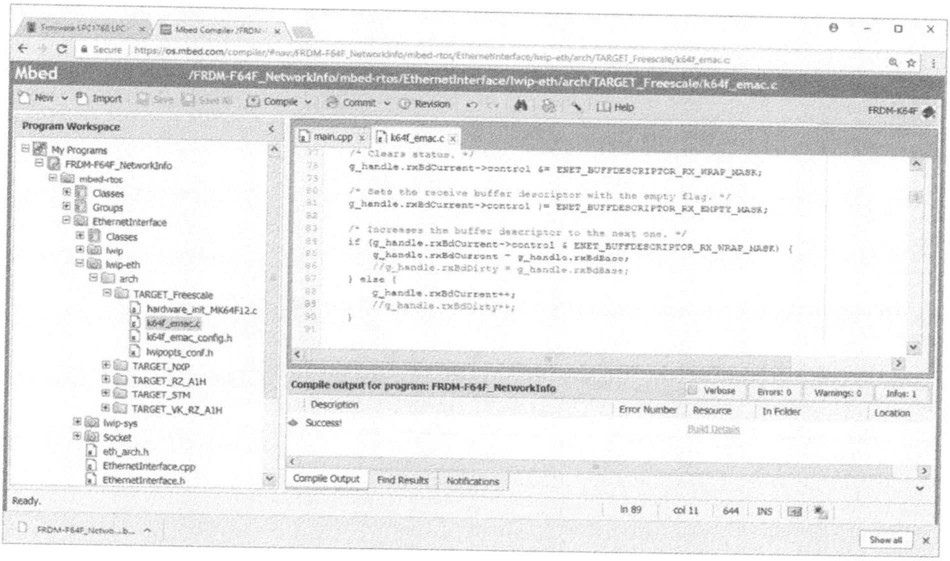

Figure 7.6 The correction of compilation errors by commenting out the two lines in the "*EthernetInterface*" library.

Further Information about Ethernet

https://docs.mbed.com/docs/mbed-os-api-reference/en/latest/APIs/communication/ethernet/

7.2 Ethernet Web Client and Web Server

World Wide Web is still the most important application on the Internet. With an Ethernet socket, it is easy to turn an FRDM-K64F board into a web client or a web server. Following is a quick example of a simple Ethernet HTTP (hyper text transfer protocol) client, i.e., web client, program. It connects to the website www.google.co.uk, then retrieves the web page information. Again, you will need both the "*EthernetInterface*" library and "*rmbed-rtos*" library.

```
***************************************************************
// Example 7.2

#include "mbed.h"
#include "EthernetInterface.h"

EthernetInterface eth;
TCPSocketConnection sock;

int main() {
  eth.init();
  eth.connect();
  printf("IP Address is %s\n", eth.getIPAddress());

  sock.connect("www.google.co.uk", 80);

  char request[] = "GET / HTTP/1.0\n\n";
  sock.send_all(request, sizeof(request)-1);

  char buffer[1024];
  int ret;
  while (true) {
      ret = sock.receive(buffer, sizeof(buffer)-1);
      if (ret <= 0)
          break;
      buffer[ret] = '\0';
      printf("Received %d chars from server:\n%s\n", ret, buffer);
  }

  sock.close();

  eth.disconnect();

  while(1) {}
}
***************************************************************
```

Following is a simple HTTP server, i.e., web server, which reads the analog input A0 (or P19 for LPC1768) and prints its value to the client. Figure 7.7 shows the web browser display of the server.

```
***************************************************************
// Example 7.3

#include "mbed.h"
#include "EthernetInterface.h"

#define PORT   80
EthernetInterface eth;
TCPSocketServer server;
TCPSocketConnection client;

#if defined(TARGET_K64F)
    AnalogIn   ain(A0);
#elif defined(TARGET_LPC1768)
    AnalogIn   ain(p19);
#endif

int main()
{
  eth.init();
  eth.connect();
  printf(" IP address: %s \r\n",eth.getIPAddress());

  server.bind(PORT);
  server.listen();

  while(true){
      int32_t status = server.accept(client);

      if (status>=0)
      {
          char msg[1024] = {};
          sprintf(msg,"A0 = %0.1f \n\r\n\r", (float) ain.read());
          client.send(msg,strlen(msg));
          client.close();
      }
   }
}
***************************************************************
```

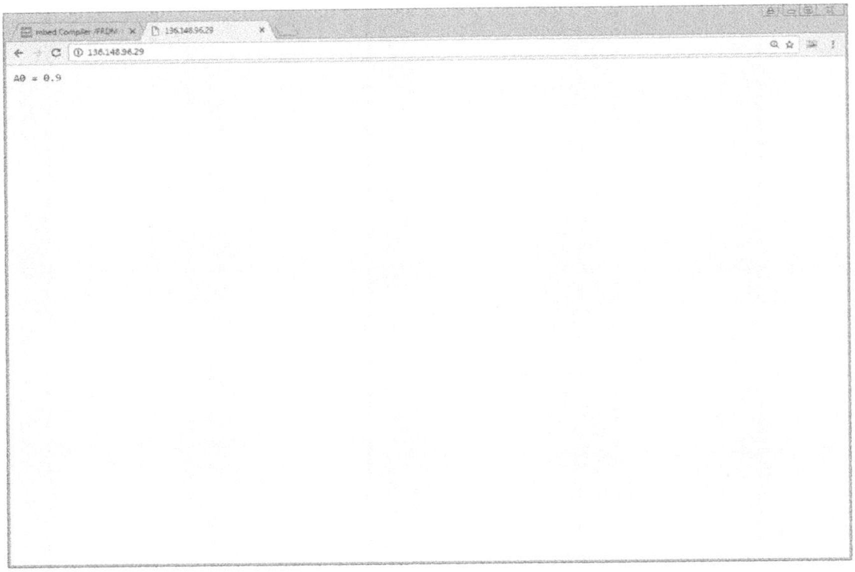

Figure 7.7 The web browser display.

Following is a slightly complicated version HTTP server, which uses a function "*web_server()*" to provide web information. The "*web_server()*" function sets up the HTTP server, accepts the HTTP client connection, prints the data sent by the client, and prints the analog input A0 value (or P19 for LPC1768) in a HTML format, including header and body, to the client.

```
*****************************************************************
// Example 7.4

#include "mbed.h"
#include "EthernetInterface.h"
#include <stdio.h>
#include <string.h>
#include "rtos.h"

#define PORT   80

EthernetInterface eth;
TCPSocketServer server;
TCPSocketConnection client;

#if defined(TARGET_K64F)
    AnalogIn   ain(A0);
#elif defined(TARGET_LPC1768)
    AnalogIn   ain(p19);
#endif
```

```
void web_server(void const *args)
{
    server.bind(PORT);
    server.listen();

    while(true){
        int32_t status = server.accept(client);

        if (status>=0)
        {
            char buffer[1024] = {};
            int n= client.receive(buffer, 1023);
            printf("Received Data:
%d\n\r\n\r%.*s\n\r",strlen(buffer),strlen(buffer),buffer);
            char Body[1024] = {};
            sprintf(Body,"<html><title></title><body><h1>A0=%0.1f
</h1></body></html>\n\r\n\r", (float) ain.read());
            char Header[256] = {};
            sprintf(Header,"HTTP/1.1 200 OK\n\rContent-Length:
%d\n\rContent-Type: text/html\n\rConnection:
Keep-Alive\n\r\n\r",strlen(Body));
            client.send(Header,strlen(Header));
            client.send(Body,strlen(Body));

            client.close();
        }
    }
}
int main() {
    EthernetInterface eth;
    eth.init();
    eth.connect();
    printf("\r\nServer IP Address is %s\r\n", eth.getIPAddress());

    web_server("");
    while(1){}
}
****************************************************************
```

Exercise 7.1

Modify the above program so that it can read the analog inputs A0, A1, A2 and display the values as a table in HTTP body message.

Further Information about the Web Client and Server

https://os.mbed.com/cookbook/HTTP-Serverhttps://os.mbed.com/cookbook/Networking

7.3 TCP Socket and UDP Socket

With Arm® Mbed™ you can also provide simple and consistent communications using TCP (File Transfer Protocol) and UDP (User Datagram Protocol) sockets. Communications using TCP sockets are connection oriented, more reliable, but more complex and slower. Communications using UDP is connectionless, and therefore much simpler, faster, but less reliable.

The following example is a simple TCP socket server. It receives data from a TCP client and echoes it back. Again, you will need both "*EthernetInterface*" library and "*rmbed-rtos*" library.

```
***************************************************************
// Example 7.5

#include "mbed.h"
#include "EthernetInterface.h"

#define PORT   7

EthernetInterface eth;
TCPSocketServer server;
TCPSocketConnection client;

int main (void) {
    eth.init();
    eth.connect();
    printf("\nServer IP Address is %s\n", eth.getIPAddress());

    server.bind(PORT);
    server.listen();

    while (true) {
        server.accept(client);
        client.set_blocking(false, 1500); // Timeout after (1.5)s

        printf("Connection from: %s\n", client.get_address());
        char buffer[256];
        while (true) {
            int n = client.receive(buffer, sizeof(buffer));
            if (n <= 0) break;

            // print received message to terminal
            buffer[n] = '\0';
            printf("Received message from Client :'%s'\n",buffer);

            // Echo received message back to client
            client.send_all(buffer, n);
            if (n <= 0) break;
```

```
        }

        client.close();
    }
}
************************************************************
```

Following is a corresponding TCP Echo Client program. In this code, you will need to change the server IP address "x.x.x.x" to correct server address.

```
************************************************************
// Example 7.6

#include "mbed.h"
#include "EthernetInterface.h"

const char* SERVER = "x.x.x.x";
const int PORT = 7;

EthernetInterface eth;
TCPSocketConnection socket;

int main() {
    eth.init();
    eth.connect();
    printf("\nClient IP Address is %s\n", eth.getIPAddress());

    while (socket.connect(SERVER, PORT) < 0) {
        wait(1);
    }
    printf("Connected to Server at %s\n",SERVER);

    // Send message to server
    char msg[] = "Hello World";
    socket.send_all(msg, sizeof(msg) - 1);

    // Receive message from server
    char buff[256];
    int n = socket.receive(buf, 256);
    buff[n] = '\0';
    printf("Received message from server: '%s'\n", buff);

    socket.close();
    eth.disconnect();

    while(true) {}
}
************************************************************
```

Exercise 7.2

Modify the above TCP client/server programs so that the server receives the message from the client, changes it to uppercase, then echoes it back to the client.

The following example is a simple UDP Echo Server; again, it receives data from a UDP client and echoes it back. It also will need both "*EthernetInterface*" library and "*rmbed-rtos*" library.

```
****************************************************************
// Example 7.7

#include "mbed.h"
#include "EthernetInterface.h"

#define PORT   7

EthernetInterface eth;
UDPSocket server;
Endpoint client;

int main (void) {
    eth.init();
    eth.connect();
    printf("\nServer IP Address is %s\n", eth.getIPAddress());

    server.bind(PORT);

    char buffer[256];
    while (true) {
        printf("\nWaiting for UDP packet...\n");
        int n = server.receiveFrom(client, buffer, sizeof(buffer));
        buffer[n] = '\0';

        server.sendTo(client, buffer, n);
    }
}
****************************************************************
```

Following is the corresponding UDP Echo Client. Again, in this code, you will need to change the server IP address "x.x.x.x" to the correct server address.

```
****************************************************************
// Example 7.8

#include "mbed.h"
#include "EthernetInterface.h"

const char* SERVER = "x.x.x.x";
const int PORT = 7;

EthernetInterface eth;
UDPSocket sock;
Endpoint echo_server;

int main() {
    eth.init();
    eth.connect();

    sock.init();
    echo_server.set_address(SERVER, PORT);

    char msg[] = "Hello World";
    sock.sendTo(echo_server, msg, sizeof(msg));

    char buffer[256];
    int n = sock.receiveFrom(echo_server, buffer, sizeof(buffer));

    buffer[n] = '\0';
    printf("Received message from server: '%s'\n", buffer);

    sock.close();
    eth.disconnect();
    while(1) {}
}
****************************************************************
```

Exercise 7.3

Modify the above UDP client/server programs so that the server reads the digital pin D0 and sends the value to the client.

Further Information about the Socket

https://docs.mbed.com/docs/mbed-os-api-reference/en/latest/APIs/communication/network_sockets/

7.4 WebSocket

The WebSocket provides full-duplex, bidirectional communications between a Web server and Web clients. Following is a WebSocket example code that simply sends a "Hello World" message every 2 seconds to a WebSocket echo server (ws://echo.websocket.org).

In this program, you will need to import three libraries:

1) "*EthernetInterface*" library (https://os.mbed.com/users/mbed_official/code/EthernetInterface/)
2) "*mbed-rtos*" library (https://os.mbed.com/users/mbed_official/code/mbed-rtos/)
3) "*WebSocketClient*" library (https://os.mbed.com/users/samux/code/WebSocketClient/).

```
*************************************************************
// Example 7.9

#include "mbed.h"
#include "EthernetInterface.h"
#include "Websocket.h"

EthernetInterface eth;

int main() {
    eth.init();
    eth.connect();
    printf("IP Address is %s\n\r", eth.getIPAddress());

    Websocket ws("ws://echo.websocket.org");
    ws.connect();

    while (1) {
        ws.send("Hello World");
        wait(2.0);
    }
}
*************************************************************
```

Exercise 7.4

Modify the above program so that it can continuously read the temperature sensor values and send it to the WebSocket server.

Following is a revised WebSocket example code that sends one "Hello World" message to a WebSocket echo server (ws://echo.websocket.org) and gets the echoed message back.

```
***********************************************************
// Example 7.10

#include "mbed.h"
#include "EthernetInterface.h"
#include "Websocket.h"

EthernetInterface eth;

int main()
{

    eth.init();
    eth.connect();

    printf("IP Address: %s\n", eth.getIPAddress());

    Websocket ws("wss://echo.websocket.org");
    ws.connect();

    char str[100];
    sprintf(str, "Hello World");
    ws.send(str);

    memset(str, 0, 100);
    wait(1.0f);

    if (ws.read(str)) {
        printf("rcv'd: %s\n", str);
    }

    ws.close();
    eth.disconnect();

    while(true);
}
***********************************************************
```

You can also build your own WebSocket server using Python or Java, as illustrated in the following Arm® Mbed™ Cookbook site (Figures 7.8 to 7.10).

https://os.mbed.com/cookbook/Websockets-Server

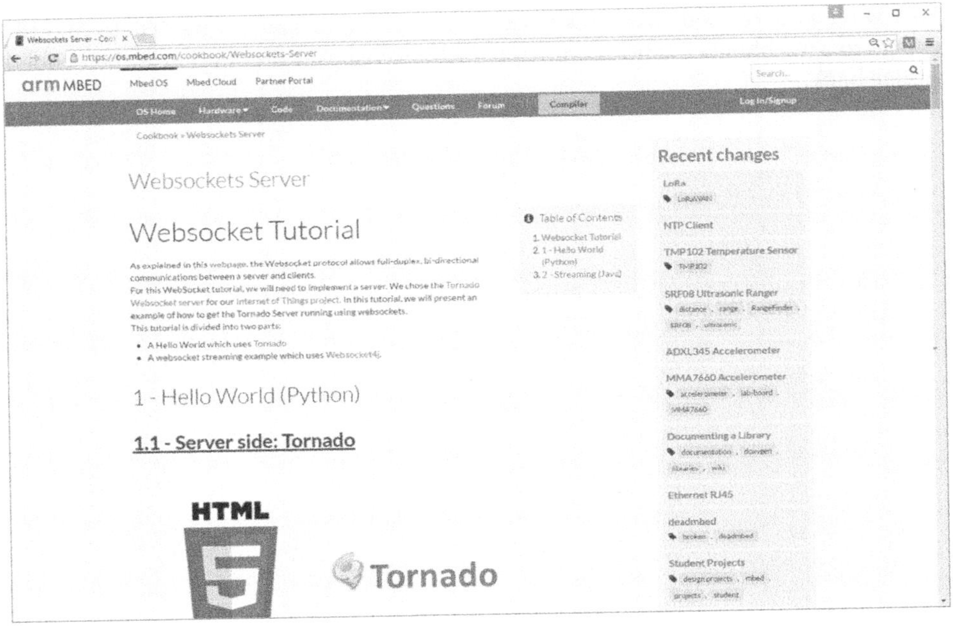

Figure 7.8 The WebSocket Server page on mbed Cookbook website.

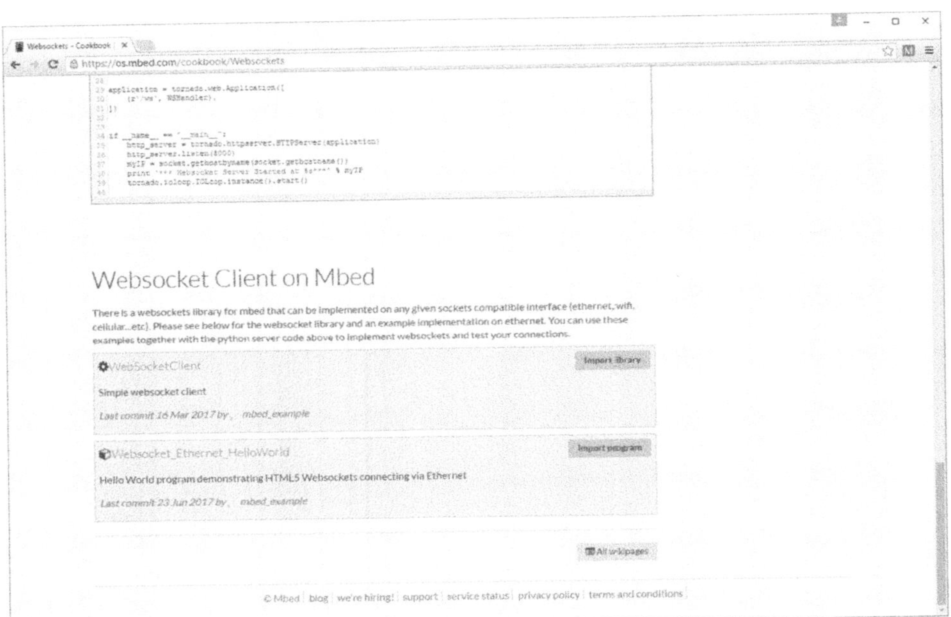

Figure 7.9 The WebSocket Client section on mbed Cookbook website.

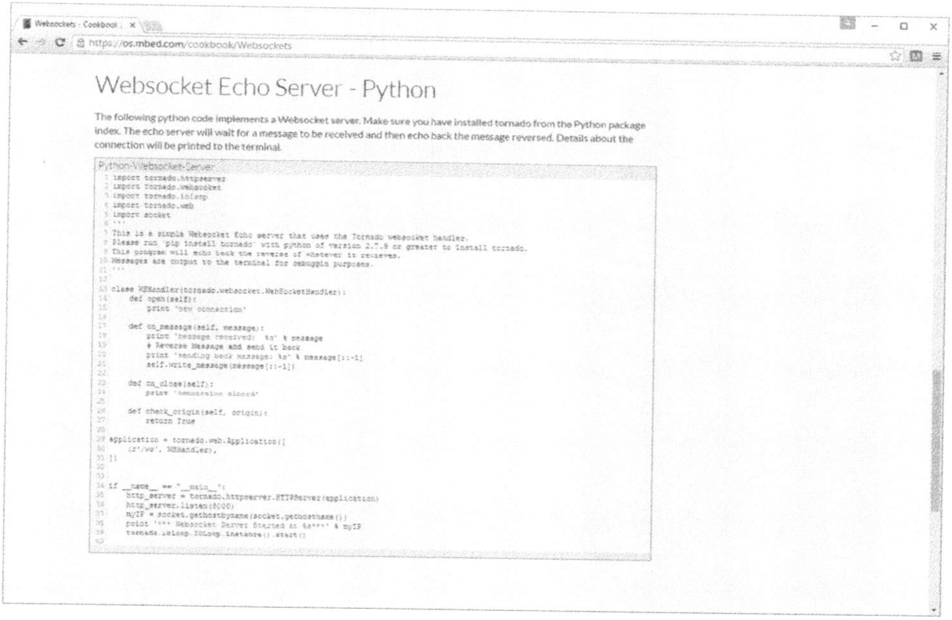

Figure 7.10 The WebSocket Echo Server -Python section on mbed Cookbook website.

Further Information about the WebSocket

https://os.mbed.com/cookbook/Websockets
https://os.mbed.com/components/HTML5-Websockets/
https://os.mbed.com/cookbook/Websockets-Server

7.5 WiFi

WiFi is another way to get the FRDM-K64F board connected to the Internet. In this example, we will use the ESP8266 WiFi module. Figure 7.11 shows the pin connections.

Here is a simple WiFi program, which connects to an access point and displays the IP address received. In this program, you will need to replace the ("*xxx*", "*ppp*") with your own access point username and password. The code should work for both FRDM-K64F and LPC 1768 boards. You will also need to import the "*ESP8266*" library:

https://os.mbed.com/users/quevedo/code/ESP8266/

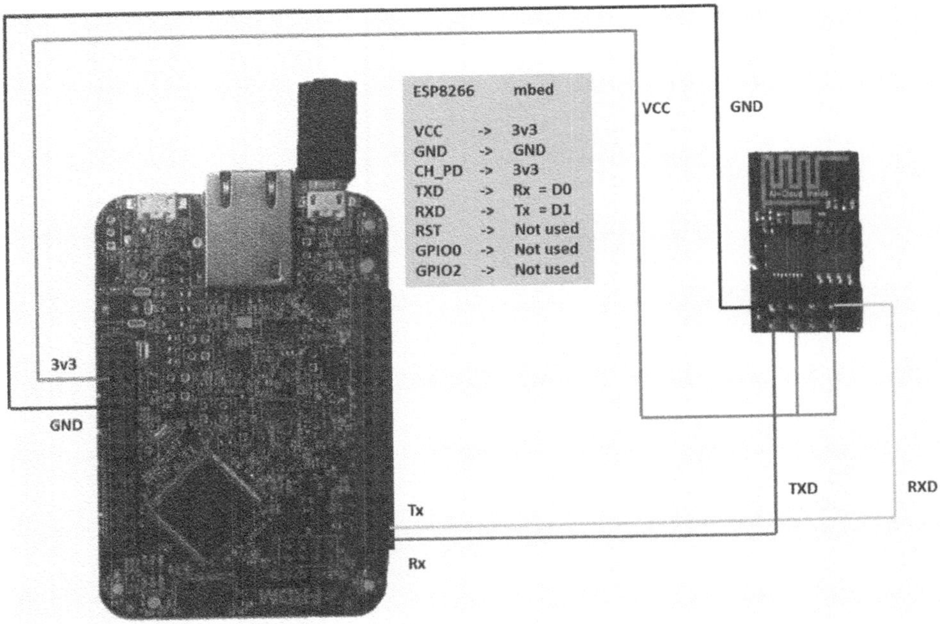

Figure 7.11 The schematic circuit diagram of the FRDM-K64F board and WiFi module (ESP8266).

```
*****************************************************************
// Example 7.11

#include "mbed.h"
#include "ESP8266.h"

Serial pc(USBTX,USBRX);
#if defined(TARGET_K64F)
    ESP8266 wifi(PTC17, PTC16, 115200); // baud rate for wifi
#elif defined(TARGET_LPC1768)
    ESP8266 wifi(P9, P10, 115200); // baud rate for wifi
#endif

char snd[255],rcv[1000];

int main () {

   pc.baud(115200);
   pc.printf("SET mode to AP\r\n");
   wifi.SetMode(1);     // set ESP mode to 1
   wifi.RcvReply(rcv, 1000);     //receive a response from ESP
   pc.printf("%s",rcv);     //Print the response onscreen
   pc.printf("Connecting to AP\r\n");
```

```
    wifi.Join("xxx", "ppp");      // Your wifi username & Password
    wifi.RcvReply(rcv, 1000);     //receive a response from ESP
    pc.printf("%s", rcv);     //Print the response onscreen
    wait(8);      //wait for response from ESP
    pc.printf("Getting IP\r\n");   //get IP address from the connected AP
    wifi.GetIP(rcv);     //receive an IP address from the AP
    pc.printf("%s", rcv);

    while (1) {

    }//While
} //main
*****************************************************************
```

Following is an improved version of the program, which brings up the ESP8266 WiFi network interface, and connects to a website:

```
*****************************************************************
// Example 7.12

#include "mbed.h"
#include "ESP8266.h"
#include "EthernetInterface.h"
#include "HTTPClient.h"

Serial pc(USBTX,USBRX);
#if defined(TARGET_K64F)
    ESP8266 wifi(PTC17, PTC16, 115200); // baud rate for wifi
#elif defined(TARGET_LPC1768)
    ESP8266 wifi(P9, P10, 115200); // baud rate for wifi
#endif

char snd[255],rcv[1000];

void getpage(void);
HTTPClient http;

int main () {

    pc.baud(115200);
    pc.printf("SET mode to AP\r\n");
    wifi.SetMode(1);     // set ESP mode to 1
    wifi.RcvReply(rcv, 1000);     //receive a response from ESP
    pc.printf("%s",rcv);     //Print the response onscreen
    pc.printf("Connecting to AP\r\n");
    wifi.Join("xxx", "ppp");      // Your wifi username & Password
```

```
    wifi.RcvReply(rcv, 1000);     //receive a response from ESP
    pc.printf("%s", rcv);     //Print the response onscreen
    wait(8);      //wait for response from ESP
    pc.printf("Getting IP\r\n");  //get IP address from the connected AP
    wifi.GetIP(rcv);     //receive an IP address from the AP
    pc.printf("%s", rcv);
    getpage();

    while (1) {

    }//While
} //main

void getpage()
{
    TCPSocketConnection sock;
    sock.connect("www.google.co.uk", 80);

    char request[] = "GET / HTTP/1.0\n\n";
    sock.send_all(request, sizeof(request)-1);

    char buffer[1024];
    int ret;
    while (true) {
        ret = sock.receive(buffer, sizeof(buffer)-1);
        if (ret <= 0)
            break;
        buffer[ret] = '\0';
        printf("Received %d chars from server:\n%s\n", ret, buffer);
    }

    sock.close();
}
*****************************************************************
```

Exercise 7.5

Modify the above program so that it can send "Hello World" to a WebSocket server.

Further Information about WiFi

https://docs.mbed.com/docs/mbed-os-api-reference/en/latest/APIs/communication/wifi/
https://os.mbed.com/users/4180_1/notebook/using-the-esp8266-with-the-mbed-lpc1768/
https://os.mbed.com/teams/ESP8266/code/mbed-os-example-esp8266/

https://github.com/armmbed/esp8266-driver/
https://os.mbed.com/teams/ESP8266/code/ESP8266_MQTT_HelloWorld/

7.6 Summary

This chapter introduces the networking and communication facilities including the Ethernet, web client, web server, TCP and UDP socket, WebSockets, and WiFi.

8

Digital Signal Processing and Control

I have not failed. I've just found 10,000 ways that won't work.

- Thomas A. Edison

Signal processing is important for many applications. With the power of modern computers, many signal processing functions can now be done digitally. In this chapter we will illustrate how to use the Arm® Mbed™-DSP library (https://developer.mbed.org/users/mbed_official/code/mbed-dsp/) for digital signal processing and control.

8.1 Low-Pass Filter

On the Arm® Mbed™ website, there is an excellent tutorial on how to design and implement a low-pass FIR (finite impulse response) filter. We will basically follow the example and extend it to high-pass filter and band-pass/stop filter.

https://os.mbed.com/handbook/Matlab-FIR-Filter

First, we need to use MATLAB software (www.mathworks.com) to create a digital filter. Digital filter design is a complex topic as it involves complicated math. MATLAB has a Signal Processing Toolbox that can make digital filter design much simpler. FIR (finite impulse response) filters and IIR (infinite impulse response) filters are the commonly used digital filters. The FIR filter is used here as it requires no feedback loop and is more stable.

Following is the MATLAB code (modified from https://os.mbed.com/handbook/Matlab-FIR-Filter) that creates a low-pass filter using *"fir1"* function. The sampling rate is 48,000 Hz, Nyquist frequency is half of the sampling frequency, 24,000 Hz, and cutoff frequency is 6000 Hz. The *"fir1"* function creates a 28th order digital filter in a normalized frequency range (0 to 1), with 1 representing the Nyquist frequency, i.e., 24,000 Hz. The normalized cutoff frequency will therefore be 6000 / 24,000 = –¼, or 0.25.

Designing Embedded Systems and the Internet of Things (IoT) with the ARM® Mbed™, First Edition. Perry Xiao

Companion website: www.wiley.com\go\xiao\designingembeddedsystemandIoTwitharmmbed

```
**************************************************************
%Example 8.1
%Modified from https://os.mbed.com/handbook/Matlab-FIR-Filter

sample_rate = 48000;

% Choose filter cutoff frequency (6 kHz)
cutoff_hz = 6000;

% Normalize cutoff frequency (wrt Nyquist frequency)
nyq_freq = sample_rate / 2;
cutoff_norm = cutoff_hz / nyq_freq;

% FIR filter order (i.e., number of coefficients - 1)
order = 28;

% Create lowpass FIR filter
fir_coeff = fir1(order, cutoff_norm);

% Analyze the filter using the Filter Visualization Tool
fvtool(fir_coeff, 'Fs', sample_rate)
**************************************************************
```

Figure 8.1 shows the FIR low-pass filter and its 29 (order +1) coefficients. We can now use these coefficients in mbed program to implement a low-pass digital filter.

```
fir_coeff =
-0.0018 -0.0016 0.0000  0.0037  0.0081  0.0085 -0.0000 -0.0174
-0.0341 -0.0334 0.0000  0.0676  0.1522  0.2229  0.2505  0.2229
 0.1522  0.0676 0.0000 -0.0334 -0.0341 -0.0174 -0.0000  0.0085
 0.0081  0.0037  0.0000 -0.0016 -0.0018
```

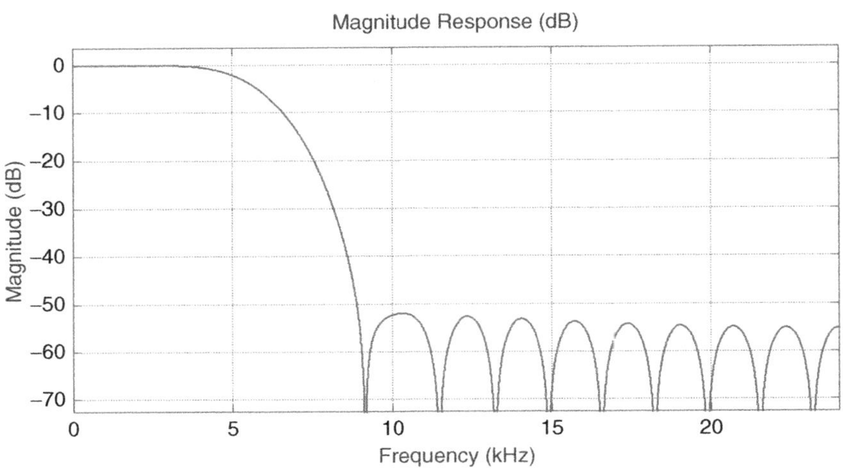

Figure 8.1 The FIR low-pass filter's coefficients (top) and the plot (bottom). Cutoff frequency is 6000 Hz, and Nyquest frequency is 24,000 Hz.

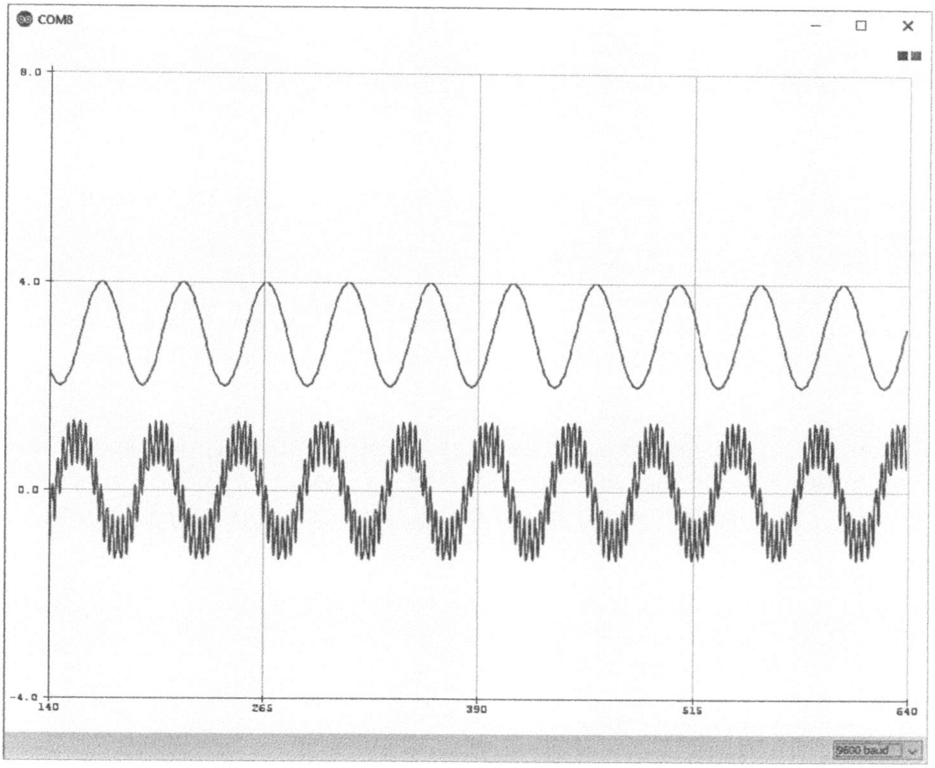

Figure 8.2 The output of the program using Arduino Serial Plotter, with the original mixed signal at the bottom and filtered signal at the top.

Following is the mbed example that uses above FIR low-pass filter coefficients. It first generates a mixed signal (32 × 20 points) using 1000 Hz sinusoid and 15,000 Hz sinusoid, then uses a FIR low-pass filter to filter out the 15,000 Hz. Finally, it prints both the original signal and filtered signal to a computer through a virtual COM port. In Figure 8.2, the filtered signal was shifted 3 V upward so that we can view the two signals separately.

In this program, you will need the mbed-DSP library:
https://os.mbed.com/users/mbed_official/code/mbed-dsp/

```
*******************************************************************
//Example 8.2
//Modified from https://os.mbed.com/handbook/Matlab-FIR-Filter

#include "mbed.h"
#include "dsp.h"

#define BLOCK_SIZE              (32)
#define NUM_BLOCKS              (20)
#define TEST_LENGTH_SAMPLES     (BLOCK_SIZE * NUM_BLOCKS)
```

```
#define SAMPLE_RATE                    (48000)

float32_t expected_output[TEST_LENGTH_SAMPLES];
float32_t           output[TEST_LENGTH_SAMPLES];

#define NUM_TAPS                29
/* FIR Coefficients buffer generated using fir1() MATLAB function:
fir1(28, 6/24) */
//Low-pass filter coefficients
const float32_t firCoeffs32[NUM_TAPS] = {
 -0.0018225230f, -0.0015879294f, +0.0000000000f, +0.0036977508f, +0.0080754303f,
 +0.0085302217f, -0.0000000000f, -0.0173976984f, -0.0341458607f, -0.0333591565f,
 +0.0000000000f, +0.0676308395f, +0.1522061835f, +0.2229246956f, +0.2504960933f,
 +0.2229246956f, +0.1522061835f, +0.0676308395f, +0.0000000000f, -0.0333591565f,
 -0.0341458607f, -0.0173976984f, -0.0000000000f, +0.0085302217f, +0.0080754303f,
 +0.0036977508f, +0.0000000000f, -0.0015879294f, -C.0018225230f
};

int main() {
    Sine_f32 sine_1KHz(  1000, SAMPLE_RATE, 1.0);
    Sine_f32 sine_15KHz(15000, SAMPLE_RATE, 0.5);
    FIR_f32<NUM_TAPS> fir(firCoeffs32);

    float32_t buffer_a[BLOCK_SIZE];
    float32_t buffer_b[BLOCK_SIZE];
    for (float32_t *sgn=output; sgn<(output+TEST_LENGTH_SAMPLES);
sgn += BLOCK_SIZE) {
      sine_1KHz.generate(buffer_a);          // Generate a 1KHz sine wave
      sine_15KHz.process(buffer_a, buffer_b);  // Add a 15KHz sine wave
      fir.process(buffer_b, sgn);        // FIR low-pass filter: 6KHz cutoff
      for (int i=0;i<BLOCK_SIZE;i++)
      {
printf("%0.3f\t%0.3f\t%0.3f\n\r",buffer_a[i],(buffer_b[i]+3),(sgn[i]+6));
        }
    }

}
*************************************************************************
```

Figure 8.2 shows the output of the program using Arduino Serial Plotter, with the original mixed signal at the bottom and the filtered signal at the top. As we can see, after filtering, only the 1000 Hz signal remains.

8.2 High-Pass Filter

Similarly, we can also implement the high-pass filter. All you need to do is to modify Example 8.1 MATLAB code, and change "*fir1*" function line from:

```
fir_coeff = fir1(order, cutoff_norm);
```

to

```
fir_coeff = fir1(order, cutoff_norm, 'high');
```

Figure 8.3 shows the corresponding FIR high-pass filter and its 29 (order +1) coefficients. We can now use these coefficients in mbed program to implement a low-pass digital filter.

```
fir_coeff =
0.0018  0.0016  -0.0000  -0.0037  -0.0080  -0.0085  -0.0000  0.0173
0.0340  0.0332  -0.0000  -0.0674  -0.1516  -0.2221  0.7487  -0.2221
-0.1516  -0.0674  -0.0000  0.0332  0.0340  0.0173  -0.0000  -0.0085
-0.0080  -0.0037  -0.0000  0.0016  0.0018
```

Modify the Example 8.2, and change the FIR coefficients using the new values, as shown below.

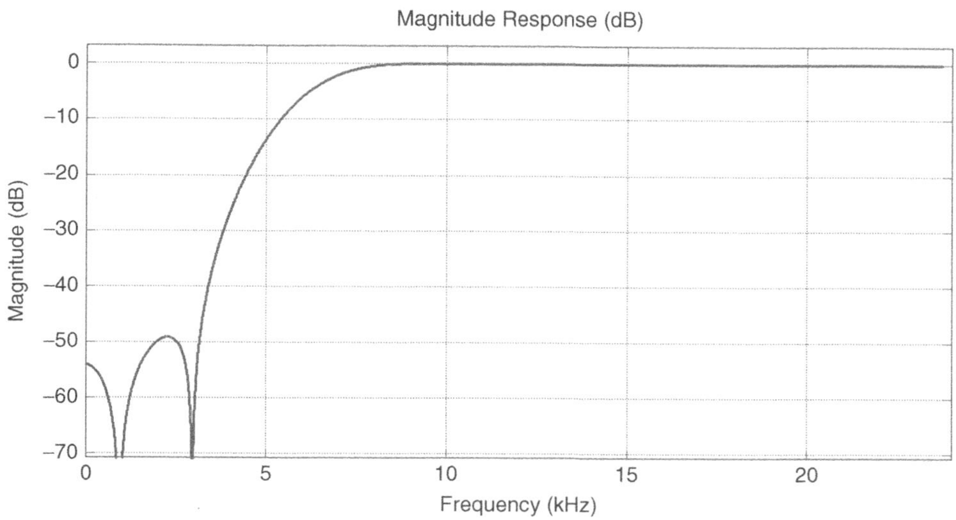

Figure 8.3 The FIR high-pass filter's coefficients (top) and the plot (bottom). Cutoff frequency is 6000 Hz, and Nyquest frequency is 24,000 Hz.

```
*****************************************************************
//Example 8.3
//Modified from https://os.mbed.com/handbook/Matlab-FIR-Filter

#include "mbed.h"
#include "dsp.h"

#define BLOCK_SIZE                (32)
#define NUM_BLOCKS                (20)
#define TEST_LENGTH_SAMPLES       (BLOCK_SIZE * NUM_BLOCKS)

#define SAMPLE_RATE               (48000)

float32_t expected_output[TEST_LENGTH_SAMPLES];
float32_t          output[TEST_LENGTH_SAMPLES];

#define NUM_TAPS              29
/* FIR Coefficients buffer generated using fir1() MATLAB function:
fir1(28, 6/24,'high') */
//high-pass filter coefficients
const float32_t firCoeffs32[NUM_TAPS] = {
  0.0018f,   0.0016f,   -0.0000f,  -0.0037f,  -0.0080f,  -0.0085f,
  -0.0000f,  0.0173f,   0.0340f,   0.0332f,   -0.0000f,  -0.0674f,
  -0.1516f,  -0.2221f,  0.7487f,   -0.2221f,  -0.1516f,  -0.0674f,
  -0.0000f,  0.0332f,   0.0340f,   0.0173f,   -0.0000f,  -0.0085f,
  -0.0080f,  -0.0037f,  -0.0000f,  0.0016f,   0.0018f
};

int main() {
   Sine_f32 sine_1KHz(  1000, SAMPLE_RATE, 1.0);
   Sine_f32 sine_15KHz(15000, SAMPLE_RATE, 0.5);
   FIR_f32<NUM_TAPS> fir(firCoeffs32);

   float32_t buffer_a[BLOCK_SIZE];
   float32_t buffer_b[BLOCK_SIZE];
   for(float32_t *sgn=output; sgn<(output+TEST_LENGTH_SAMPLES); sgn +=
BLOCK_SIZE)
  {
     sine_1KHz.generate(buffer_a);         // Generate a 1KHz sine wave
     sine_15KHz.process(buffer_a, buffer_b); // Add a 15KHz sine wave
     fir.process(buffer_b, sgn);      // FIR lcw-pass filter: 6KHz cutoff
     for (int i=0;i<BLOCK_SIZE;i++)
     {
       printf("%0.3f\t%0.3f\t%0.3f\n\r",buffer_a[i],(buffer_b[i]+3),
(sgn[i]+6));
     }
  }

}
*****************************************************************
```

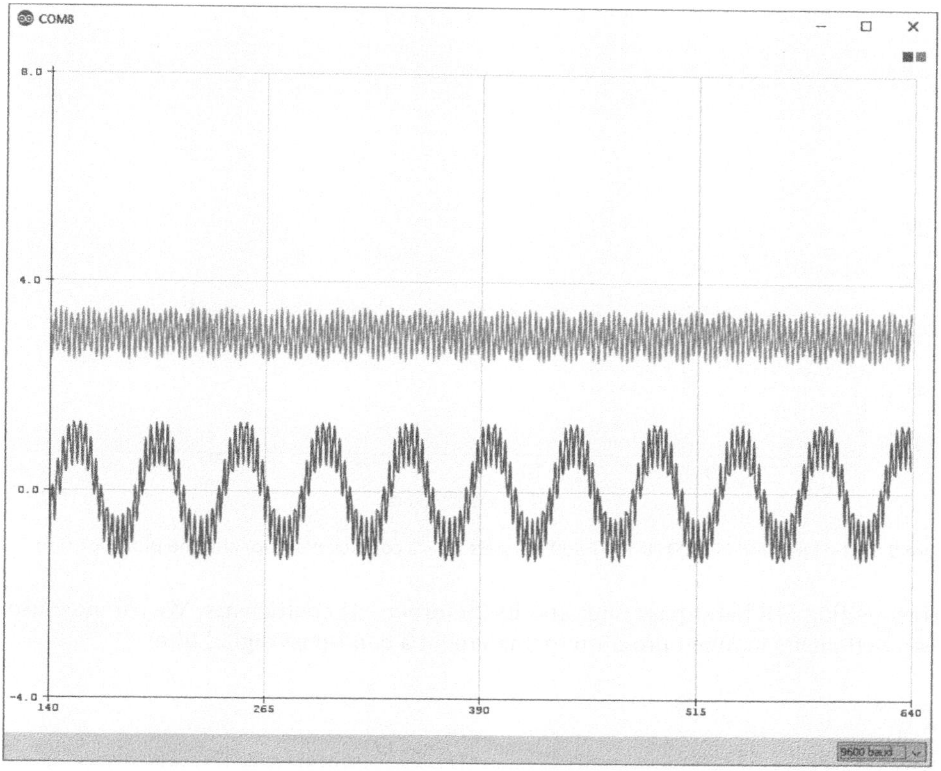

Figure 8.4 The output of the program using Arduino Serial Plotter, with original mixed signal at the bottom and filtered signal at the top.

Figure 8.4 shows the output of the program using Arduino Serial Plotter, with the original mixed signal at the bottom and filtered signal at the top. As we can see this time, after filtering, only 15,000 Hz signal remained.

8.3 Band-Pass Filter

For the band-pass filter, modify Example 8.1 MATLAB code and change "*fir1*" function line from:

```
fir_coeff = fir1(order, cutoff_norm);
```

to

```
fir_coeff = fir1(order, [0.5 0.7]);
```

In this case, only the frequencies between 0.5 × 24,000 (12,000 Hz) to 0.7 × 24,000 (16,800 Hz) are allowed to pass; other frequencies are blocked. Figure 8.5 shows the

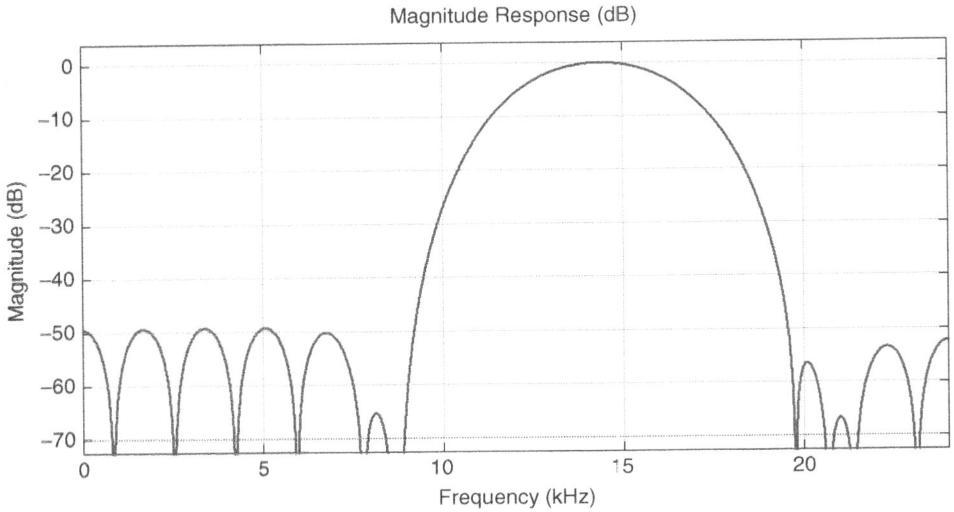

Figure 8.5 The FIR band (12000 Hz to 16800 Hz) pass filter's coefficients (top) and the plot (bottom).

corresponding FIR band-pass filter and its 29 (order +1) coefficients. We can now use these coefficients in mbed program to implement a band-pass digital filter.

```
fir_coeff =
-0.0011   -0.0030   0.0033    0.0010    0.0000   -0.0024   -0.0171    0.0332
0.0207   -0.0974    0.0400    0.1292   -0.1494   -0.0622    0.2069   -0.0622
-0.1494   0.1292    0.0400   -0.0974    0.0207    0.0332   -0.0171   -0.0024
0.0000    0.0010    0.0033   -0.0030   -0.0011
```

Modify Example 8.2 and change the FIR coefficients using the new values, as shown below.

```
*****************************************************************
//Example 8.4
//Modified from https://os.mbed.com/handbook/Matlab-FIR-Filter

#include "mbed.h"
#include "dsp.h"

#define BLOCK_SIZE             (32)
#define NUM_BLOCKS             (20)
#define TEST_LENGTH_SAMPLES    (BLOCK_SIZE * NUM_BLOCKS)
```

```
#define SAMPLE_RATE                    (48000)


float32_t expected_output[TEST_LENGTH_SAMPLES];
float32_t          output[TEST_LENGTH_SAMPLES];

#define NUM_TAPS               29
/* FIR Coefficients buffer generated using fir1() MATLAB function:
fir1(28, [0.5 0.7]) */
//band-pass filter coefficients
const float32_t firCoeffs32[NUM_TAPS] = {
-0.0011f, -0.0030f, 0.0033f, 0.0010f, 0.0000f, -0.0024f, -0.0171f, 0.0332f,
0.0207f, -0.0974f, 0.0400f, 0.1292f, -0.1494f, -0.0622f, 0.2069f, -0.0622f,
-0.1494f, 0.1292f, 0.0400f, -0.0974f, 0.0207f, 0.0332f, -0.0171f, -0.0024f,
0.0000f,  0.0010f,  0.0033f,  -0.0030f,  -0.0011f,
};
int main() {
  Sine_f32 sine_1KHz(  1000, SAMPLE_RATE, 1.0);
  Sine_f32 sine_15KHz(15000, SAMPLE_RATE, 0.5);
  FIR_f32<NUM_TAPS> fir(firCoeffs32);

  float32_t buffer_a[BLOCK_SIZE];
  float32_t buffer_b[BLOCK_SIZE];
  for(float32_t *sgn=output; sgn<(output+TEST_LENGTH_SAMPLES); sgn
+= BLOCK_SIZE)
  {
     sine_1KHz.generate(buffer_a);        // Generate a 1KHz sine wave
     sine_15KHz.process(buffer_a, buffer_b); // Add a 15KHz sine wave
     fir.process(buffer_b, sgn);    // FIR low-pass filter: 6KHz cutoff
     for (int i=0;i<BLOCK_SIZE;i++)
     {
       printf("%0.3f\t%0.3f\t%0.3f\n\r",buffer_a[i],(buffer_b[i]+3),
(sgn[i]+6));
     }
  }

}
*************************************************************
```

Figure 8.6 shows the output of the program using Arduino Serial Plotter, with the original mixed signal at the bottom and filtered signal at the top. As we can see this time, after the band-pass filtering, only 15,000 Hz signal remained.

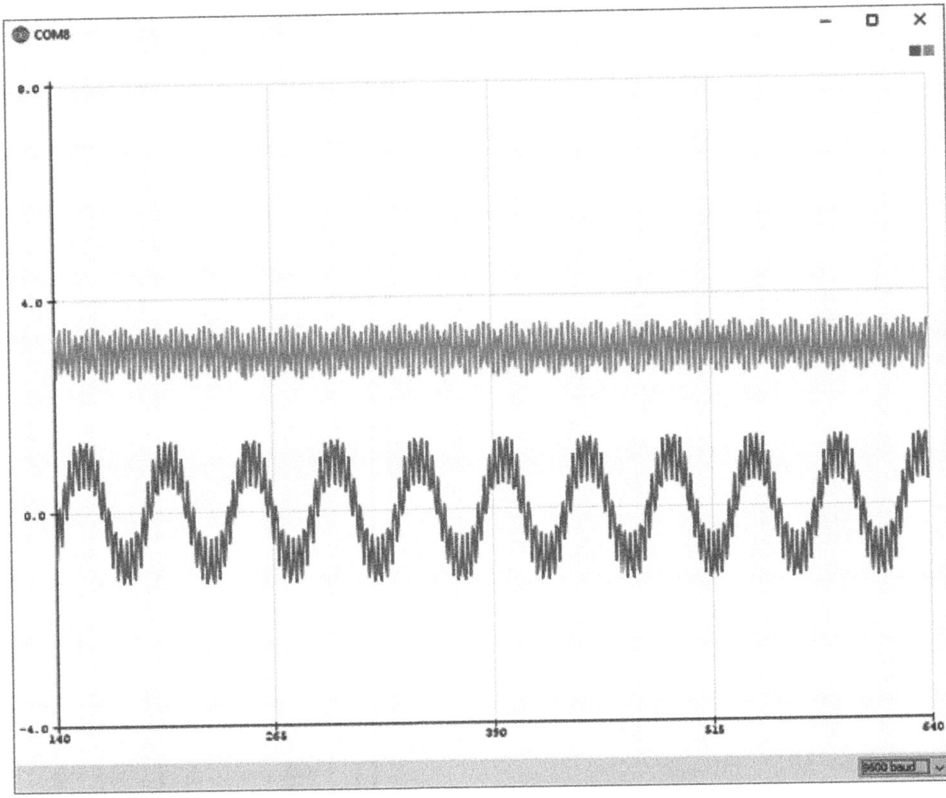

Figure 8.6 The output of the program using Arduino Serial Plotter, with the original mixed signal at the bottom and filtered signal at the top.

8.4 Band-Stop Filter and Notch Filter

For the band-stop filter, modify the Example 8.1 MATLAB code, and change *"fir1"* function line from:

```
fir_coeff = fir1(order, cutoff_norm);
```

to

```
fir_coeff = fir1(order, [0.5 0.7],'stop');
```

In this case, the frequencies between 0.5 × 24,000 (12,000 Hz) to 0.7 × 24000 (16,800 Hz) will be blocked and other frequencies are allowed. When the band becomes narrow enough, band-stop filter will become notch filter. Figure 8.7 shows

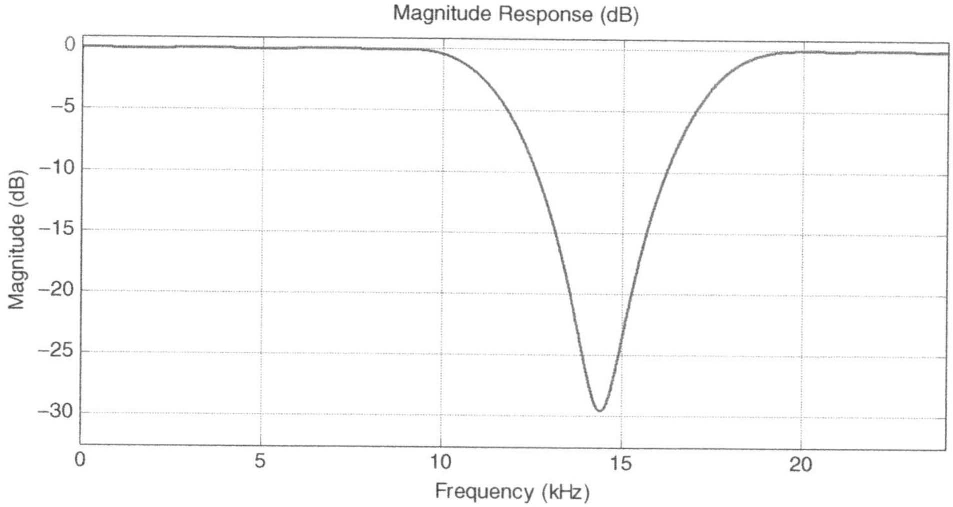

Figure 8.7 The FIR band (12,000 Hz to 16,800 Hz) stop filter's coefficients (top) and the plot (bottom).

the corresponding FIR band-stop filter and its 29 coefficients, which can be used in mbed program to implement a band-stop digital filter.

```
fir_coeff =
0.0011  0.0029  -0.0032  -0.0010  -0.0000  0.0023  0.0165  -0.0320
-0.0200  0.0939  -0.0385  -0.1245  0.1440  0.0599  0.7974  0.0599
0.1440  -0.1245  -0.0385  0.0939  -0.0200  -0.0320  0.0165  0.0023
-0.0000  -0.0010  -0.0032  0.0029  0.0011
```

Modify the Example 8.2, and change the FIR coefficients using the new values, as shown below.

```
*****************************************************************
//Example 8.5
//Modified from https://os.mbed.com/handbook/Matlab-FIR-Filter
#include "mbed.h"
#include "dsp.h"

#define BLOCK_SIZE              (32)
#define NUM_BLOCKS              (20)
#define TEST_LENGTH_SAMPLES     (BLOCK_SIZE * NUM_BLOCKS)

#define SAMPLE_RATE             (48000)
```

```
float32_t expected_output[TEST_LENGTH_SAMPLES];
float32_t            output[TEST_LENGTH_SAMPLES];

#define NUM_TAPS              29
/* FIR Coefficients buffer generated using fir1() MATLAB function:
fir1(28, [0.5 0.7],'stop'); */
//band-stop filter coefficients
const float32_t firCoeffs32[NUM_TAPS] = {
0.0011f, 0.0029f, -0.0032f, -0.0010f, -0.0000f, 0.0023f, 0.0165f, -0.0320f,
-0.0200f, 0.0939f, -0.0385f, -0.1245f, 0.1440f, 0.0599f, 0.7974f, 0.0599f,
0.1440f, -0.1245f, -0.0385f, 0.0939f, -0.0200f, -0.0320f, 0.0165f, 0.0023f,
-0.0000f, -0.0010f, -0.0032f, 0.0029f,  0.0011f,
};
int main() {
   Sine_f32 sine_1KHz(  1000, SAMPLE_RATE, 1.0);
   Sine_f32 sine_15KHz(15000, SAMPLE_RATE, 0.5);
   FIR_f32<NUM_TAPS> fir(firCoeffs32);

   float32_t buffer_a[BLOCK_SIZE];
   float32_t buffer_b[BLOCK_SIZE];
   for(float32_t *sgn=output; sgn<(output+TEST_LENGTH_SAMPLES); sgn
   += BLOCK_SIZE)
  {
    sine_1KHz.generate(buffer_a);      // Generate a 1KHz sine wave
    sine_15KHz.process(buffer_a, buffer_b); // Add a 15KHz sine wave
    fir.process(buffer_b, sgn);    // FIR low-pass filter: 6KHz cutoff
    for (int i=0;i<BLOCK_SIZE;i++)
    {
      printf("%0.3f\t%0.3f\t%0.3f\n\r",buffer_a[i],(buffer_b[i]+3),
(sgn[i]+6));
    }
  }

}
***********************************************************
```

Figure 8.8 shows the output of the program using Arduino Serial Plotter, with original mixed signal at the bottom and filtered signal at the top. As we can see, after filtering, 15,000 Hz signal was stopped, and only 1000 Hz signal remained.

Further Information about the Digital Filter:

http://uk.mathworks.com/help/signal/ref/fir1.html
https://en.wikipedia.org/wiki/Digital_filter

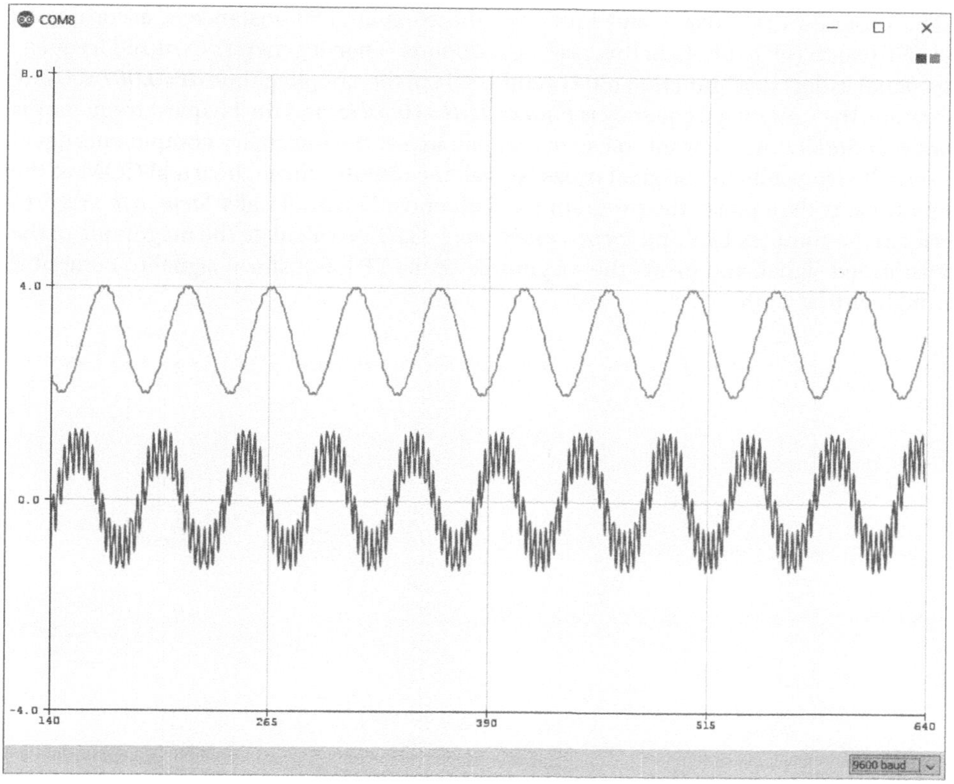

Figure 8.8 The output of the program using Arduino Serial Plotter, with original mixed signal at the bottom and filtered signal at the top.

8.5 Fast Fourier Transform (FFT)

Fast Fourier Transform (FFT) and inverse FFT have many important applications. In this section, we will show how to perform an FFT and inverse FFT using the mbed-DSP library:

https://os.mbed.com/users/mbed_official/code/mbed-dsp/

The following example illustrates how to use *"arm_cfft_f32()"* to perform complex FFT. The *"arm_cfft_f32()"* function can only be used for data lengths of [16, 32, 64, ..., 4096], but can be use both for FFT and inverse FFT. Check the above mbed-DSP library for the details of the function.

```
arm_cfft_f32(S, samples, 0, 1);                //FFT
arm_cfft_f32(S, samples, 1, 1);                //Inverse FFT
```

The program first creates and initializes the complex FFT instance *S*, according to the FFT length (FFT_LEN), in this case, is 512 points. Then it generates a mixed frequencies signal using "*sin()*" function (30 Hz and 100 Hz), the sampling time ***dt=0.001*** second; therefore, the sampling frequency is ***Fmax=1/dt=1000Hz***, and the Nyquest frequency is ***Fmax/2=500Hz***. As we want to use real signal, we set the imaginary components equal to zero. It also prints the original mixed signal to computer through virtual COM at the same time. It then pause the program for 5 seconds. Finally, it calls "*arm_cfft_f32()*" to perform the complex FFT and "*arm_cmplx_mag_f32()*" to calculate the magnitude of the transformed signal, and prints the magnitude of the FFT transform signal to computer through virtual COM.

```
*****************************************************************
//Example 8.6

#include "mbed.h"
#include "arm_const_structs.h"

const int FFT_LEN   = 512;

const static arm_cfft_instance_f32 *S;

float samples[FFT_LEN*2];
float magnitudes[FFT_LEN];

int main()
{
  int32_t i = 0;

  // Init arm_ccft_32
  switch (FFT_LEN)
  {
  case 16:
      S = & arm_cfft_sR_f32_len16;
      break;
  case 32:
      S = & arm_cfft_sR_f32_len32;
      break;
  case 64:
      S = & arm_cfft_sR_f32_len64;
      break;
  case 128:
      S = & arm_cfft_sR_f32_len128;
      break;
  case 256:
      S = & arm_cfft_sR_f32_len256;
      break;
```

```
    case 512:
        S = & arm_cfft_sR_f32_len512;
        break;
    case 1024:
        S = & arm_cfft_sR_f32_len1024;
        break;
    case 2048:
        S = & arm_cfft_sR_f32_len2048;
        break;
    case 4096:
        S = & arm_cfft_sR_f32_len4096;
        break;
    }
    double dt=0.001;        //time interval
    double f1=30;           //frequency 1
    double f2=100;          //frequency 2

    for(i = 0; i< FFT_LEN*2; i+=2)
    {
      samples[i] = sin(2*3.1415926*f1*dt*i) + 0.5*sin(2*3.1415926*f2*dt*i) ;
      samples[i+1]  = 0;
      printf("%f\r\n",samples[i] );
    }
    wait(5);

    arm_cfft_f32(S, samples, 0, 1);                        //FFT
    arm_cmplx_mag_f32(samples, magnitudes, FFT_LEN);       //FFT Magnitudes

    for(i = 0; i< FFT_LEN/2; i++)
    {
        printf("%f\r\n",magnitudes[i]);
    }

    while(1)
    {

    }
}
***************************************************************
```

Again, we can use Arduino Serial Plotter to view the results. As shown in Figure 8.9, the mixed original signal is at the top, and magnitude of the correspond FFT transformed signal at the bottom. As the FFT transformed signal always contains duplicated peaks mirrored in the middle, we only need to look at the first half of the plot, when we can clearly see two frequency peaks (30 Hz and 100 Hz). The peaks values are also proportional to the original signal amplitude (1.0 and 0.5).

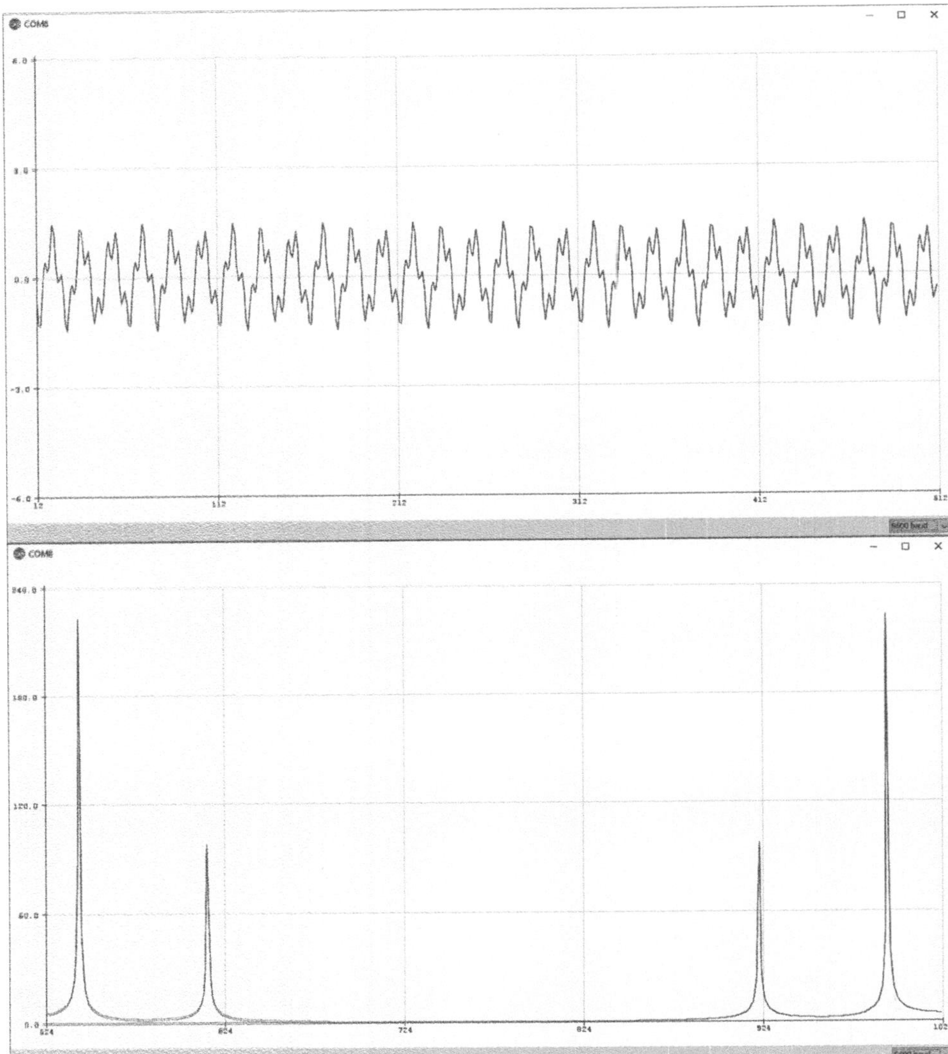

Figure 8.9 The output of the program using Arduino Serial Plotter, with original mixed signal (top) and the corresponding magnitude of the complex FFT transformed signal (bottom).

A very interesting application with FFT is that we can modify the FFT transformed signal, such as applying a low-pass filter or a high-pass filter, then perform the inverse FFT. In the following example, after FFT, the low-frequency components (<50 Hz) was removed by setting them to zero, this is equivalent to applying a high-pass filter. The inverse FFT is then performed, by using "*arm_cfft_f32(S, samples, 1, 1)*" function.

```
*******************************************************************
//Example 8.7

#include "mbed.h"
#include "arm_const_structs.h"

const int FFT_LEN   = 512;

const static arm_cfft_instance_f32 *S;

float samples[FFT_LEN*2];
float magnitudes[FFT_LEN];

int main()
{
  int32_t i = 0;

  // Init arm_ccft_32
  switch (FFT_LEN)
  {
  case 16:
      S = & arm_cfft_sR_f32_len16;
      break;
  case 32:
      S = & arm_cfft_sR_f32_len32;
      break;
  case 64:
      S = & arm_cfft_sR_f32_len64;
      break;
  case 128:
      S = & arm_cfft_sR_f32_len128;
      break;
  case 256:
      S = & arm_cfft_sR_f32_len256;
      break;
  case 512:
      S = & arm_cfft_sR_f32_len512;
      break;
  case 1024:
      S = & arm_cfft_sR_f32_len1024;
      break;
  case 2048:
      S = & arm_cfft_sR_f32_len2048;
      break;
  case 4096:
      S = & arm_cfft_sR_f32_len4096;
      break;
  }
```

```
double dt=0.001;        //time interval
double f1=30;           //frequency 1
double f2=100;           //frequency 2

for(i = 0; i< FFT_LEN*2; i+=2)
{
  samples[i] = sin(2*3.1415926*f1*dt*i) + 0.5*sin(2*3.1415926*f2*dt*i) ;
  samples[i+1] = 0;
  printf("%f\r\n",samples[i] );
}
wait(5);

arm_cfft_f32(S, samples, 0, 1);                               //FFT
arm_cmplx_mag_f32(samples, magnitudes, FFT_LEN);     //FFT Magnitudes

for(i = 0; i< FFT_LEN; i++)
{
    printf("%f\r\n",magnitudes[i]);
}
wait(5);

double Fmax=1/dt;                      //maximum frequency
double df=Fmax/(FFT_LEN*2);          //delta frequency
double Fcut=50/df;                  //set cutoff frequency as 50 Hz

//high-pass filter
for(i = 0; i< FFT_LEN*2; i+=2)  //set frequencies <50 Hz to zero
{
    if ((i<Fcut*2)||(i>(FFT_LEN*2-Fcut*2))){
        samples[i]       = 0 ;
        samples[i+1]     = 0;
    }
}

arm_cmplx_mag_f32(samples, magnitudes, FFT_LEN);  //FFT magnitude

for(i = 0; i< FFT_LEN; i++)
{
    printf("%f\r\n",magnitudes[i]);
}
wait(5);
arm_cfft_f32(S, samples, 1, 1);                          //inverse FFT
for(i = 0; i< FFT_LEN*2; i+=2)
  {
    printf("%f\r\n",samples[i] );
}
```

```
    while(1)
    {

    }
}
***********************************************************
```

Figures 8.10 and 8.11 show the corresponding four consequent outputs of the program. There was a 5-second delay between each output. Figure 8.10 (top) shows the original mixed frequency signal, and Figure 8.10 (bottom) shows its corresponding FFT frequency domain signal. Figure 8.11 (top) shows the FFT frequency domain signal with lower frequency components (<50 Hz) removed, and Figure 8.11 (bottom) shows the

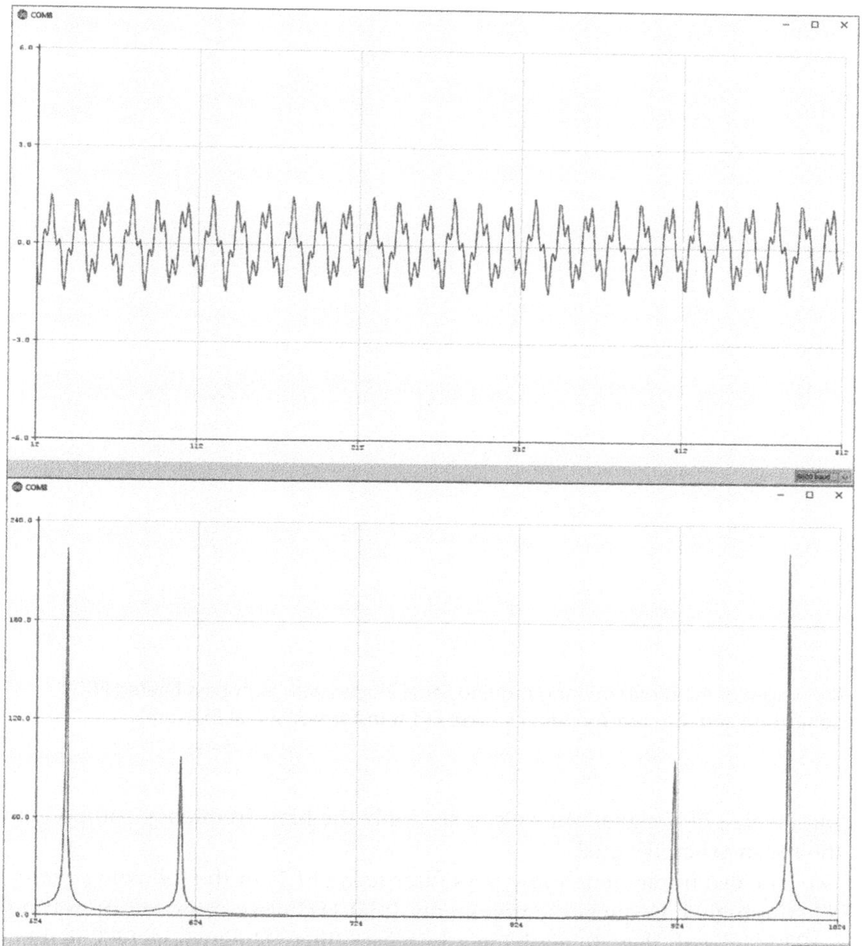

Figure 8.10 The output of the program using Arduino Serial Plotter, with original mixed signal (top) and the corresponding magnitude of the complex FFT transformed signal (bottom).

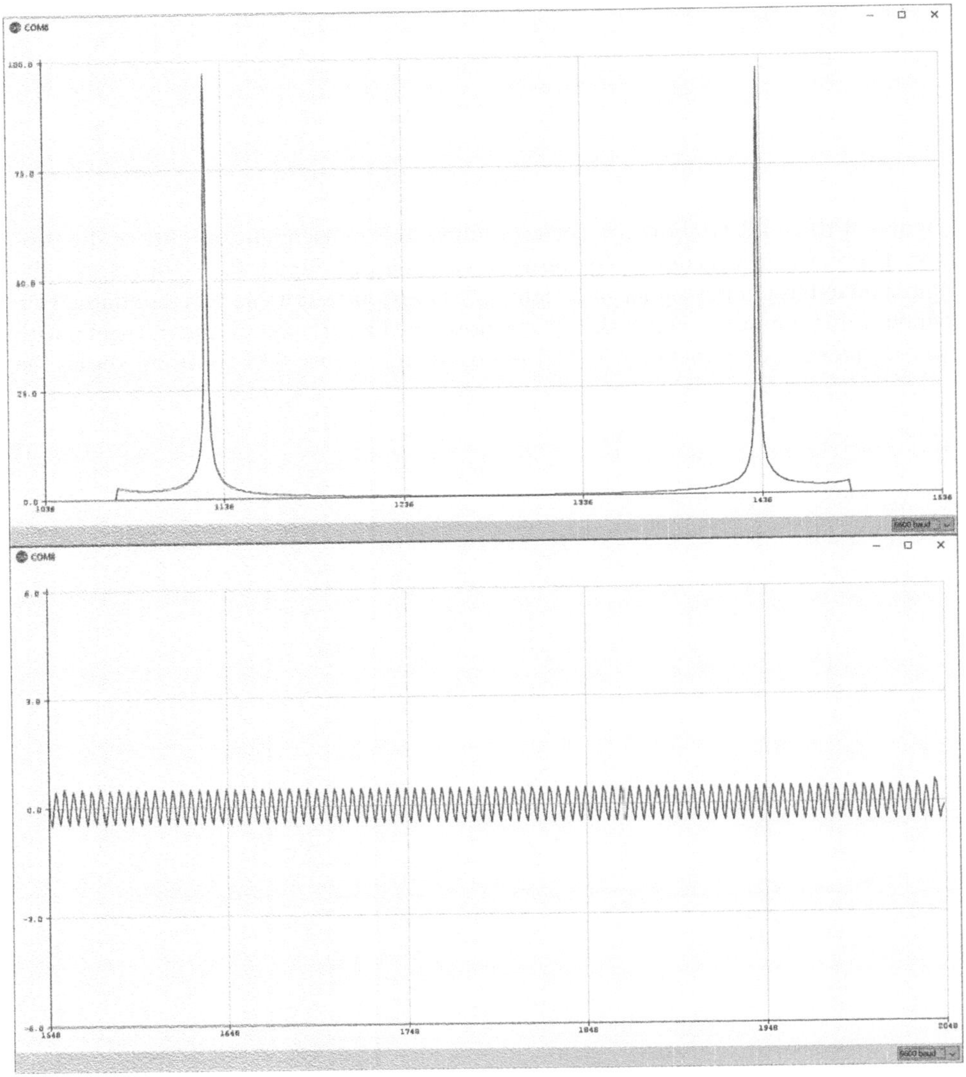

Figure 8.11 The output of the program using Arduino Serial Plotter, with high-pass filtered FFT transformed signal (top) and the corresponding inverse FFT transformed signal (bottom).

corresponding inverse FFT signal. As we can see, only the high-frequency component remains in the reconstructed signal.

Similarly, we can also implement a low-pass filter using FFT. In the following example, after FFT, the higher frequency components (>50 Hz) were removed by setting them to zero. This is equivalent to applying a low-pass filter. The inverse FFT is then performed, by using "*arm_cfft_f32(S, samples, 1, 1)*" function.

```
************************************************************************
//Example 8.8

#include "mbed.h"
#include "arm_const_structs.h"

const int FFT_LEN   = 512;

const static arm_cfft_instance_f32 *S;

float samples[FFT_LEN*2];
float magnitudes[FFT_LEN];

int main()
{
  int32_t i = 0;

  // Init arm_ccft_32
  switch (FFT_LEN)
  {
  case 16:
      S = & arm_cfft_sR_f32_len16;
      break;
  case 32:
      S = & arm_cfft_sR_f32_len32;
      break;
  case 64:
      S = & arm_cfft_sR_f32_len64;
      break;
  case 128:
      S = & arm_cfft_sR_f32_len128;
      break;
  case 256:
      S = & arm_cfft_sR_f32_len256;
      break;
  case 512:
      S = & arm_cfft_sR_f32_len512;
      break;
  case 1024:
      S = & arm_cfft_sR_f32_len1024;
      break;
  case 2048:
      S = & arm_cfft_sR_f32_len2048;
      break;
  case 4096:
      S = & arm_cfft_sR_f32_len4096;
      break;
  }
```

```
double dt=0.001;       //time interval
double f1=30;          //frequency 1
double f2=100;          //frequency 2

for(i = 0; i< FFT_LEN*2; i+=2)
{
  samples[i] = sin(2*3.1415926*f1*dt*i) + 0.5*sin(2*3.1415926*f2*dt*i) ;
  samples[i+1] = 0;
  printf("%f\r\n",samples[i] );
}
wait(5);

arm_cfft_f32(S, samples, 0, 1);                              //FFT
arm_cmplx_mag_f32(samples, magnitudes, FFT_LEN);    //FFT Magnitudes

for(i = 0; i< FFT_LEN; i++)
{
    printf("%f\r\n",magnitudes[i]);
}
wait(5);
double Fmax=1/dt;                              //maximum frequency
double df=Fmax/(FFT_LEN*2);                   //delta frequency
double Fcut=50/df;                           //set cutoff frequency as 50 Hz

//low-pass filter
for(i = 0; i< FFT_LEN*2; i+=2)     //set frequencies >50 Hz to zero
{
  if (((i>Fcut*2)&&(i<FFT_LEN))||((i>FFT_LEN) &&(i<(FFT_LEN*2-Fcut*2))))
  {
      samples[i]       = 0 ;
      samples[i+1]     = 0;
  }
}

arm_cmplx_mag_f32(samples, magnitudes, FFT_LEN);    //FFT magnitudes

for(i = 0; i< FFT_LEN; i++)
{
    printf("%f\r\n",magnitudes[i]);
}
wait(5);
arm_cfft_f32(S, samples, 1, 1);                         //inverse FFT
for(i = 0; i< FFT_LEN*2; i+=2)
{
    printf("%f\r\n",samples[i] );
}
```

```
    while(1)
    {

    }
}
*************************************************************
```

Figure 8.12 (top) show the FFT frequency domain signal with higher frequency components (>50 Hz) removed, and Figure 8.12 (bottom) the corresponding inverse FFT signal. As we can see, only the low-frequency component remains in the reconstructed signal.

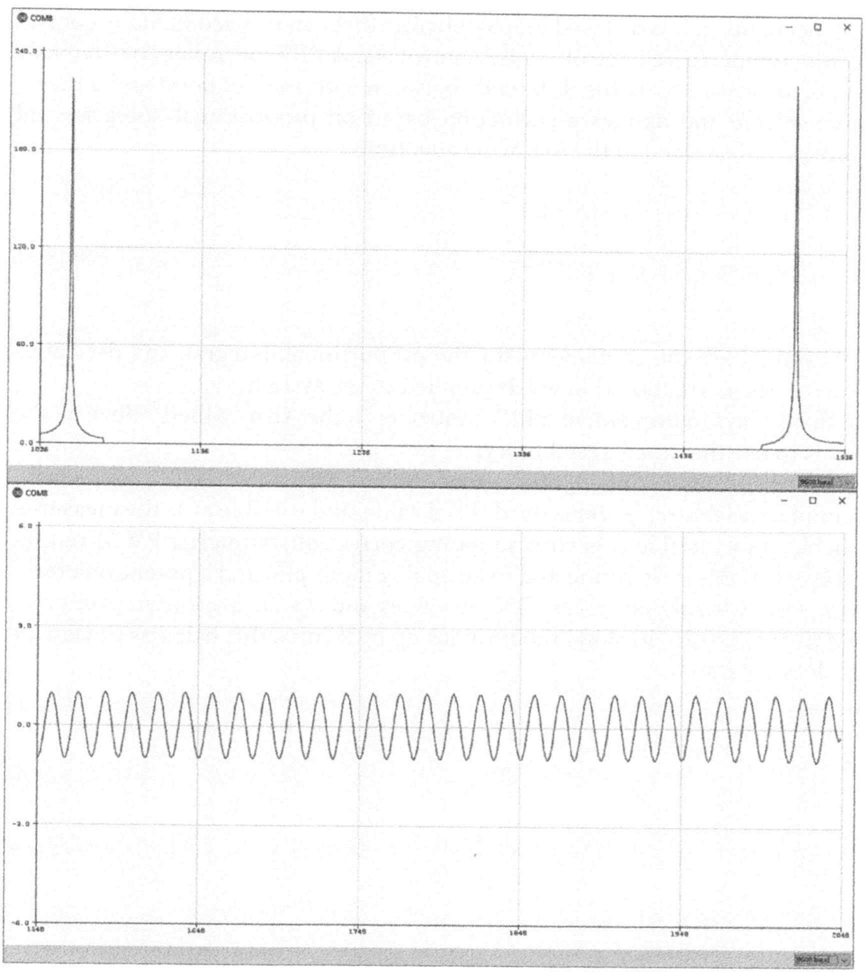

Figure 8.12 The output of the program using Arduino Serial Plotter, with low-pass filtered FFT transformed signal (top) and the corresponding inverse FFT transformed signal (bottom).

Further Information about FFT:

https://os.mbed.com/users/jcobb/code/audio_FFT/file/5b7b619f59cd/main.cpp/
https://os.mbed.com/users/tony1tf/code/KL25Z_FFT_Demo_tony/file/b8c9dffbbe7e/main.cpp/
https://os.mbed.com/users/cpm219/code/fft_test_k22f/
http://paulbourke.net/miscellaneous/dft/
https://rosettacode.org/wiki/Fast_Fourier_transform

8.6 PID Controller

The PID (proportional–integral–derivative) controller is one of the most commonly used control mechanism. It is a closed loop controller that can be used in many control systems, such as temperature control, cruise control etc. A PID controller continuously calculates an error value *e(t)* as the difference between a *desired set point* and a measured *process variable* and applies a correction based on proportional, integral, and derivative terms, as described in the following equations.

$$e(t) = \text{Set point} - \text{Process variable}$$
$$u(t) = K_p e(t) + K_i \int_0^t e(\tau)dt + K_d \frac{de(t)}{dt} \qquad \text{Eq. (8.1)}$$

where K_p, K_i, and K_d are the coefficients for the proportional, integral, and derivative terms, and $u(t)$ is the correction that will be applied to the system.

There are many ways to implement a PID controller in the Arm® Mbed™ device. The simplest way is to use the mbed-DSP library:

https://os.mbed.com/users/mbed_official/code/mbed-dsp/

In this example, variable *set_point* is the desired value, and variable *pv* is the measured process variable, and variable *u* is corresponding correction through a PWM output pin. In this case, variable *pv* is connected to an analog input pin, and a potentiometer is used to change the variable *pv* values. The *pv* values and *u* values are then printed to virtual COM port. The *(2 + u)* shifts the *u* values up by 2 volts, this helps us to view *pv* plots and *u* plots separately.

```
************************************************************
// Example 8.9

#include "mbed.h"
#include "dsp.h"

#if defined(TARGET_K64F)
   AnalogIn pv(A0);
   PwmOut u(D9);
#elif defined(TARGET_LPC1768)
```

```
    AnalogIn pv(p19);
    PwmOut u(p21);
#endif
Serial pc(USBTX, USBRX);

arm_pid_instance_f32 pid;
float set_point = 0.8;

int main()
{
    //Set the initial duty cycle to 0%
    u = 0.0;

    //Initialize the PID instance structure
    pid.Kp = 1.0;
    pid.Ki = 0.002;
    pid.Kd = 5.0;
    arm_pid_init_f32(&pid, 1);

    while(1) {
        float out = arm_pid_f32(&pid, set_point - pv.read());
        //Range limits the output
        if (out < 0.0)
            out = 0.0;
        else if (out >= 1.0)
            out = 1.0;

        //Set the new output duty cycle
        u = out;
        pc.printf("%0.3f\t%0.3f\n\r",pv.read(), (2+u));
        wait(0.1);
    }
}
*************************************************************
```

Figure 8.13 shows the output of the program using Arduino Serial Plotter, with the variable *pv* value at the bottom and the variable *u* value at the top. The results show that as soon as the variable *pv* value changes, the variable *u* value changes accordingly. The apparent oscillations of the *u* value as it changes can be reduced by adjusting the PID K_p, K_i, and K_d coefficients.

We can modify the above example to make it more flexible, as shown in the following code. The K_p, K_i, and K_d coefficients can be taken from the virtual COM port from computer, if available. The K_p, K_i, and K_d coefficients are sent as three numbers separated by |, as shown in Figure 8.14.

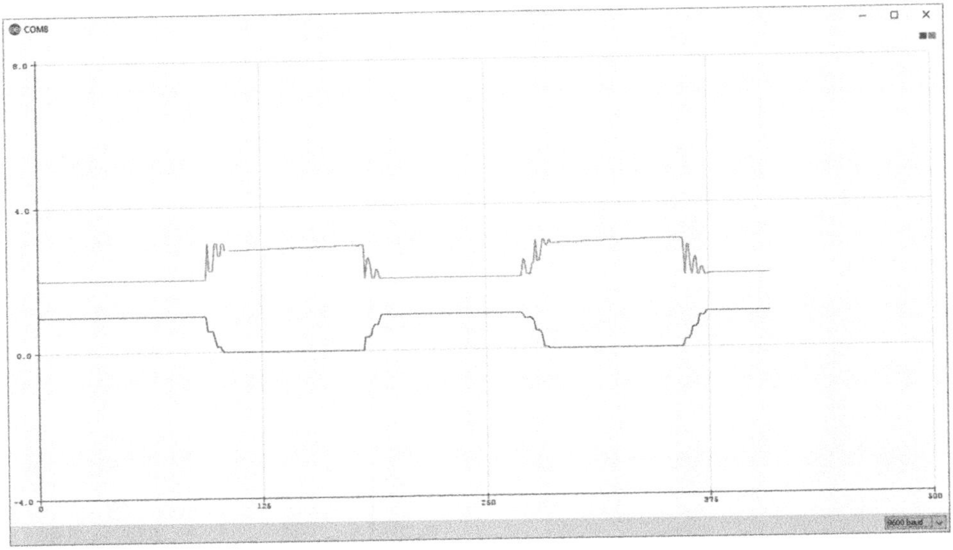

Figure 8.13 The output of the program using Arduino Serial Plotter, with the variable *pv* value at the bottom and the variable *u* value at the top.

```
************************************************************
// Example 8.10

#include "mbed.h"
#include "dsp.h"

#if defined(TARGET_K64F)
   AnalogIn pv(A0);
   PwmOut u(D9);
#elif defined(TARGET_LPC1768)
   AnalogIn pv(p19);
   PwmOut u(p21);
#endif
Serial pc(USBTX, USBRX);

arm_pid_instance_f32 pid;
float set_point = 0.8;

int main()
{
    //Set the initial duty cycle to 0%
    u = 0.0;
```

```
    //Initialize the PID instance structure
    pid.Kp = 1.0;
    pid.Ki = 0.002;
    pid.Kd = 5.0;
    arm_pid_init_f32(&pid, 1);


    while(1) {
        if(pc.readable()) {
            char buff [256]="";
            pc.gets(buff, 256);
            pc.printf("%s\n\r", buff);
            sscanf (buff,"%f|%f|%f",&pid.Kp,&pid.Ki,&pid.Kd);
            arm_pid_init_f32(&pid,1);
        }
        float out = arm_pid_f32(&pid, set_point - pv.read());
        //Range limit the output
        if (out < 0.0)
            out = 0.0;
        else if (out >= 1.0)
            out = 1.0;

        //Set the new output duty cycle
        u = out;
        pc.printf("%0.3f\t%0.3f\n\r",pv.read(), (2+u));
        wait(0.1);
    }
}
***********************************************************
```

Further Information about the PID:

https://en.wikipedia.org/wiki/PID_controller
https://os.mbed.com/users/aberk/code/PID/
https://os.mbed.com/questions/1904/mbed-DSP-Library-PID-Controller/
https://os.mbed.com/teams/FRDM-K64F-Code-Share/code/PID/

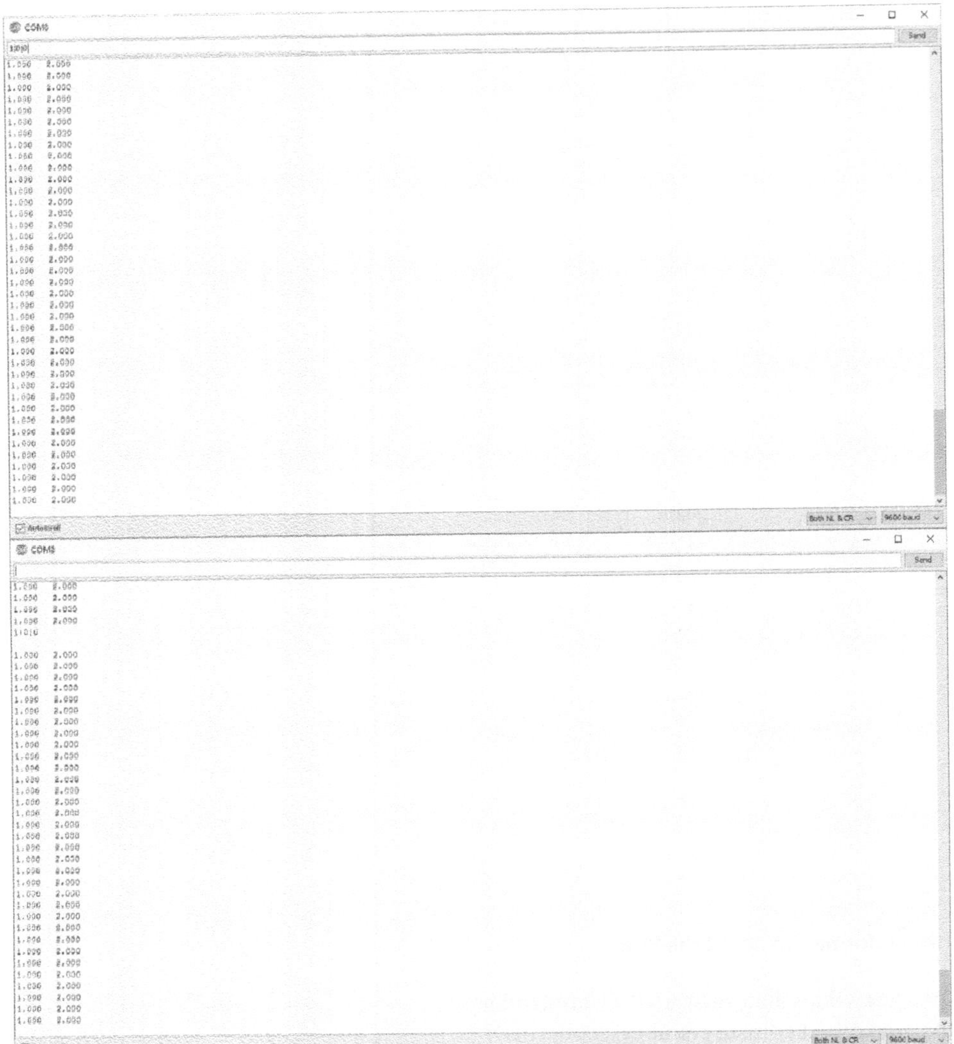

Figure 8.14 The output of the program using Arduino Serial Monitor, with the K_p, K_i, and K_d coefficients (1|0|0) sent from computer to mbed device through virtual COM port (top) and the corresponding responses of the virtual COM port (bottom).

8.7 Summary

This chapter illustrates how to use the Arm® Mbed™-DSP library to perform digital signal processing, e.g., low-pass filter, high-pass filter, band-pass filter, band-stop filter and notch filter. It also illustrates how to use the Arm® Mbed™-DSP library to perform FFT (fast Fourier transform) and inverse FFT, as well as how to implement a PID controller.

9

Debugging, Timer, Multithreading, and Real-Time Programming

Never, never, never give up.

- Winston Churchill

9.1 Debugging

Debugging is an important process to get rid of the errors in the code. In programming terms, "bugs" mean errors, and "debugging" means removing errors. The use of the terms can be dated back to the 1940s, when Admiral Grace Hopper was working on a Mark II Computer at Harvard University, her associates discovered a moth stuck in a relay and thereby impeding operation, whereupon she remarked that they were "debugging" the system.

Although the full debugging capabilities, which allows you to set break points, step into the code etc., are not available for the online compiler, many techniques allow you to get debugging information about your code.

There are generally two types of errors, compile-time errors and run-time errors. The compile-time errors are normally due to incorrect syntax and misuse of variables and functions. They can be relatively easily corrected, as otherwise your code would not be able to be compiled. Run-time errors are more difficult to eliminate. However, the Arm® Mbed™ provides a mechanism called *light of death*, which will flash LEDs like the siren lights, when a runtime error is encountered. Following is a typical example that will result in light of death.

Designing Embedded Systems and the Internet of Things (IoT) with the ARM® Mbed™, First Edition. Perry Xiao

Companion website: www.wiley.com\go\xiao\designingembeddedsystemandIoTwitharmmbed

```
************************************************************
// Example 9.1

#include "mbed.h"

PwmOut pout(D3);
int main() {
    while(1) {
        for(float p = 0.0f; p < 1.0f; p += 0.1f) {
            pout = p;
            wait(0.1);
        }
    }
}
************************************************************
```

The Arm® Mbed™ also contains some features for reporting runtime errors, for example:

- **printf()**—Print a formatted message to the USB serial port (stdout default).
- **error()**—Print a formatted message to the USB serial port, then die with "Siren Lights."

Following is an example on how to use the above methods to report errors.

```
************************************************************
// Example 9.2

#include "mbed.h"

DigitalIn button(p21);
AnalogIn pot(p20);

int main() {
    while(pot > 0.0) {
        printf("Pot value = %f", pot.read());
        wait(0.1);
    }
    error("Loop unexpectedly terminated");
}
************************************************************
```

You can also use different LEDs to indicate the flow of your code, as shown in the following example.

```
****************************************************************
// Example 9.3

#include "mbed.h"

AnalogIn ain(A0);
DigitalOut led1(LED1); //use for debug
DigitalOut led2(LED2); //use for debug
DigitalOut led3(LED3); //use for debug

int main() {
    while(1) {
        if(ain>2.0)) {
            led1=1;
            led2=0;
            led3=0;
        }else if(ain>1.0)) {
            led1=0;
            led2=1;
            led3=0;
        }else{
            led1=0;
            led2=0;
            led3=1;
        }
    }
}
****************************************************************
```

Further Information about Debugging:

https://os.mbed.com/handbook/Debugging

9.2 Timer, Timeout, Ticker, and Time

Timer is very useful for measuring small time changes, as illustrated in the following example. It first starts the Timer, does some calculations, then stops the Timer and reads the time elapsed in seconds. Any number of Timer objects can be created and can be started and stopped independently.

```
*************************************************************
// Example 9.4

#include "mbed.h"

Timer t;

int main() {
   t.start();
   int x=10,y=20;
   y=x+y;
   t.stop();
   printf("Time=%f seconds\n",t.read());
}
*************************************************************
```

Figure 9.1 show the Arduino Serial Monitor output of the program. In this example, the calculation only takes 3 microseconds.

Exercise 9.1

Modify the above program so that it can measure the time for running a *"for"* loop 10,000 times.

The Timeout interface is used to set up an interrupt to call a function after a specified delay. In the following example, a function *"fun()"* has been assigned to a Timeout event, which will interrupt the main loop after 2 seconds. A Timeout event will only occur once.

Any number of Timeout objects can be created, allowing multiple outstanding interrupts at the same time.

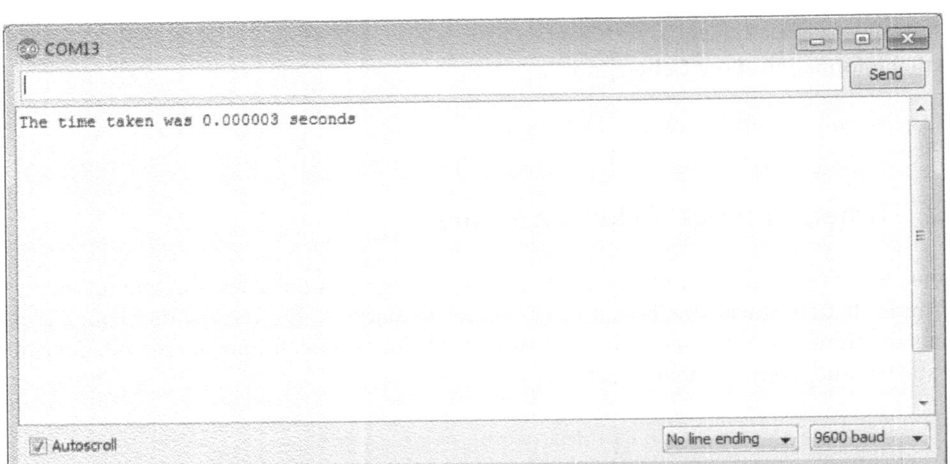

Figure 9.1 The Arduino Serial Monitor output of Timer example.

```
************************************************************
// Example 9.5

#include "mbed.h"

Timeout tout;

void fun() {
  printf("Timeout print......\r\n");
}

int main() {
  tout.attach(&fun, 2.0); //set up Timeout to call fun() after 2 seconds
  while(1) {
      printf("Main loop......\r\n");
      wait(0.2);
   }
}
************************************************************
```

Figure 9.2 show the Arduino Serial Monitor output of the program. As we expected, Timeout event only happened once, after 2 seconds of program running.

The Ticker interface is used to set up a recurring interrupt to repeatedly call a function at a specified rate. In the following example, a function "*fun()*" has been assigned to a Ticker event, which will interrupt the main loop every 2 seconds. A Ticker event will occur repeatedly.

Any number of Ticker objects can be created, allowing multiple outstanding interrupts at the same time. The function can be a static function or a member function of a particular object.

Figure 9.2 The Arduino Serial Monitor output of Timeout example.

```
****************************************************************
// Example 9.6
#include"mbed.h"

Tickertk;

void fun() {
    printf("Timeout print......\r\n");
}

int main() {
    tk.attach(&fun, 2.0);  //set up Ticker to call fun() every 2 seconds
    while(1) {
        printf("Main loop......\r\n");
        wait(0.2);
    }
}
****************************************************************
```

Figure 9.3 show the Arduino Serial Monitor output of the program. In this case, Ticker event happened every 2 seconds.

The Arm® Mbed™ also has a Time function. Following is a simple program to set and get date and time:

```
****************************************************************
// Example 9.7

#include "mbed.h"

int main() {
    set_time(1256729737);  // Set RTC time to Wed, 28 Oct 2009 11:35:37

    while (true) {
        time_t seconds = time(NULL);

        printf("Time as seconds since January 1, 1970 = %d\n", seconds);

        printf("Time as a basic string = %s", ctime(&seconds));

        char buffer[32];
        strftime(buffer, 32, "%I:%M %p\n", localtime(&seconds));
        printf("Time as a custom formatted string = %s", buffer);

        wait(1);
    }
}
****************************************************************
```

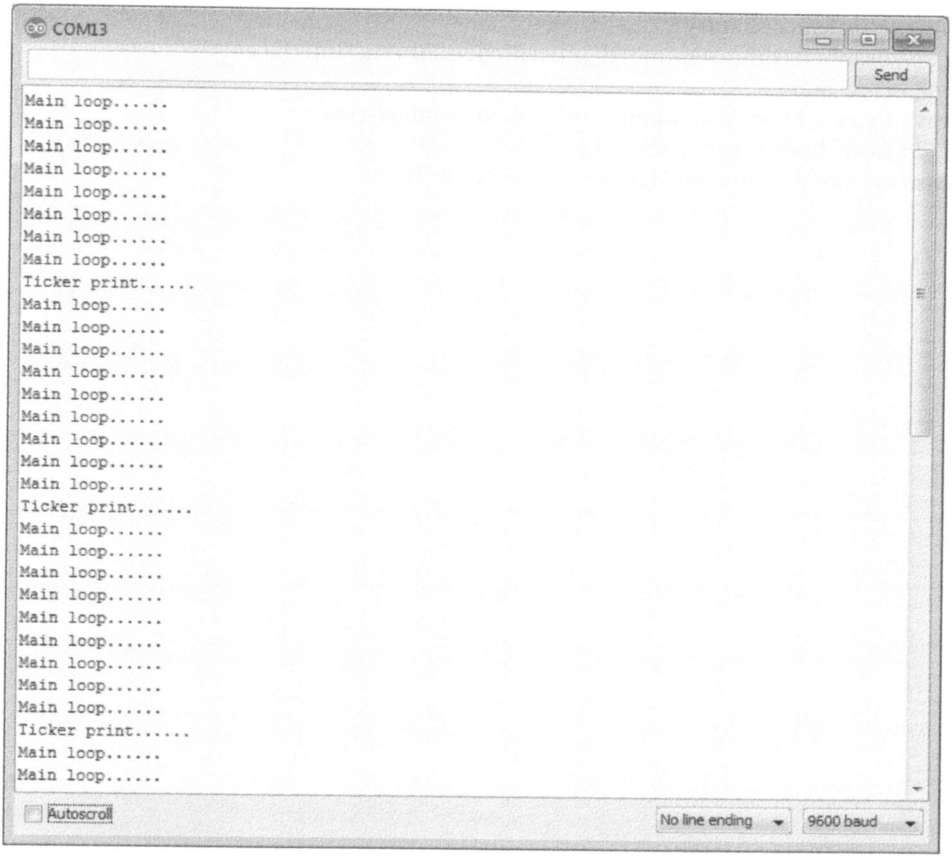

Figure 9.3 The Arduino Serial Monitor output of Ticker example.

Further Information about Time:

https://docs.mbed.com/docs/mbed-os-api-reference/en/latest/APIs/tasks/Timer/
https://docs.mbed.com/docs/mbed-os-api-reference/en/latest/APIs/tasks/Timeout/
https://docs.mbed.com/docs/mbed-os-api-reference/en/latest/APIs/tasks/Ticker/
https://docs.mbed.com/docs/mbed-os-api-reference/en/latest/APIs/tasks/Time/
https://docs.mbed.com/docs/mbed-os-api-reference/en/latest/APIs/tasks/events/
https://docs.mbed.com/docs/mbed-os-api-reference/en/latest/APIs/tasks/wait/

9.3 Network Time Protocol (NTP)

Network Time Protocol (NTP) is a networking protocol for clock synchronization between computer systems over packet-switched, variable-latency data networks. Following is a simple example to show how to get time information from the Internet using NTP. It uses the following libraries:

"*EthernetInterface*" library
https://os.mbed.com/users/mbed_official/code/EthernetInterface/
"*mbed-rtos*" library
https://os.mbed.com/users/mbed_official/code/mbed-rtos/
"*NTPClient*" library
https://os.mbed.com/users/donatien/code/NTPClient/

```
**************************************************************
// Example 9.8
// Modified from
// https://developer.mbed.org/users/donatien/code/NTPClient_HelloWorld/

#include "mbed.h"
#include "EthernetInterface.h"
#include "NTPClient.h"

EthernetInterface eth;
NTPClient ntp;

int main()
{
    eth.init();
    eth.connect();

    if (ntp.setTime("0.pool.ntp.org") == 0)
    {
      time_t ctTime;
      ctTime = time(NULL);
      printf("Time is set to (UTC): %s\r\n", ctime(&ctTime));
    }

    eth.disconnect();

    while(1) { }
}
**************************************************************
```

Figure 9.4 show the corresponding output of the program.

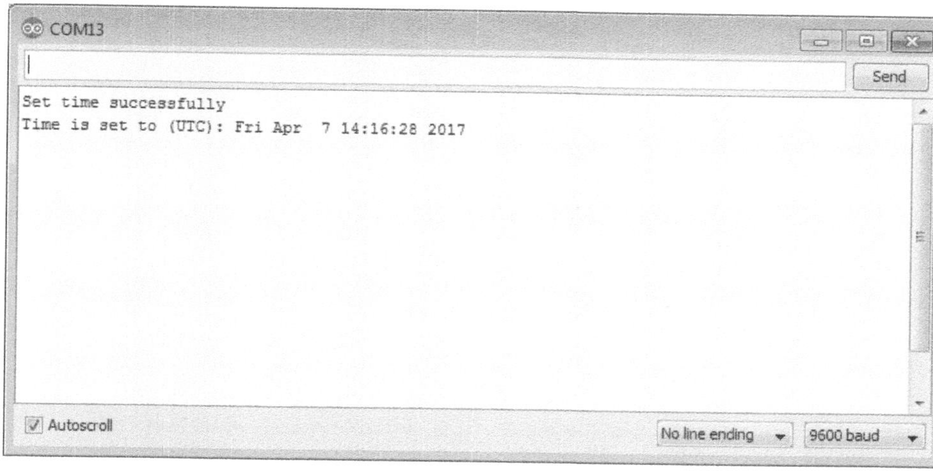

Figure 9.4 The Arduino Serial Monitor output of the NTP example.

Further Information about NTP:

https://os.mbed.com/users/donatien/notebook/ntp-client/
https://os.mbed.com/cookbook/NTP-Client

9.4 Multithreading and Real-Time Programming

Multithreading is a powerful feature that comes along with the Arm® Mbed™ OS 5. It allows you to run the tasks in parallel. For example, you can use one thread for communication, and one thread for control. You will see more multithreading examples in Chapter 12 with a multithreaded web server, and multi-thread smart lighting.

Multithreading is provided through Real-Time Operating System (RTOS), one of the key features of the new mbed OS 5. This much-requested feature is now incorporated in the core of the mbed operating system. RTOS provides native thread support to the OS and applications, simplifying development and integration of complex and robust application components like networking stacks. The RTOS requires very limited system overhead.

To create a multithread program in the Arm® Mbed™ is very simple. Just import the "*mbed-rtos*" library, create a function that describes what you would like to do, and call that function in a thread. The following program uses main loop to print out a message and uses a separate thread to call the "*fun_1()*" function to print out another message.

```
************************************************************
// Example 9.9

#include "mbed.h"
#include "rtos.h"

void fun_1(void const *args) {
    while (true) {
        printf("Thread 1 ... ... \n\r");
        Thread::wait(200);
    }
}

int main() {
    Thread thread(fun_1);

    while (true) {
        printf("Main Loop Thread ... ... \n\r");
        Thread::wait(100);
    }
}
************************************************************
```

Figure 9.5 shows the Arduino Serial Monitor output of the example. As you can see, the main loop thread executed every 100 ms, and the separate thread executed every 200 ms.

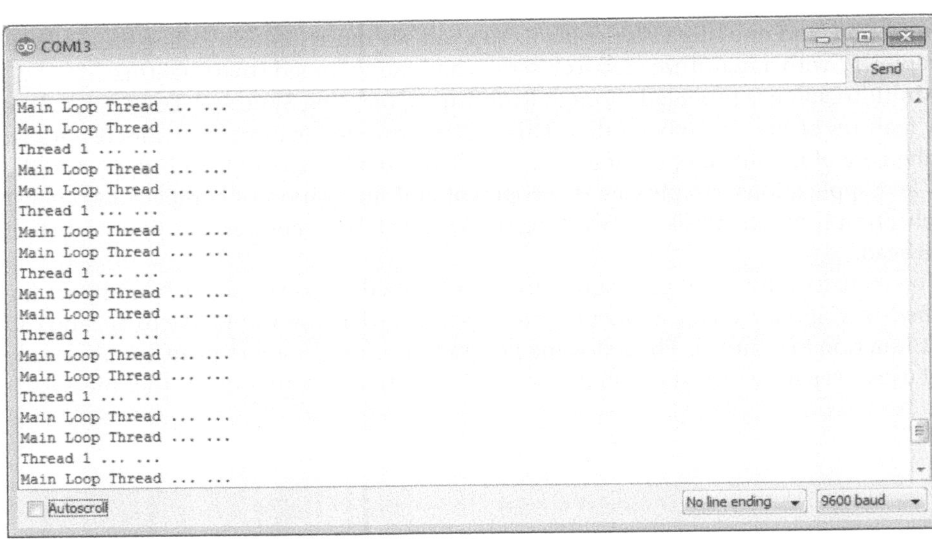

Figure 9.5 The Arduino Serial Monitor output of the multithread example.

Following is an improved version of the above example, in which two functions were created to print two messages and two threads were created in the "*main()*" function to call these two functions.

```
************************************************************
// Example 9.10

#include "mbed.h"
#include "rtos.h"

void fun_1(void const *args) {
    while (true) {
        printf("Thread 1 … … \n\r");
        Thread::wait(200);
    }
}
void fun_2(void const *args) {
    while (true) {
        printf("Thread 2 … … \n\r");
        Thread::wait(500);
    }
}

int main() {
    Thread thread1(fun_1);
    Thread thread2(fun_2);

    while (true) {
    }
}
************************************************************
```

Figure 9.6 shows the Arduino Serial Monitor output of the example. As you can see, the two separate threads run simultaneously.

Following is another multithreaded example, in which two functions were created to do some calculations and two threads were created in the "*main()*" function to call these two functions. The two functions can both use global variable *mode* to share information between them. As shown in Figure 9.7, function "*fun_2()*" modified the value of *mode* and function "*fun_1()*" could pick up the corresponding changes.

Figure 9.6 The Arduino Serial Monitor output of the improved multithread example.

```
**************************************************************
// Example 9.11

#include "mbed.h"
#include "rtos.h"

int mode =0;

void fun_1(void const *args) {
    while (true) {
        printf("mode: %d\n\r", mode);
        Thread::wait(200);
    }
}
void fun_2(void const *args) {
    while (true) {
        mode++;
        Thread::wait(1000);
    }
}

int main() {
    Thread thread1(fun_1);
    Thread thread2(fun_2);

    while (true) {
    }
}
**************************************************************
```

Figure 9.7 The Arduino Serial Monitor output of the above multithread example.

We can also combine the above multithreaded code with the web server code (Example 8,8) to create a multithreaded web server.

```
***********************************************************
// Example 9.12

#include "mbed.h"
#include "EthernetInterface.h"
#include <stdio.h>
#include <string.h>
#include "rtos.h"

#define PORT   80

EthernetInterface eth;

TCPSocketServer server;
bool serverIsListened = false;

TCPSocketConnection client;
bool clientIsConnected = false;

void web_thread(void const *args)
{

    //set up tcp socket
    if(server.bind(PORT)< 0) {
        serverIsListened = false;
```

```
    } else {
        printf("tcp server bind succeeded.\n\r");
        serverIsListened = true;
    }

    server.listen();

    //listening for http GET request
    while (serverIsListened) {
        //blocking mode(never timeout)
        if(server.accept(client)<0) {
            printf("failed to accept connection.\n\r");
        } else {
            printf("connection success!\n\rIP: %s\n\r",client.
get_address());
            clientIsConnected = true;

            while(clientIsConnected) {
                char buffer[1024] = {};
                if(client.receive(buffer, 1023)<1){
                    break;
                }
                else{
                        printf("Received
Data: %d\n\r\n\r%.*s\n\r",strlen(buffer),strlen(buffer),buffer);
                        if(buffer[0] == 'G' && buffer[1] == 'E' &&
buffer[2] == 'T' && buffer[3] == ' ' && buffer[4] == '/' ) {
                            printf("GET request incoming.\n\r");
                            //set up http response header & data
                            char Body[1024] = {};
                            sprintf(Body,"<html></title>
<body>Hello World %d </body></html>\n\r\n\r",strlen(buffer));
                            char Header[256] = {};
                            sprintf(Header,"HTTP/1.1 200 OK\n\
rContent-Length: %d\n\rContent-Type: text\n\rConnection:
Close\n\r\n\r",strlen(Body));
                            client.send(Header,strlen(Header));
                            client.send(Body,strlen(Body));
                            clientIsConnected = false;
                        }

                }
            }
            printf("close connection.\n\rtcp server is listening...\n\r");
            client.close();

        }
```

```
    }
}
int main() {
    EthernetInterface eth;
    eth.init(); //Use DHCP
    eth.connect();
    printf("\r\nServer IP Address is %s\r\n", eth.getIPAddress());

    Thread thread(web_thread);
    while(1){}
}
*****************************************************************
```

Further Information about Multithreading and Real-Time Programming:

https://os.mbed.com/handbook/RTOS
https://docs.mbed.com/docs/mbed-os-api-reference/en/latest/APIs/tasks/rtos/
https://os.mbed.com/blog/entry/Introducing-mbed-OS-5/

9.5 Summary

This chapter introduces how to debug, how to use Timer, Timeout, Ticker, and Time, as well as how to get time and date information from the Internet using NTP (Network Time Protocol). It also introduces multithread programming and real-time programming.

10

Libraries and Programs

Start by doing what's necessary; then do what's possible; and suddenly you are doing the impossible.

- Francis of Assisi

10.1 Import Libraries and Programs

The Arm® Mbed™ developer website (used to be https://developer.mbed.org and now has changed to https://os.mbed.com) is a wonderful resource for your program. There are a lot of programs and libraries available. A very good way to learn is to import existing programs into your online compiler workspace (Figure 10.1). Just click the "*Import!*" button from your online compiler. An "*Import Wizard*" will be displayed. From the "*Programs*" tab, search what you are looking for from "*mbed.org*", then double-click the program (or click the "*Import!*" button) to import! From the "*Bookmarked*" tab, you can also import programs from a website and from the "*Upload*" tab, you can also import programs from your local computer!

Similarly, you can also import libraries into your program, as shown in the following screenshot, Figure 10.2. Just click the "*Import!*" button from your online compiler, then from the "*Libraries*" tab, search what you are looking for from "*mbed.org*" and then double-click the library to import!

10.2 Export Your Program

You can also export your project to an offline third-party compiler software, also called toolchains. From your online compiler, in the "*Program Workspace,*" select the project you would like to export, right-click it to display a pop-up menu, and select "*Export Program...*" (Figure 10.3).

A pop-up "*Export program*" window will appear (Figure 10.4); just make sure you select the correct export target and export toolchains (Figure 10.5). There are a number of popular toolchains supported, such as Keil uVersion4, GCC, IAR Systems, and Kinetic Design Studio. You can also re- import the programs that you exported; just follow the steps of importing from your local computer, described in section 10.1.

Designing Embedded Systems and the Internet of Things (IoT) with the ARM® Mbed™, First Edition. Perry Xiao

Companion website: www.wiley.com\go\xiao\designingembeddedsystemandIoTwitharmmbed

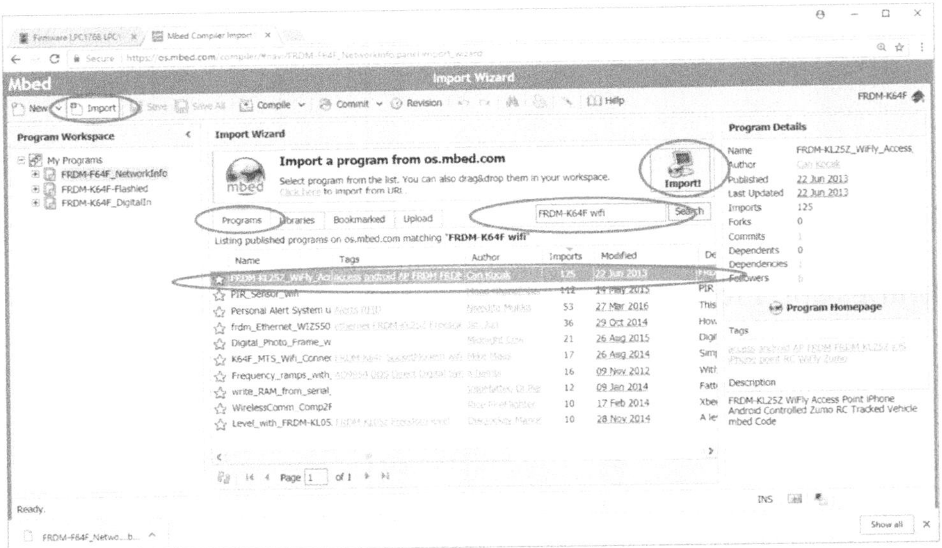

Figure 10.1 Import a program from the online compiler.

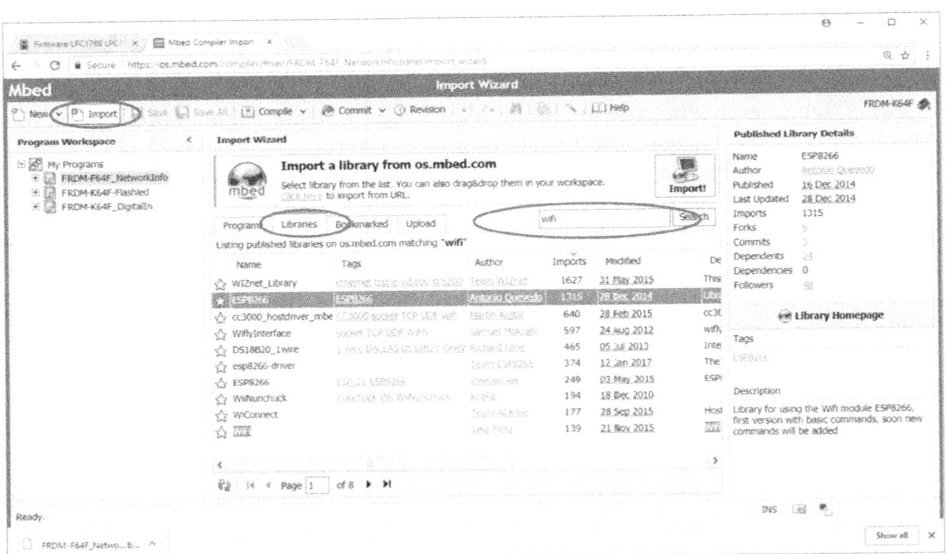

Figure 10.2 Import a library from the online compiler.

10.3 Write Your Own Library

A library is a collection of code that is designed to provide certain functions, or to handle certain hardware components.

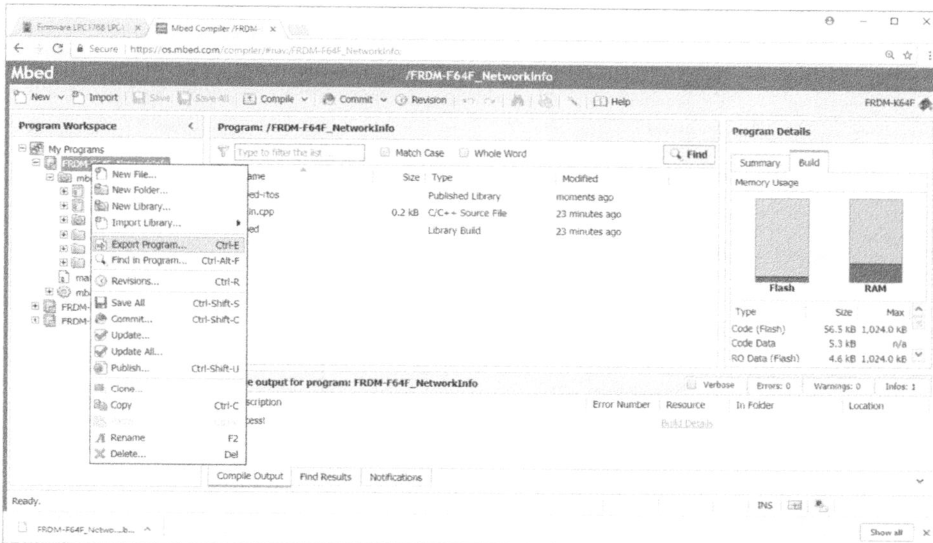

Figure 10.3 Export a program from the online compiler.

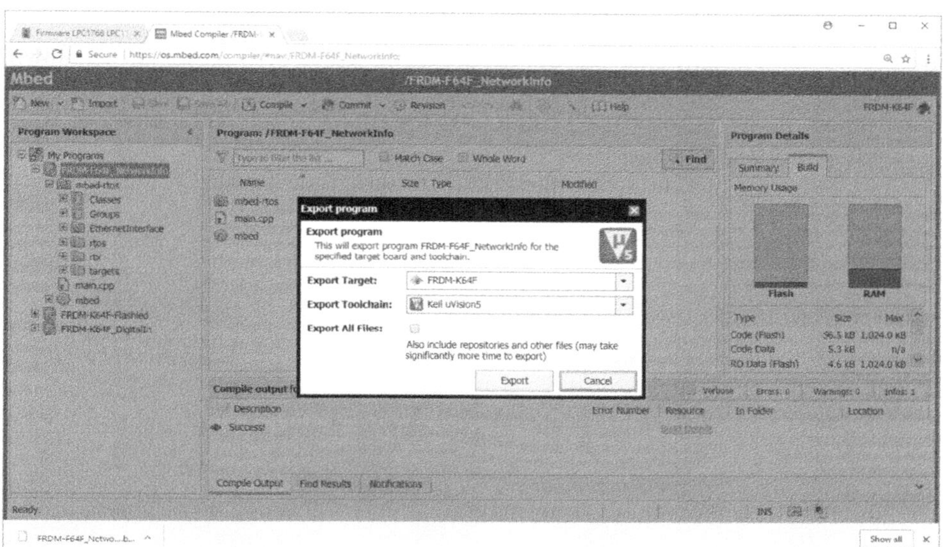

Figure 10.4 Export program pop-up window.

Libraries are crucial in project development, as libraries allow users to share and reuse the code, so you don't need to keep reinventing the wheel. For example, as you see in Chapter 6, if you want to use the combo accelerometer and magnetometer sensor, or SD card etc., you just need to import the corresponding libraries that deal with the sensor and SD card; you don't need to rewrite the code from scratch!

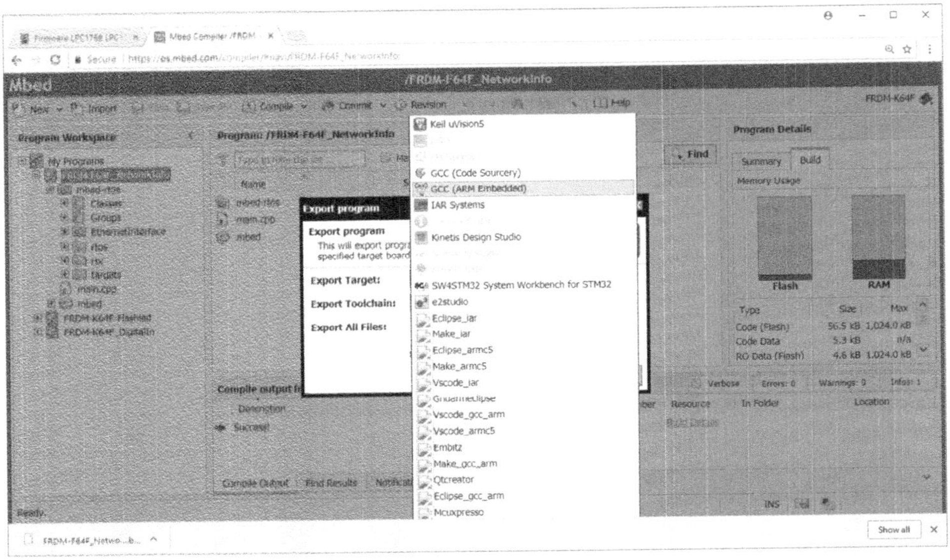

Figure 10.5 Export program toolchains option.

To create a library, right-click on your program and select *"New Library..."* as shown in Figure 10.6.

Enter library name, in this case, I call it *"PXMathLibrary"*; it'll add a folder to your program (Figure 10.7).

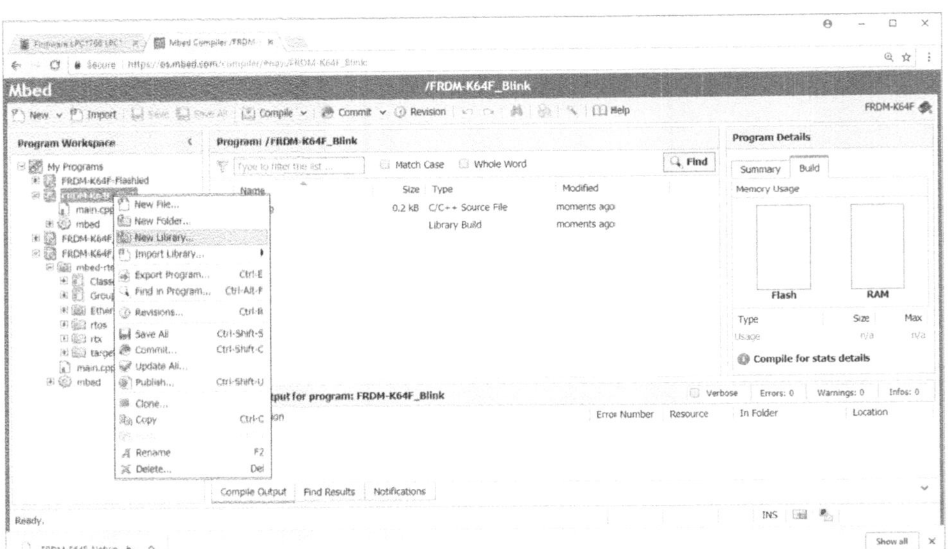

Figure 10.6 Create a new library from mbed.org.

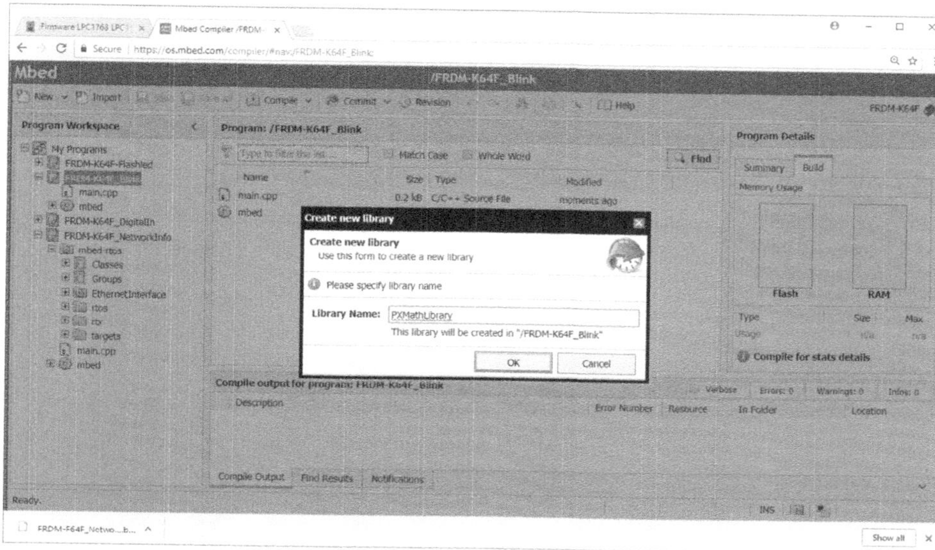

Figure 10.7 New library name pop-up window.

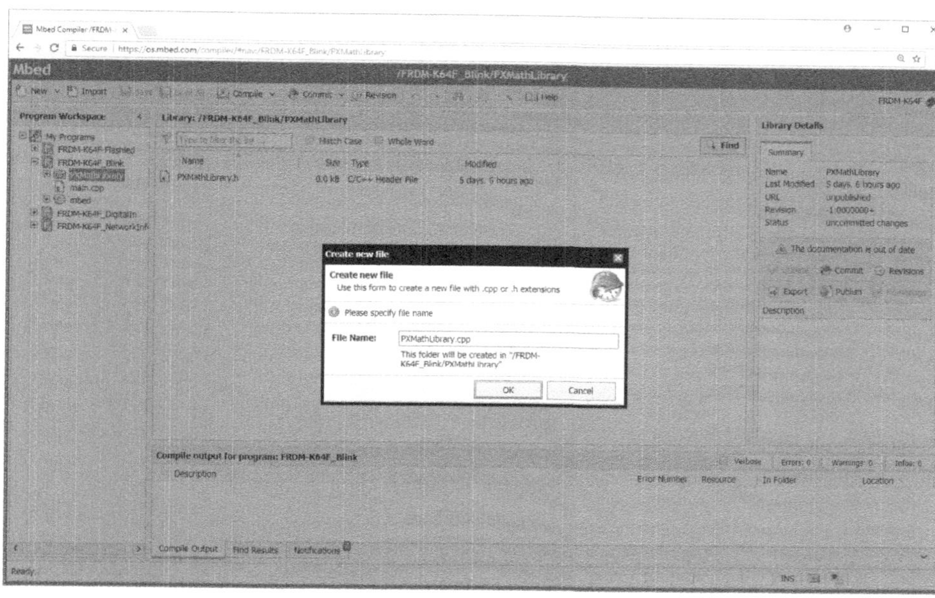

Figure 10.8 Create a new file in the library.

Then add two files into the folder, "*PXMathLibrary.h*" and "*PXMathLibrary.cpp*" see Figure 10.8. Figure 10.9 shows the corresponding contents of the two files. In this library, it gets two analogue input pins, then uses a function called "***mean()***" to calculate the average value of the two analogue pins, and display all the values to the computer serial port.

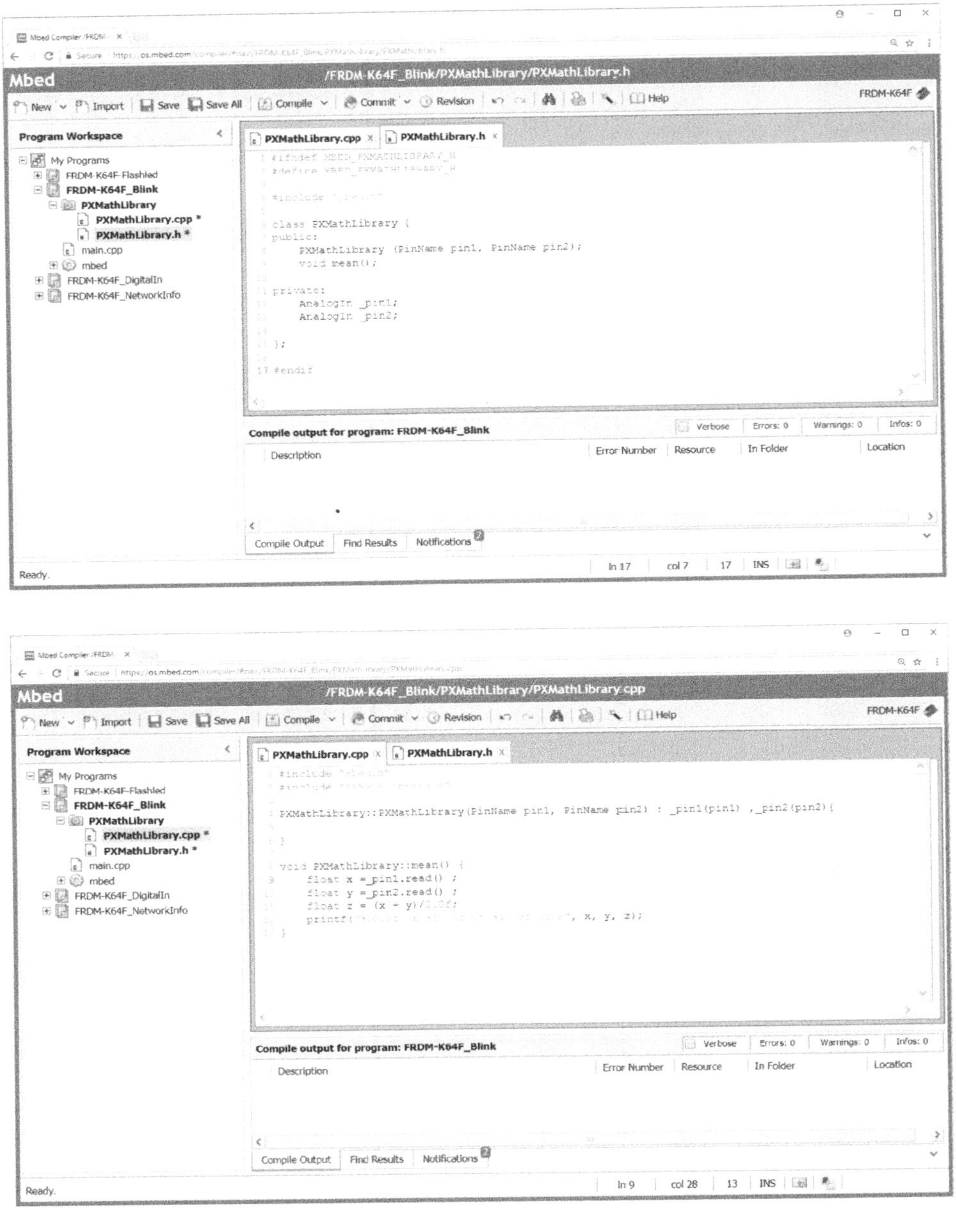

Figure 10.9 The new library program header file page (top) and CPP file page (bottom).

Finally, you can call you library and use the "*mean()*" function, as shown in Figure 10.10. In this example, the two analogue input pins are "*A0*" and "*A1*".

Figure 10.11 shows the "Tera Term" screenshot of the three values send to serial port, the first two values are the analogue input pins "*A0*" and "*A1*", and the third is their average value.

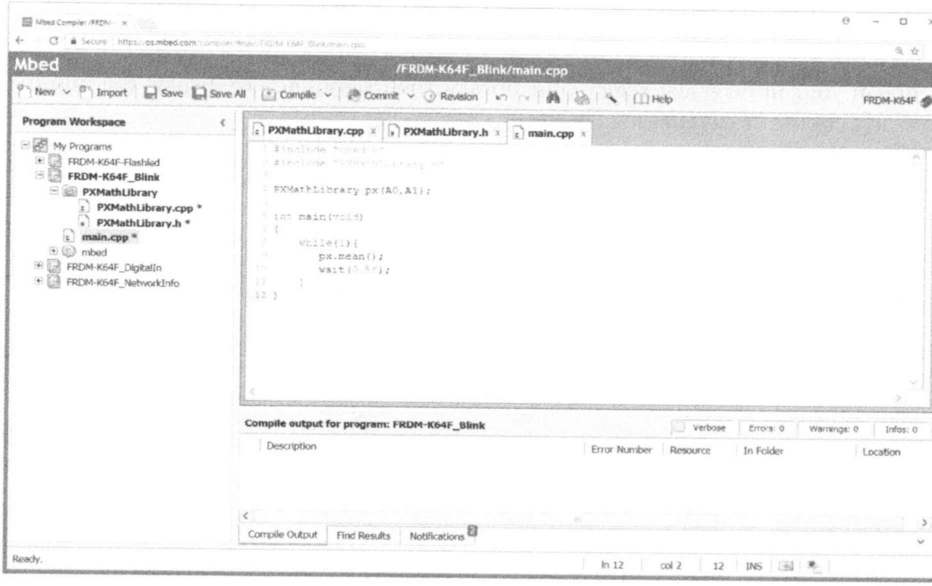

Figure 10.10 The main.ccp file page.

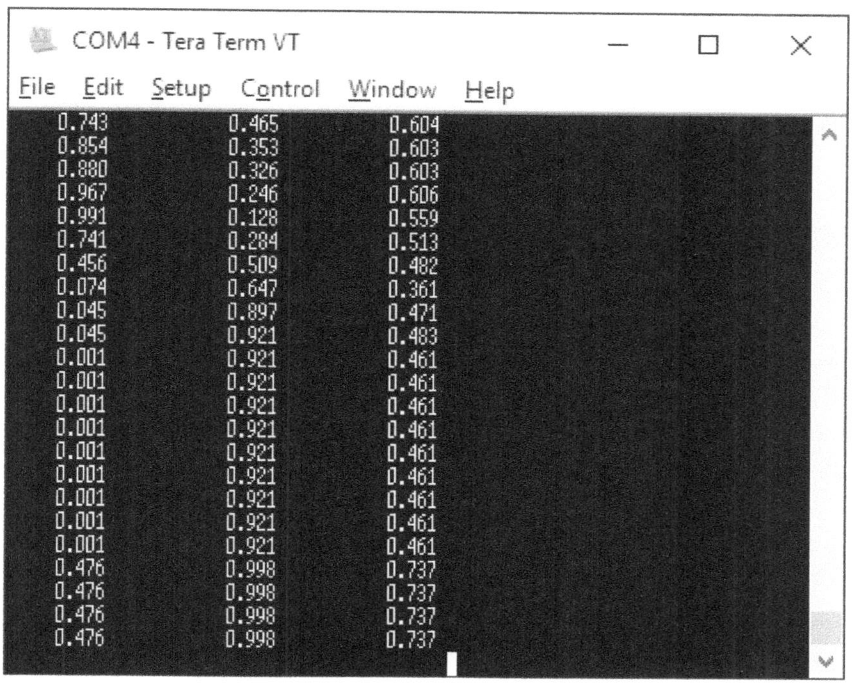

Figure 10.11 The Tera Term outputs.

10.4 Publish Your Library

To publish your library, just select your library, then right click to display the pop-up menu, and choose "***Publish***", see Figure 10.12.

A "***Revision Commit***" window will appear, and type in the commit message, as shown in Figure 10.13. After clicking the "OK" button, a "***Publish Repository***" window will

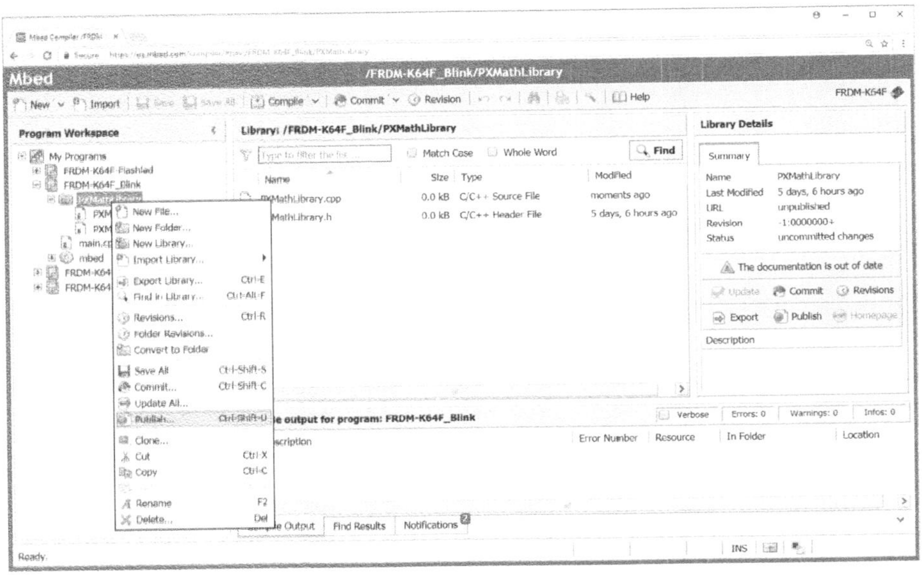

Figure 10.12 Publish a library from the Online Compiler.

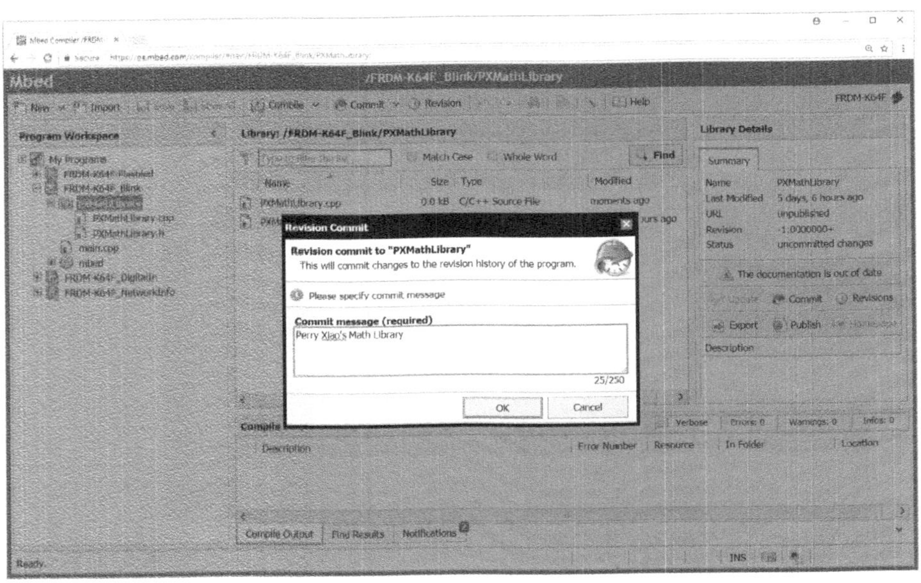

Figure 10.13 The Revision commit pop up window.

appear, see Figure 10.14, make sure all the information is correct, and click the "OK" button. A confirmation window will show the URL that your library is published on, see Figure 10.15. That is it!

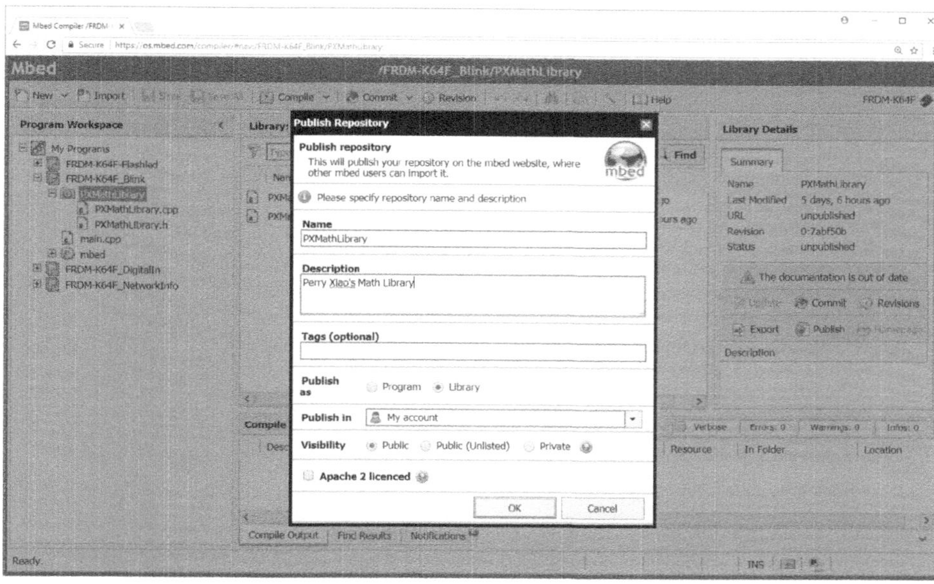

Figure 10.14 The Publish Repository popup window.

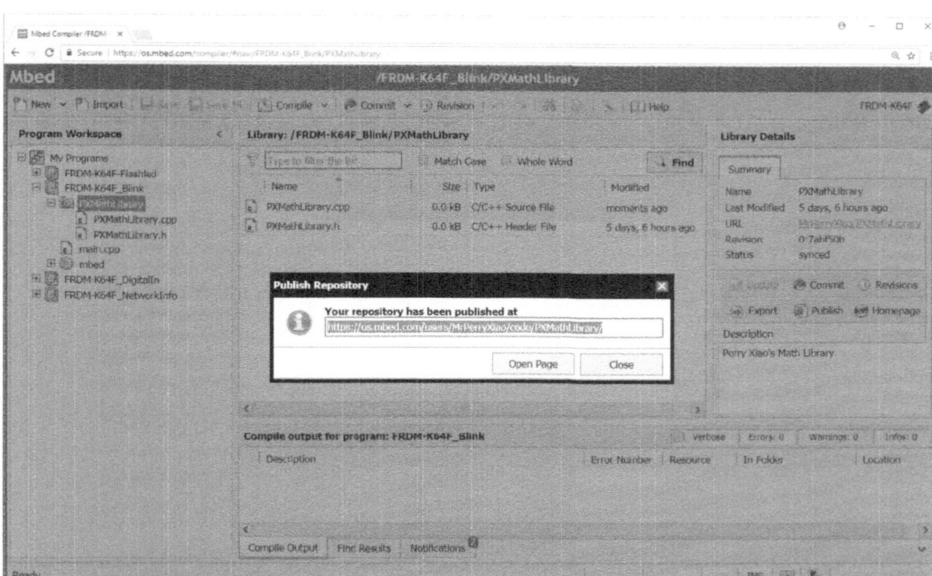

Figure 10.15 The Publish Repository confirmation popup window.

10.5 Publish Your Program

You can also publish your program, in a similar way as you publish your library, see steps shown in Figure 10.16, Figure 10.17, Figure 10.18 and Figure 10.19.

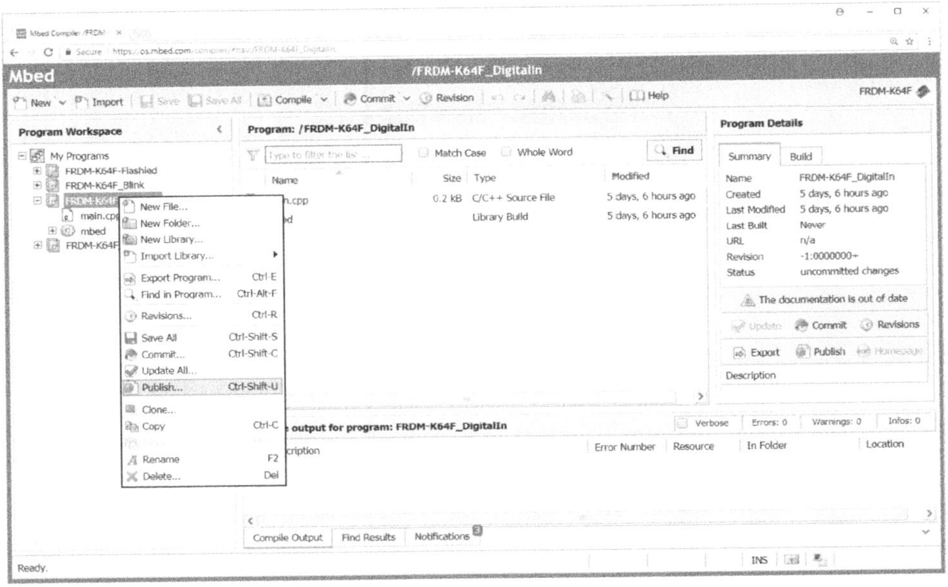

Figure 10.16 Publish a program from the online compiler.

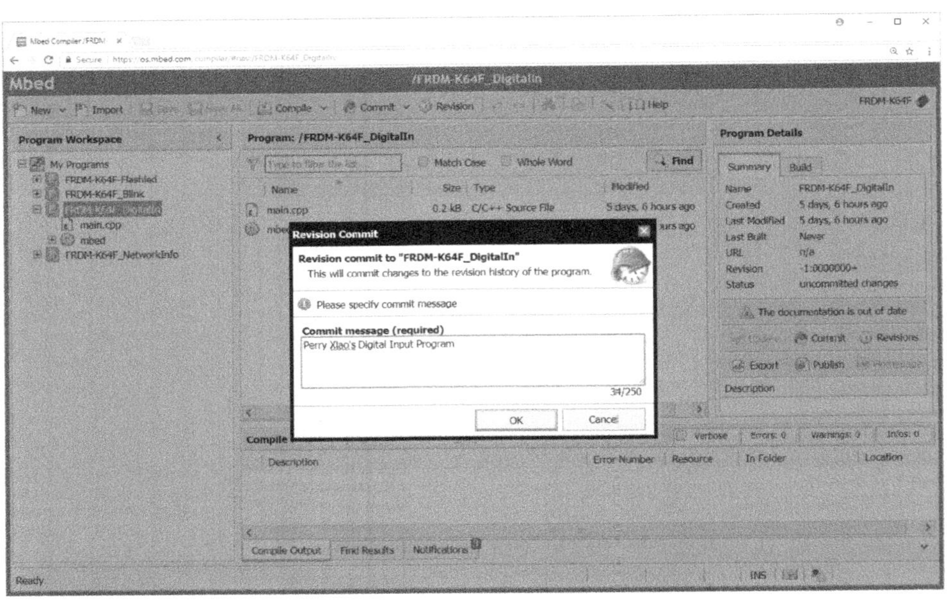

Figure 10.17 The Revision Commit pop-up window.

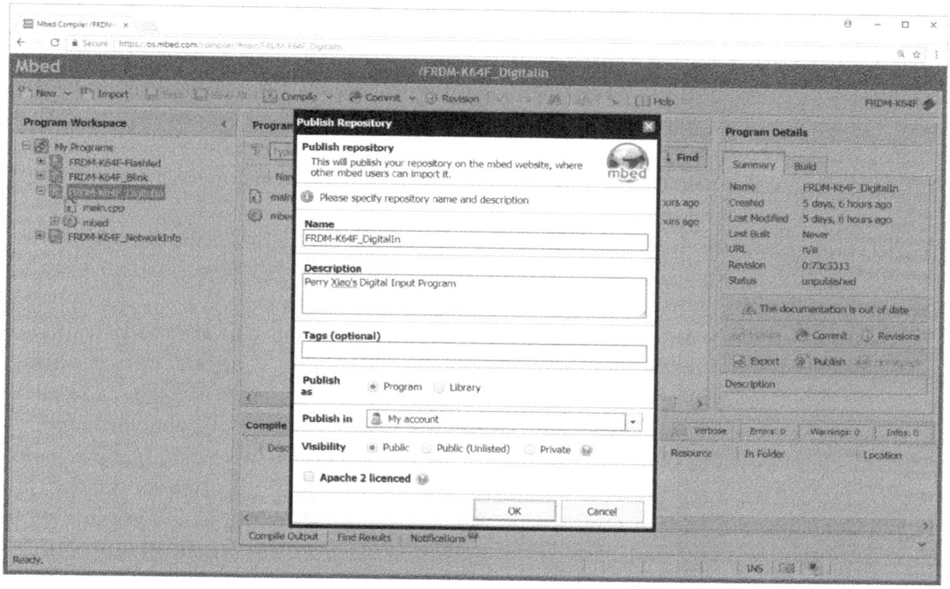

Figure 10.18 The Publish Repository pop-up window.

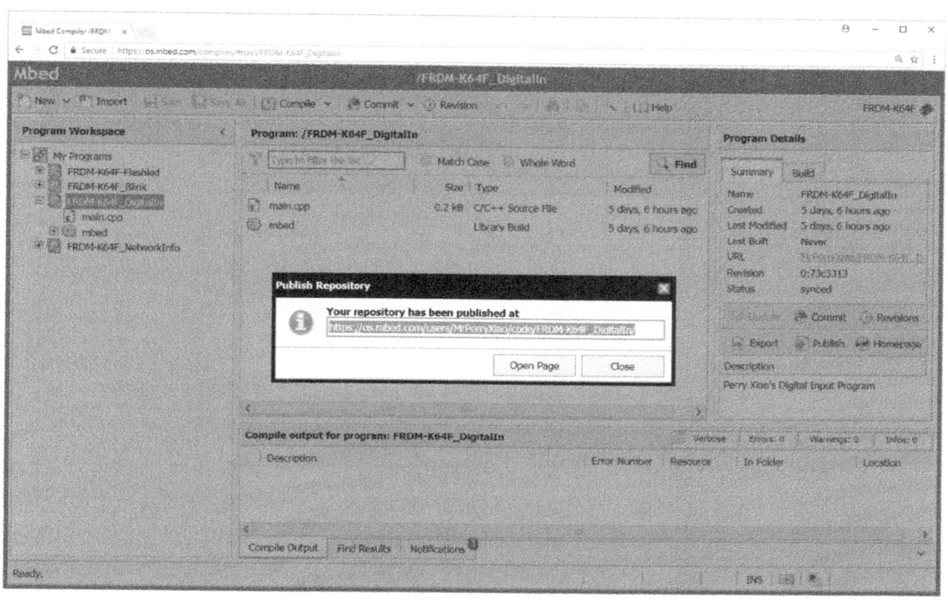

Figure 10.19 The Publish Repository confirmation window.

10.6 Version Control

Version control is very important in programming, especially for a large project, where a lot of Things could go wrong. The Arm® Mbed™ online compiler comes along with a powerful, built-in version control. With version control. you can easily view the history of different versions, compare the differences of different versions of your code, and go back to the previous version if necessary.

To use version control, from your online compiler, select the program, then click the "*Revision*" button. A "*Revision History*" window will be displayed (Figure 10.20). On the top of the list is you current working set program.

You continue to work on your program (Figure 10.21); when you are ready to finish the current version and move on to next version, just click the "*Commit*" button on the top. A "*Revision Commit*" window will pop up. Type in the "*Commit message*" and click "*OK*" button (Figure 10.22). Figure 10.23 shows the finished version of your program.

You can keep working on your program (Figure 10.24). In this case, *if(din.is_connected())* is added to your program. When you are ready to make another version, again, just click the "*Commit*" button on the top. "*Revision Commit*" window will pop up, type in the "*Commit message*" and click "*OK*" button (Figure 10.25). Again, Figure 10.26 shows all the versions of your program.

You simply repeat this process throughout your development, and whenever you would like to view the previous versions, just click the "*Revision*" button, and all the revision history are available from the "*Revision History*" window. Figure 10.27 shows a further updated program, with *else* structure added. Figures 10.28 and 10.29 show all the versions of your program before and after the *else* structure is added.

You can then compare versions, merge versions, and even switch back to the previous version. Figure 10.30 shows how to select the previous version, and Figure 10.31 shows the previous version program code, which is exactly the same as shown in Figure 10.21.

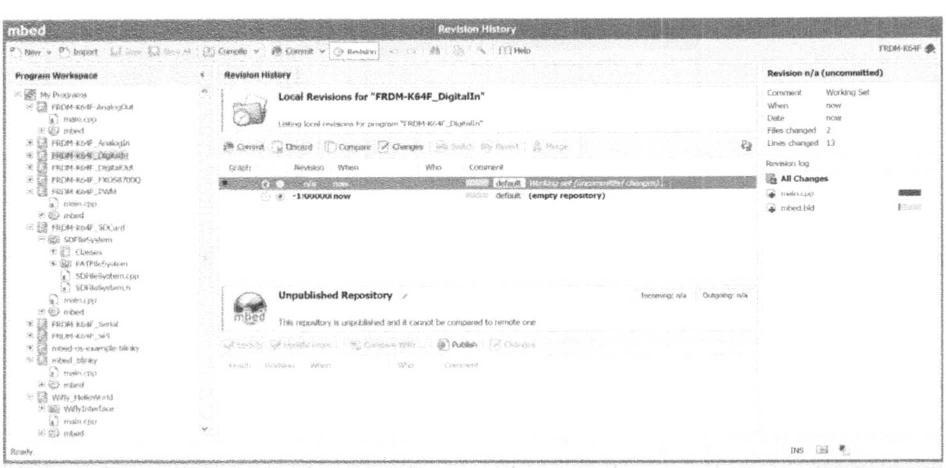

Figure 10.20 Revision History in your program page.

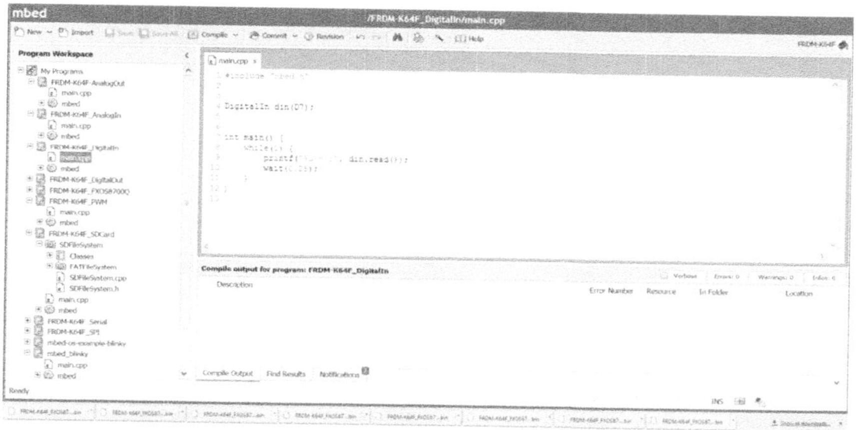

Figure 10.21 Your current program page.

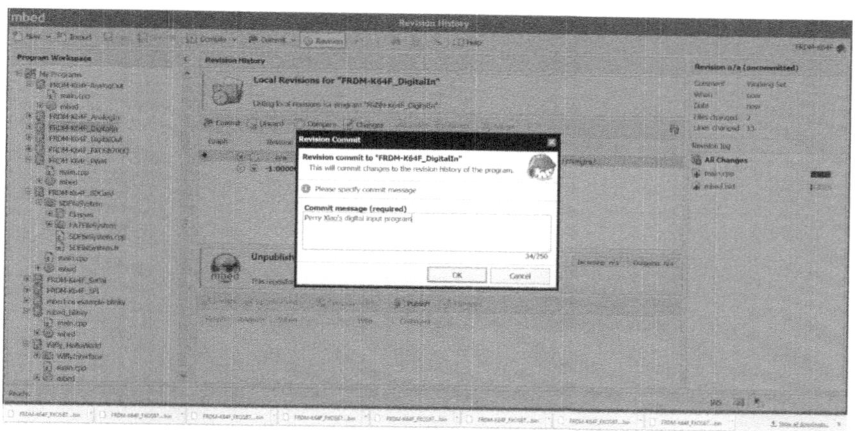

Figure 10.22 Revision Commit pop-up window.

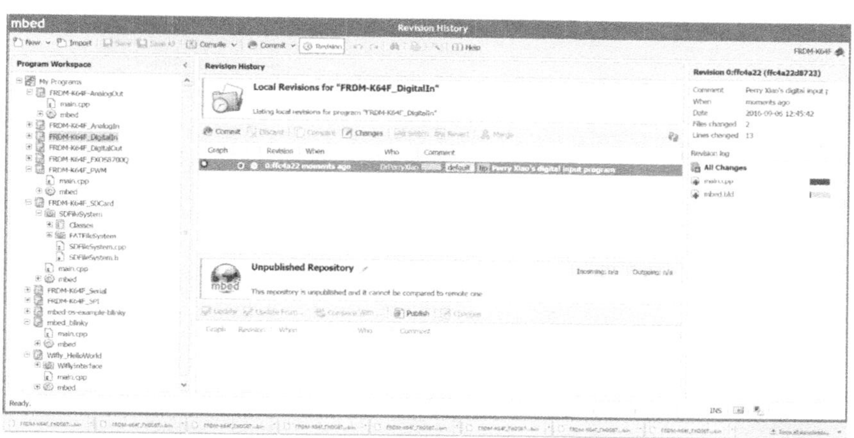

Figure 10.23 The finished version of your program shown in Revision History.

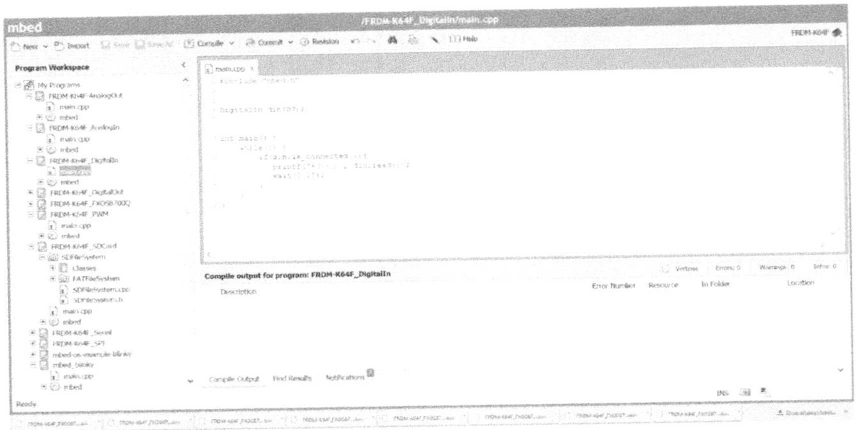

Figure 10.24 Your updated program with *if (din.is_connected())* added.

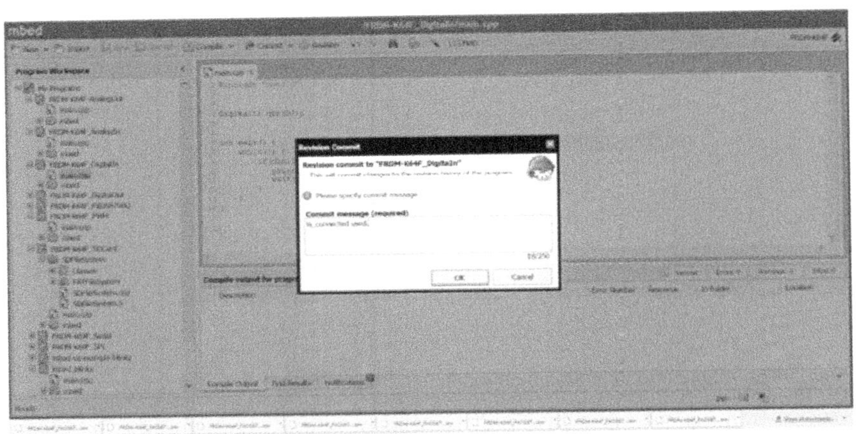

Figure 10.25 The Revision Commit pop-up window.

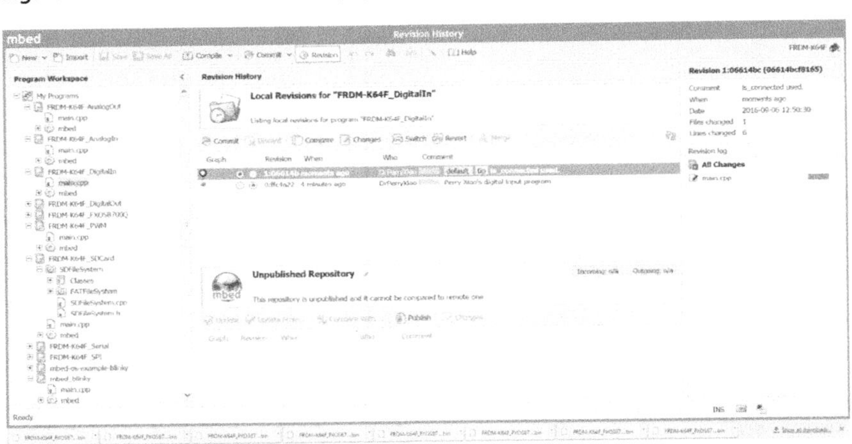

Figure 10.26 The updated versions of your program shown in Revision History.

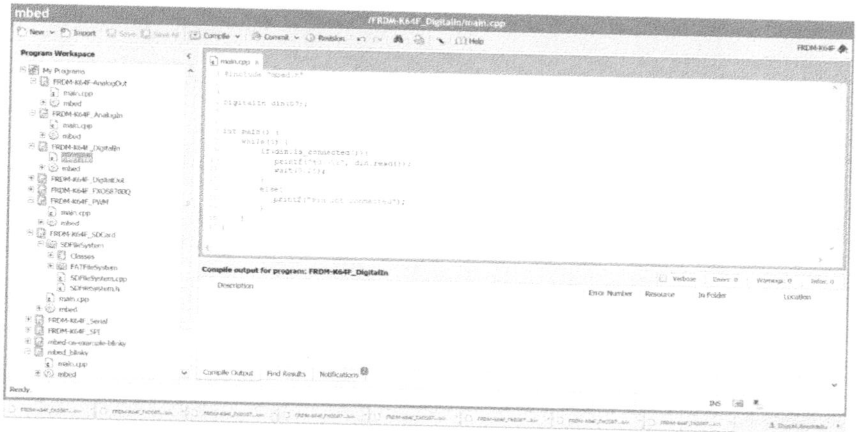

Figure 10.27 Your further updated program with *else* structure added.

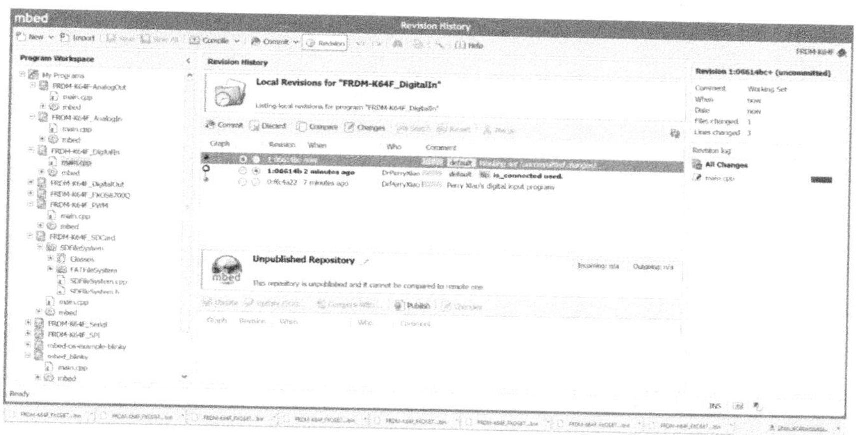

Figure 10.28 The versions of your program before *else* structure is added.

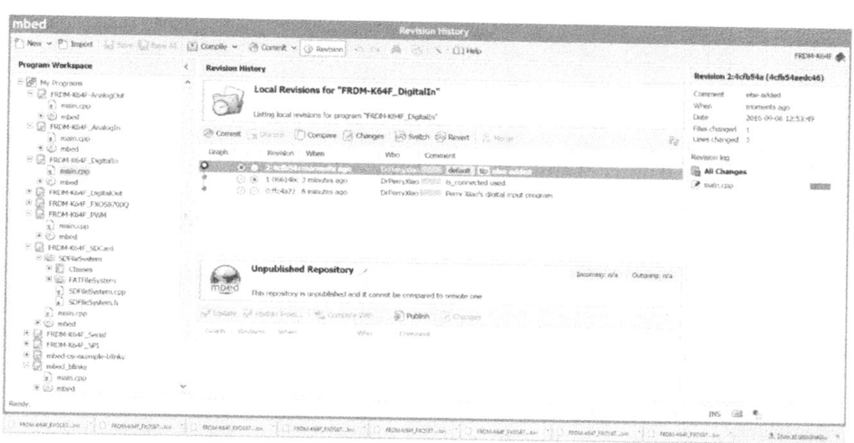

Figure 10.29 The versions of your program after *else* structure is added.

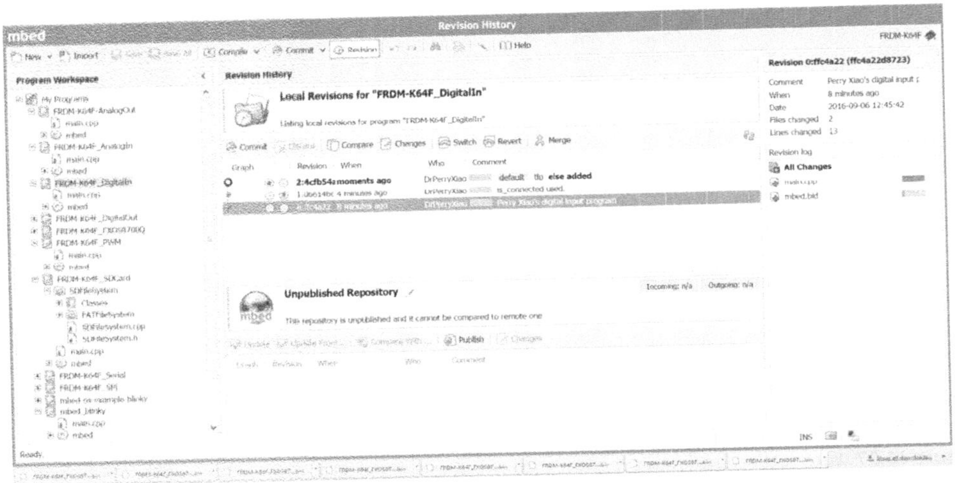

Figure 10.30 Select the previous version.

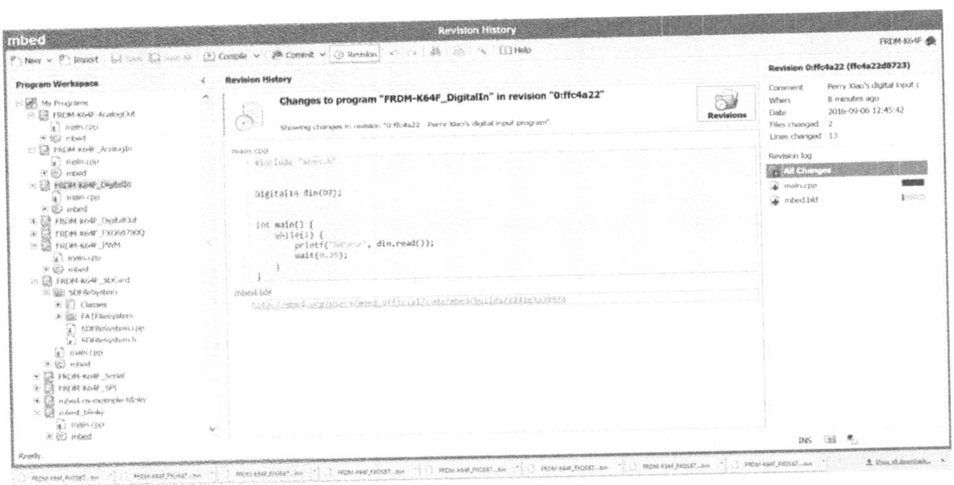

Figure 10.31 The program code of previous version.

Further Information about Version Control:

https://docs.mbed.com/docs/mbed-os-handbook/en/latest/collab/versions/
https://os.mbed.com/docs/v5.6/tools/collab-online-comp.html
https://os.mbed.com/docs/v5.6/tools/version-control.html

10.7 Collaborations

The Arm® Mbed™ online compiler also allows multiple users to work on the same program, i.e., collaborations. To add users to your program, first you need to publish your program, then from the "*Program Details*" panel on the right hand side, click "*Homepage*"

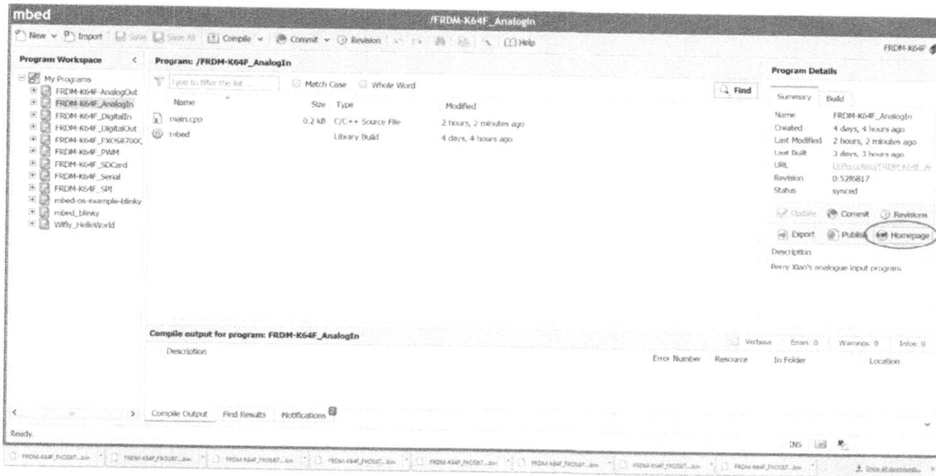

Figure 10.32 The *Homepage* button on the right-hand side of your program page.

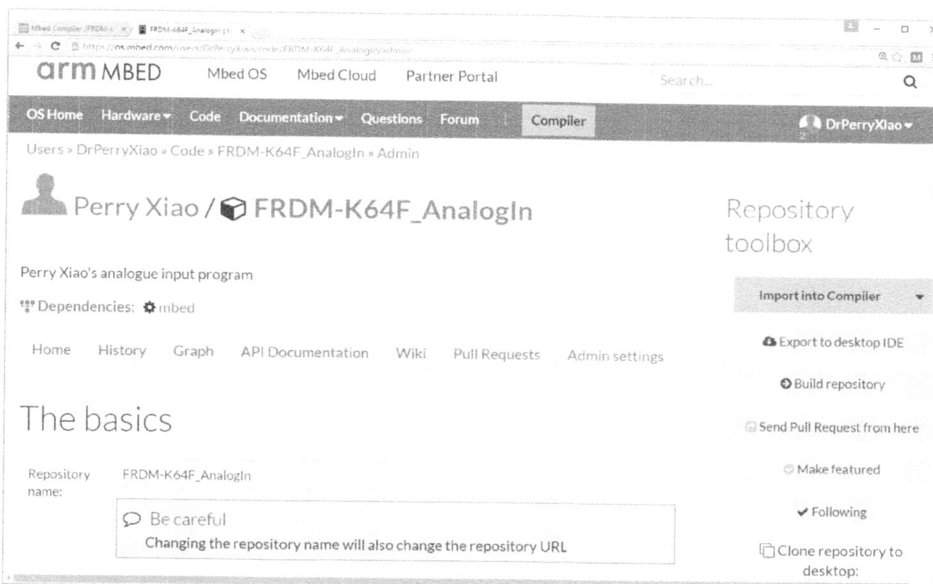

Figure 10.33 The *"Admin settings"* tab on your program repository page.

button, to go to your program homepage, also called your repository homepage (Figure 10.32).

Select the "*Admin settings*" tab (Figure 10.33), and somewhere in the middle of the page, there is "*Privacy Settings.*"

In the "*Privacy Settings*" (Figure 10.34), you can add one or more developers to your program. In this case, two developers are added, "Perry Xiao" is the original developer, and "Johnny English" is the additional developer. Remember to click the "*Save changes*" button to save the changes!

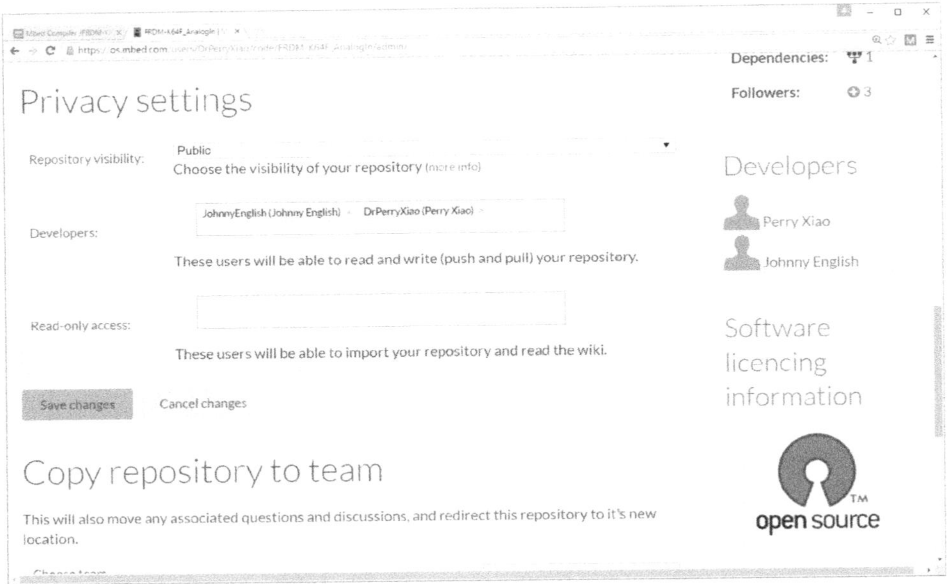

Figure 10.34 The *"Privacy Settings"* on your Program Repository page.

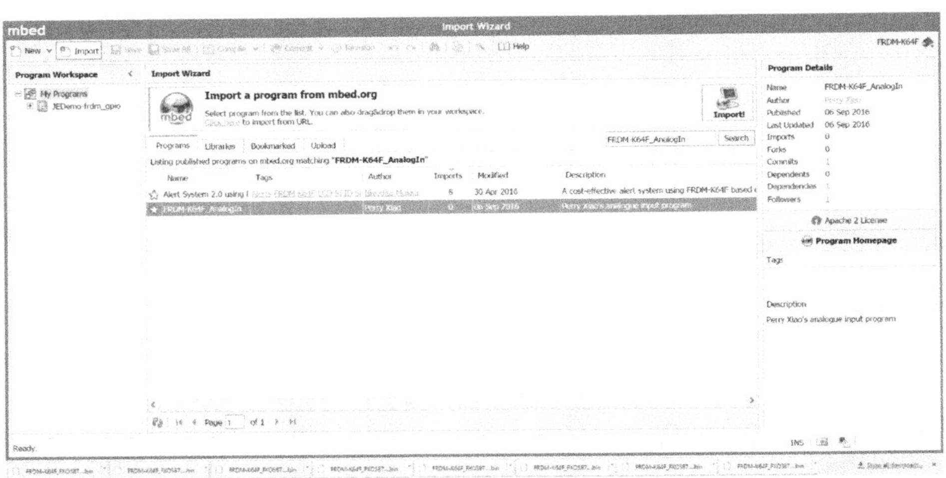

Figure 10.35 The additional developer can import your program into his workspace.

The additional developer will then be able to import the program (Figure 10.35) into his own program workspace, modify it, save it, commit to a version (Figure 10.36), etc. You will then be able to see the additional developer's version in your own Revision History page (Figure 10.37).

When ready, the additional user can publish the program back to the original repository homepage (Figure 10.38). Figure 10.39 shows the publication confirmation pop-up window.

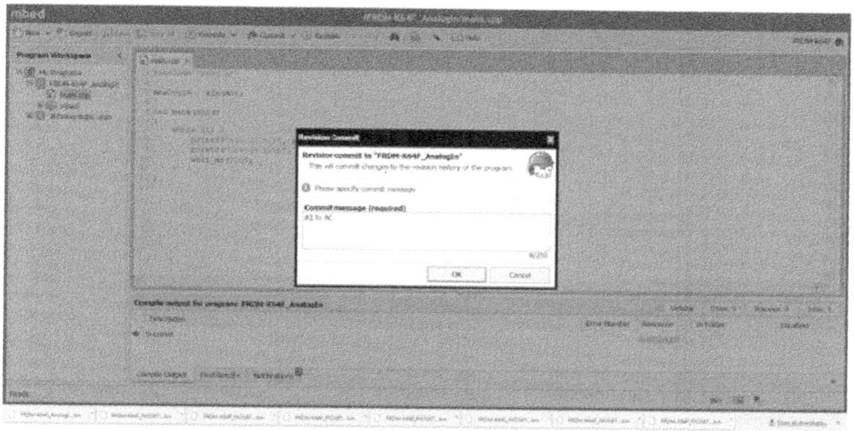

Figure 10.36 The additional developer can also commit to a revision of the program.

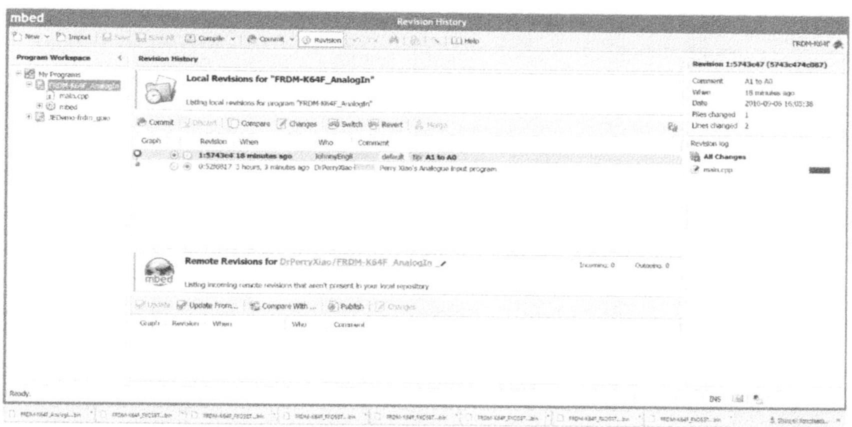

Figure 10.37 The revision history in your program shows additional developer's revision.

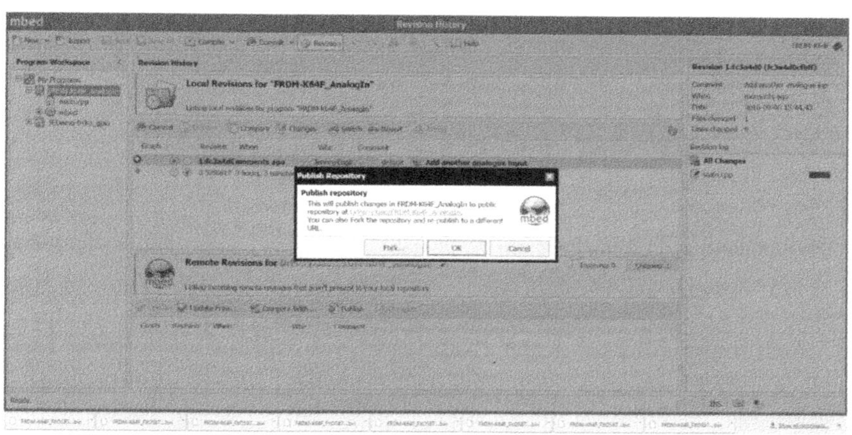

Figure 10.38 The additional developer can also publish the program.

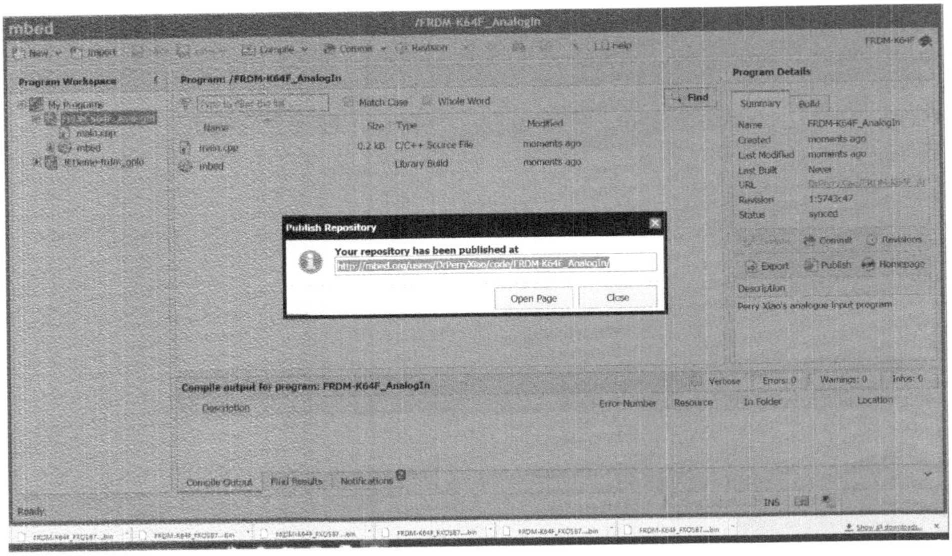

Figure 10.39 The additional developer publication confirmation pop-up window.

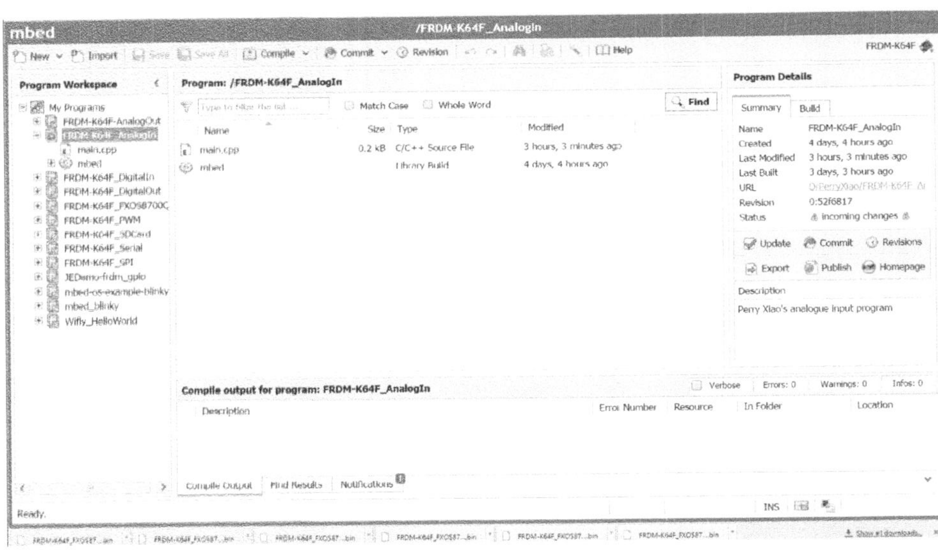

Figure 10.40 The *"Update"* button in the *"Program Details"* tab.

After that, when the original developer (you) log in, you will see an *"Update"* sign appear on the corresponding program. After you click the *"Update"* button in the *"Program Details"* tab, the program will be updated to the latest version (Figure 10.40). By clicking *"Revision"* button, you will also be able to see the revision history (Figure 10.41).

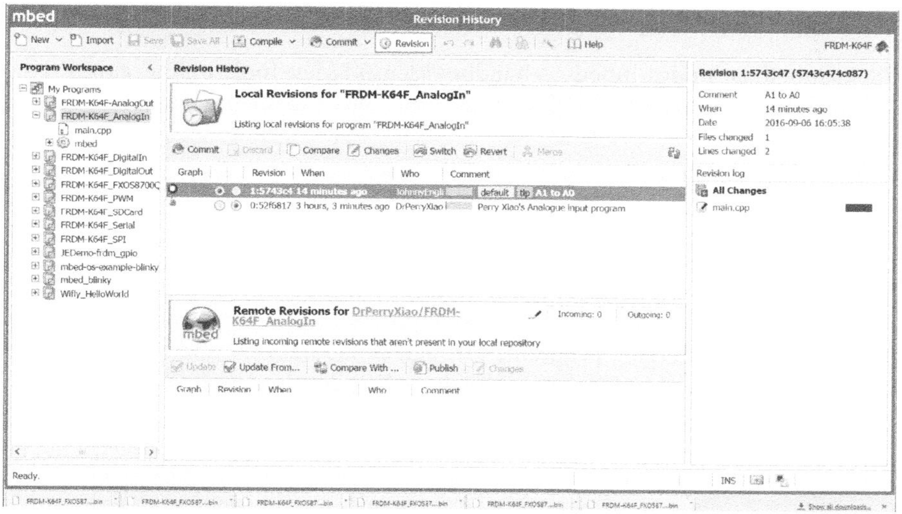

Figure 10.41 The updated Revision History.

Further Information about Collaborations:
https://os.mbed.com/docs/v5.6/tools/collaborate.html

10.8 Update Your Library and Program

Whenever your library or your program has a newer version available, a green cycle arrow will appear on your library or program icon (Figure 10.42). To update, just select your library or program and click the "*Update*" button on the right-hand "*Program Details*" panel.

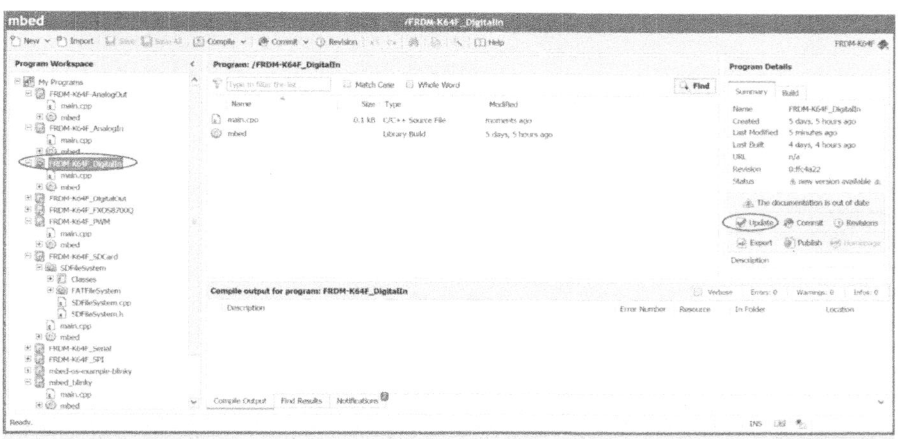

Figure 10.42 The "*Update*" button and green arrows on the right-hand "*Program Details*" panel.

Further Information about Arm® Mbed™ Libraries and Projects:

https://docs.mbed.com/docs/mbed-os-handbook/en/5.1/dev_tools/online_comp/
https://docs.mbed.com/docs/mbed-os-handbook/en/5.1/getting_started/blinky_compiler/

10.9 Summary

This chapter introduces how to import libraries and programs, how to export programs, how to write your own libraries, how to publish libraries and programs, how to perform version control, how to develop programs through collaborations, and how to update your libraries and programs in the Arm® Mbed™ online development environment.

Part III

The IoT Starter Kit and IoT Projects

In This Part

11

Arm® Mbed™ Ethernet IoT Starter Kit

It is never too late to be what you might have been.

- George Eliot

The Arm® Mbed™ Ethernet IoT Starter Kit comes with two components, FRDM-K64F development board and mbed application shield. As introduced in Chapter 1, section 1.4.4, the mbed application shield has many useful features, such as 128×32 LCD, joystick, RGB LED, two potentiometers, a speaker, three-axis accelerometer, and LM75B temperature sensor. These features are great for developing IoT applications. To use the kit, you will need to mount the mbed application shield on the top of the FRDM-K64F development board, and make sure all pins are fully pushed in (Figure 11.1). The NXP LPC1768 and its mbed application board, as described in Chapter 1, section 1.4.1, provide many similar features to the IoT Starter Kit, so for the purpose of backward compatibility, most of the examples here are made to also work on the NXP LPC1768 and its mbed application board.

11.1 128×32 LCD

The mbed application shield's onboard 128×32 LCD (liquid crystal display) is connected by pins D11, D13, D12, D7, and D10 to the FRDM-K64F board. Following is an example LCD program, which prints text "Hello World" at (0,3) position on the LCD and prints a counting variable at (0,15) position on the LCD. In this program, you will need to import the "*C12832*" LCD library (https://developer.mbed.org/users/chris/code/C12832/).

Designing Embedded Systems and the Internet of Things (IoT) with the ARM® Mbed™, First Edition. Perry Xiao

Companion website: www.wiley.com\go\xiao\designingembeddedsystemandIoTwitharmmbed

Figure 11.1 The Arm® Mbed™ Ethernet IoT Starter Kit.

```
*****************************************************************
// Example 11.1
#include"mbed.h"
#include"C12832.h"

#if defined(TARGET_K64F)              //FRDM-K64F IoT Starter Kit
    C12832lcd(D11,D13,D12,D7,D10);
#elif defined(TARGET_LPC1768)         //LPC1768 + Application board
    C12832 lcd(p5, p7, p6, p8, p11);
#endif

intmain()
{
    intj=0;
    lcd.cls();
    lcd.locate(0,3);
    lcd.printf("Hello World");

    while(true){
        lcd.locate(0,15);
        lcd.printf("Counting:%d",j);
        j++;
        wait(1.0);
    }
}
*****************************************************************
```

Exercise 11.1

Modify the above program so that it displays your name and your telephone number on the LCD.

11.2 Joystick

The mbed application shield's onboard joystick is connected by pins A2, A3, A4, A5, and D4 to the FRDM-K64F board.

Following is an example program for the Arm® Mbed™ application board that uses the joystick button. It reads the joystick inputs and print correspondingly to a computer through a serial port.

```
*******************************************************************
// Example 11.2
#include "mbed.h"

#if defined(TARGET_K64F)              //FRDM-K64F IoT Starter Kit
    DigitalIn up(A2);
    DigitalIn down(A3);
    DigitalIn left(A4);
    DigitalIn right(A5);
    DigitalIn centre (D4);

#elif defined(TARGET_LPC1768)    //LPC1768 + Application board
    DigitalIn up(p15);
    DigitalIn down(p12);
    DigitalIn left(p13);
    DigitalIn right(p16);
    DigitalIn center(p14);

#endif

int main()
{

    while (1) {
        while (1)  {
        if(up){
            printf("up\n\r");
        }
        if(down){
            printf("down\n\r");
        }
```

```
        if(left){
            printf("left\n\r");
        }
        if(right){
            printf("right\n\r");
        }
        if(centre){
            printf("center\n\r");
        }
        wait(0.2);
    }     }
}

***************************************************************
```

Exercise 11.2

Modify the above program so that it displays up, down, left, right, and press down on LCD when the joystick button is pressed.

11.3 Two Potentiometers

The mbed application shield's two onboard potentiometers (pot 1 and pot 2) are connected at pin A0 and A1 of the FRDM-K64F board. Following is a sample code to display the two potentiometers values on LCD.

```
***************************************************************
// Example 11.3
#include "mbed.h"
#include "C12832.h"

#if defined(TARGET_K64F)           //FRDM-K64F IoT Starter Kit
    C12832 lcd(D11, D13, D12, D7, D10);
    AnalogIn pot1 (A0);
    AnalogIn pot2 (A1);
#elif defined(TARGET_LPC1768)      //LPC1768 + Application board
    C12832 lcd(p5, p7, p6, p8, p11);
    AnalogIn pot1 (p19);
    AnalogIn pot2 (p20);
#endif

int main()
{
```

```
    while(1) {
        lcd.cls();
        lcd.locate(0,3);
        lcd.printf("P1 : %10.2f", (float)pot1);
        lcd.locate(0,15);
        lcd.printf("P2 : %10.2f", (float)pot2);
        wait(0.01);
    }
}
****************************************************************
```

Exercise 11.3

Modify the above program so that it displays the sum and differences of two potentiometers on LCD screen.

11.4 Speaker

The mbed application shield's onboard speaker is connected at pin D6 of the FRDM-K64F board. Following is an example code to play the sounds on speaker, ranging from 2000 Hz to 12,000 Hz, with 100 Hz interval.

```
****************************************************************
// Example 11.4

#include "mbed.h"

#if defined(TARGET_K64F)          //FRDM-K64F IoT Starter Kit
    PwmOut speaker(D6);
#elif defined(TARGET_LPC1768)     //LPC1768 + Application board
    PwmOut speaker(p26);
#endif

int main()
{
    for (int i=0; i<100; i++) {
        float f=i*100+2000;       //frequency 2000 Hz to 12000 Hz
        float T=1.0/f;            //Period
        speaker.period(T);
        speaker =0.5;
        wait(0.02);
    }
}
****************************************************************
```

Table 11.1 The Frequency of Music Notes.

C	D	E	F	G	A	B
261.63 Hz	293.66 Hz	329.63 Hz	349.23 Hz	392.00 Hz	440.00 Hz	493.88 Hz

In music, the frequency of notes can be calculated by the following formula:

$$f(n) = 2^{\frac{n-49}{12}} \times 440\text{Hz}$$

Where n is the nth key in piano. The note middle C is the 40th key in a standard piano, and has a frequency of 261.63 Hz. Table 11.1 shows the frequencies of seven basic notes.

Following is an example code to play middle C note (261.63 Hz) on a speaker.

```
************************************************************
// Example 11.5

#include "mbed.h"

#if defined(TARGET_K64F)              //FRDM-K64F IoT Starter Kit
    PwmOut speaker(D6);
#elif defined(TARGET_LPC1768)    //LPC1768 + Application board
    PwmOut speaker(p26);
#endif

int main()
{
    float f=261.63;         //frequency of middle C
    float T=1.0/f;              //Period
    speaker.period(T);
    speaker =0.5;
    wait(0.02);
}
************************************************************
```

Exercise 11.4

Use the information in Table 11.1 to modify the above program so that it plays other music notes.

Exercise 11.5

Write a program that plays the song "Twinkle Twinkle, Little Star."

11.5 Three-Axis Accelerometer

An accelerometer is an electromechanical device that will measure acceleration forces. The mbed application shield's onboard three-axis accelerometer, which uses I2C for communications, is connected by pins D14 and D15 (SDA, SCL) to the FRDM-K64F board. Following is an example to get the acceleration information from the *X*, *Y*, and *Z* axes. In this program, you will need to import the *MMA7660* accelerometer library (https://developer.mbed.org/users/Sissors/code/MMA7660/).

```
*************************************************************
// Example 11.6

#include "mbed.h"
#include "C12832.h"
#include "MMA7660.h"

#if defined(TARGET_K64F)          //FRDM-K64F IoT Starter Kit
    C12832 lcd(D11,D13,D12,D7,D10);
    MMA7660 MMA(D14,D15);               // I2C (SDA,SCL)
#elif defined(TARGET_LPC1768)    //LPC1768 + Application board
    C12832 lcd(p5, p7, p6, p8, p11);
    MMA7660 MMA(p28,p27);               // I2C (SDA,SCL)
#endif

int main()
{
    lcd.cls();
    while(1){
        lcd.locate(0,3);
        lcd.printf("x=%.2f y=%.2f z=%.2f",MMA.x(),MMA.y(),MMA.z());
        wait(0.1);
    }
}
*******************************************************************
```

Exercise 11.6

Modify the above program so that if *X*-axis values are larger than a certain value, it switches on red RGB LED; if *Y*-axis values are larger than a certain value, it switches on green RGB LED; and if *Z*-axis values are larger than a certain value, it switches on blue RGB LED.

11.6 LM75B Temperature Sensor

The mbed application shield's onboard LM75B temperature sensor, which also uses I2C for communications, is connected by pins D14 and D15 (SDA, SCL) to the FRDM-K64F board. Following is an example for the onboard LM75B temperature sensor. In this program, you will need to import the *LM75B* library (https://developer.mbed.org/users/chris/code/LM75B/).

```
**************************************************************
// Example 11.7

#include "mbed.h"
#include "LM75B.h"
#include "C12832.h"

#if defined(TARGET_K64F)            //FRDM-K64F IoT Starter Kit
    C12832lcd(D11,D13,D12,D7,D10);
    LM75B sensor(D14,D15);              // I2C (SDA,SCL)
#elif defined(TARGET_LPC1768)    //LPC1768 + Application board
    C12832 lcd(p5, p7, p6, p8, p11);
    LM75B sensor(p28,p27);              // I2C (SDA,SCL)
#endif

int main ()
{

       while (1) {
         lcd.cls();
         lcd.locate(0,3);
         lcd.printf("Temp = %.1f\n", sensor.read());
         wait(1.0);
       }
}
**************************************************************
```

Exercise 11.7

Modify the above program so that you can change the temperature display either in Celsius or Fahrenheit by pressing the joystick.

11.7 RGB LED

The mbed application shield's RGB LED is connected by pins D5, D8, and D9 to the FRDM-K64F board.

Following is an example program for the onboard RGB LED. It uses PWM to gradually light up RGB LED one by one. For RGB LED, value 1 means off, and 0 means fully on.

```
************************************************************
// Example 11.8
#include "mbed.h"

#if defined(TARGET_K64F)              //FRDM-K64F IoT Starter Kit
    PwmOut r (D5);
    PwmOut g (D8);
    PwmOut b (D9);
#elif defined(TARGET_LPC1768)         //LPC1768 + Application board
    PwmOut r (p23);
    PwmOut g (p24);
    PwmOut b (p25);
#endif

int main()
{
    r.period(0.001);
    while(1) {
        for(float i = 0.0; i < 1.0 ; i += 0.01) {
            r = 1.0 - i;
            g=1;
            b=1;
            wait (0.01);
        }
        for(float i = 0.0; i < 1.0 ; i += 0.01) {
            r=1;
            g = 1.0 - i;
            b=1;
            wait (0.01);
        }
        for(float i = 0.0; i < 1.0 ; i += 0.01) {
            r=1;
            g=1;
            b = 1.0 - i;
            wait (0.01);
        }
    }
}
************************************************************
```

Exercise 11.8

Modify the above program so that it displays red and blue colors at different intensities, depending on the inputs of two potentiometers.

Further Information about Ethernet IoT Starter Kit:

https://os.mbed.com/platforms/IBMethernetKit/
https://os.mbed.com/components/mbed-Application-Shield/

11.8 Summary

This chapter provides example codes for the Arm® Mbed™ Ethernet IoT Starter Kit, illustrating the usages of 128×32 LCD, joystick, RGB LED, two potentiometers, *speaker*, three-axis accelerometer, and LM75 temperature sensor.

12

IoT Projects with Arm® Mbed™

Reach for the stars.

- Christa McAuliffe

12.1 Temperature Monitoring over the Internet

Temperature measurement is one of the most fundamental, most commonly performed measurements. It can be temperature of a room, temperature of a person, or temperature of a device. Being able to monitor the temperature remotely, over the Internet, has many potential important applications. For example, many old people live alone. If they fall ill, it might be some time before it is discovered that they are in crisis. If we can remotely monitor their body temperature, then when they are sick, especially with a life-threatening illness, we can alert the doctors, healthcare providers, and relatives instantaneously. In this project, you will use the LM75B temperature sensor on the application shield as the temperature sensor, and Ethernet as the means to connect to the Internet. Figure 12.1 shows the schematic diagram of the project.

Hardware Required

- Arm® Mbed™ Ethernet IoT Starter Kit (FRDM-K64F + mbed application shield)
- Mini USB cable and Ethernet cable

Software Required

- An Internet browser

Procedure

Connect the Arm® Mbed™ Ethernet IoT Starter Kit to a computer using a mini USB cable, and connect it to the Internet using an Ethernet cable. There are many ways to monitor temperature over the Internet. The simplest way is to turn FRDM-K64F into a web server. The following example illustrates how to set up a web server, read the temperature sensor data, and print it on the LCD as well as on the web page. You need to import four libraries to run this code:

- *"LM75B"* library—for temperature sensor
  https://developer.mbed.org/users/chris/code/LM75B/

Designing Embedded Systems and the Internet of Things (IoT) with the ARM® Mbed™, First Edition. Perry Xiao

Companion website: www.wiley.com\go\xiao\designingembeddedsystemandIoTwitharmmbed

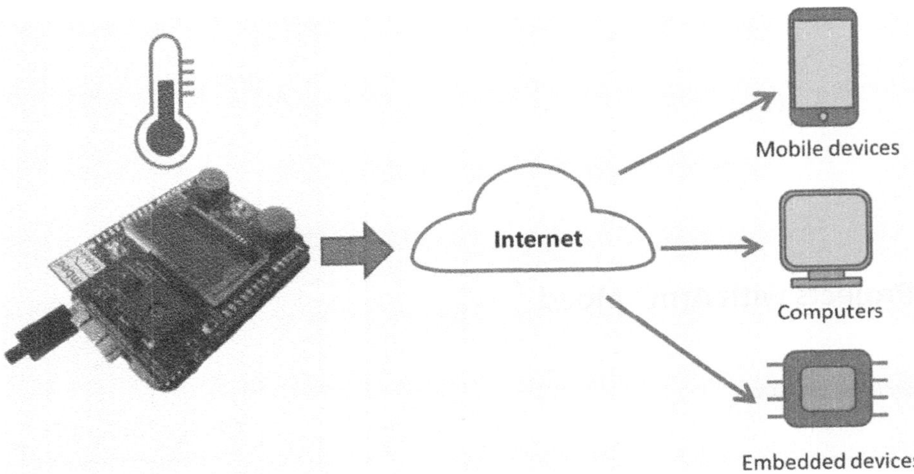

Figure 12.1 The schematic diagram of the temperature monitoring project over the Internet.

- "*C12832*" library—for LCD
  https://developer.mbed.org/users/chris/code/C12832/
- "*EthernetInterface*" library—for Ethernet connection
  https://os.mbed.com/users/mbed_official/code/EthernetInterface/
- "*mbed-rtos*" library—for EthernetInterface and multithreading
  https://os.mbed.com/users/mbed_official/code/mbed-rtos/

```
*****************************************************************
// Example 12.1

#include "mbed.h"
#include "EthernetInterface.h"
#include "rtos.h"
#include <stdio.h>
#include <string.h>
#include "LM75B.h"
#include "C12832.h"

#define PORT   80

bool serverIsListened = false;

TCPSocketConnection client;
bool clientIsConnected = false;
int mode=0;

C12832 lcd(D11, D13, D12, D7, D10);
LM75B sensor(D14,D15);
```

```
EthernetInterface eth;
TCPSocketServer server;
float temp=0;

void web_thread(void const *args){
  if(server.bind(PORT)< 0) {
       serverIsListened = false;
  } else {
      serverIsListened = true;
  }

  server.listen();

  //listening for http GET request
  while (serverIsListened) {
    if(server.accept(client)<0) {
        printf("failed to accept connection.\n\r");
    } else {
        printf("connection success!\n\rIP: %s\n\r",client.get_address());
        clientIsConnected = true;

        while(clientIsConnected) {
            char buffer[1024] = {};
            if(client.receive(buffer, 1023)<1){
                break;
            }
            else{
                printf("Received
Data: %d\n\r\n\r%.*s\n\r",strlen(buffer),strlen(buffer),buffer);
                if(buffer[0] == 'G' && buffer[1] == 'E' && buffer[2]
== 'T' && buffer[3] == ' ' && buffer[4] == '/' ) {
                    printf("GET request incoming.\n\r");
                    //set up http response header & data
                    char Body[1024] = {};
                    sprintf(Body,"Temp = %f \n\r\n\r",temp);
                    char Header[256] = {};
                    sprintf(Header,"HTTP/1.1 200 OK\n\rContent-Length:
%d\n\rContent-Type: text\n\rConnection: Close\n\r\n\r",strlen(Body));
                    client.send(Header,strlen(Header));
                    client.send(Body,strlen(Body));
                    clientIsConnected = false;
                  }
              }
          }
          printf("close connection.\n\r tcp server is listening...\n\r");
           client.close();
      }
```

```
    }
}

int main (void)
{
    eth.init(); //Use DHCP
    eth.connect();
    printf("\r\nServer IP Address is %s\r\n", eth.getIPAddress());

    Thread thread(web_thread);
    while (1) {
        lcd.cls();
        lcd.locate(0,3);
        temp=sensor.read();
        lcd.printf("Temp = %.1f\n", temp);

        printf("Temp = %.1f\n\r", temp);
        wait(1.0);
    }
}
*********************************************************************
```

In this case, you can write a simple Java TCP client to send commands to the starter kit; see the following example code and make sure you use the correct server IP address "x.x.x.x". Change the mode value to 0, 1, and 2, to switch the light off, on, and to automatic mode. You can compile and execute the code using an online Java compiler, such as this one (Figure 12.2):

http://www.tutorialspoint.com/compile_java_online.php

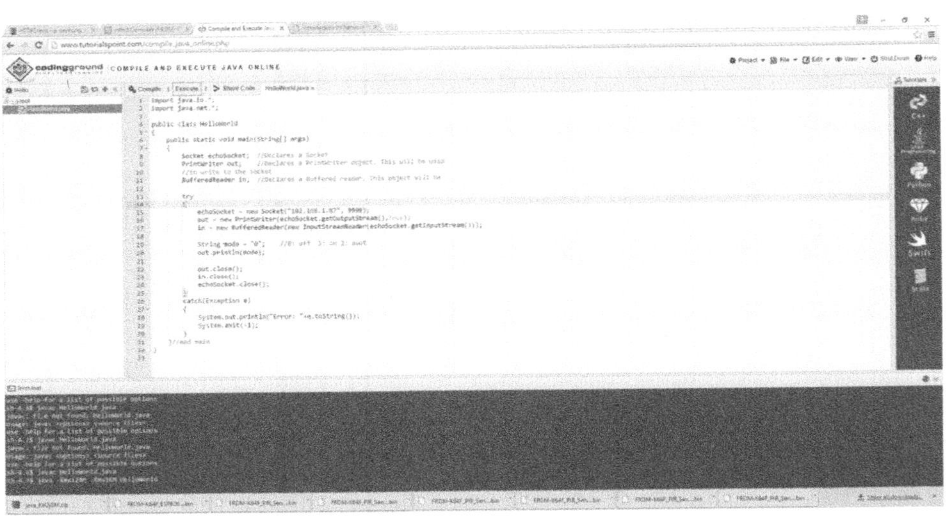

Figure 12.2 Online Java compiler.

```
*************************************************************
//   Example 12.2

import java.io.*;
import java.net.*;

public class TCPClient
{
   public static void main(String[] args)
   {
     Socket echoSocket;     //Declares a Socket
     PrintWriter out;       //Declares a PrintWriter object.
     BufferedReader in;
     try
     {
        echoSocket = new Socket("x.x.x.x", 9999);
        out = new PrintWriter(echoSocket.getOutputStream(),true);
        in = new BufferedReader(new
InputStreamReader(echoSocket.getInputStream()));

        String mode = "0";     //0: off  1: on 2: auto
        out.println(mode);

        out.close();
        in.close();
        echoSocket.close();
     }
     catch(Exception e)
     {
        System.out.println("Error: "+e.toString());
        System.exit(-1);
     }
   }//end main
}
*************************************************************
```

Alternatively, you can also set up your Arm® Mbed™ Ethernet IoT Starter Kit as a HTTP client, and use the POST method to update the temperature values to a remote web server. Following is a simple example to post some data to a web server (http://httpbin.org/post). In this code, you will need to import "*HTTPClient*" library:

https://os.mbed.com/users/donatien/code/HTTPClient/

```
*******************************************************************
// Example 12.3

#include "mbed.h"
#include "EthernetInterface.h"
#include "HTTPClient.h"

EthernetInterface eth;
HTTPClient http;
char str[512];

int main()
{
  eth.init();
  eth.connect();

  HTTPMap map;
  HTTPText inText(str, 512);
  map.put("Hello", "World");
  map.put("test", "1234");
  printf("\nTrying to post data...\n\r");
  ret = http.post("http://httpbin.org/post", map, &inText);
  if (!ret)
  {
   printf("Executed POST successfully - read %d characters\n",
strlen(str));
   printf("Result: %s\n\r", str);
  }
  else
  {
   printf("Error - ret = %d -
HTTP return code = %d\n\r", ret, http.getHTTPResponseCode());
  }

  eth.disconnect();

  while(1) { }
}
*******************************************************************
```

Exercise 12.1

Using the above program as an example, modify the example code 12.1 so that it sends the temperature reading to a web server using the POST method.

You can also send the temperature values to an email address. Following is a simple email example. Make sure you use the correct server, port, username, password, sender address, and receiver address when you are running it. In this example, you will need to import the "*SimpleSMTPClient*" library.

https://os.mbed.com/users/sunifu/code/SimpleSMTPClient/

```
****************************************************************
//  Example 12.4

#include "mbed.h"
#include "EthernetInterface.h"
#include "SimpleSMTPClient.h"

#define DOMAIN "gmail.com"
#define SERVER "smtp.gmail.com"
#define PORT "587" //25 or 587,465(OutBound Port25 Blocking )
#define USER "xxxx"
#define PWD "xxxx"
#define FROM_ADDRESS "xxxx@xxxx"
#define TO_ADDRESS "xxx@xxx"

#define SUBJECT "Test Mail"

int main()
{
    EthernetInterface eth;
    eth.init();
    eth.connect();

    SimpleSMTPClient smtp;
    int ret;
    char msg[]="Hello World";

    smtp.setFromAddress(FROM_ADDRESS);
    smtp.setToAddress(TO_ADDRESS);
    smtp.setMessage(SUBJECT,msg);

    ret = smtp.sendmail(SERVER, USER, PWD, DOMAIN,PORT,SMTP_AUTH_NONE);

    if (ret) {
        printf("Email Sending Error\r\n");
    } else {
        printf("Email Sending OK\r\n");
    }

    return 0;
}
****************************************************************
```

Exercise 12.2

Using the above program as an example, modify the example code 12.1 so that it sends the temperature reading out by email.

A much more elegant way is to use MQTT (Message Queuing Telemetry Transport) protocol. Following is an example program that connects to a MQTT broker (iot.eclipse.org) at port *1883*, creates a topic called "PX-Sensor," and publishes the data (Hello World) four times.

```
*****************************************************************
//  Example 12.5

#define MQTTCLIENT_QOS2 1
#include "MQTTEthernet.h"
#include "MQTTClient.h"

int arrivedcount = 0;

void messageArrived(MQTT::MessageData& md)
{
   MQTT::Message &message = md.message;
   ++arrivedcount;
}

int main(int argc, char* argv[])
{
   MQTTEthernet ipstack = MQTTEthernet();
   char* topic = "PX-Sensor";

   MQTT::Client<MQTTEthernet, Countdown> client =
MQTT::Client<MQTTEthernet, Countdown>(ipstack);

   char* hostname = "iot.eclipse.org";
   int port = 1883;

   int rc = ipstack.connect(hostname, port);

   MQTTPacket_connectData data = MQTTPacket_connectData_initializer;
   data.MQTTVersion = 3;
   data.clientID.cstring = "PX-Sensor";
   data.username.cstring = "testuser";
   data.password.cstring = "testpassword";
   if ((rc = client.connect(data)) != 0)
       printf("From MQTT connect: %d\n\r", rc);
```

```
    if ((rc = client.subscribe(topic, MQTT::QOS2, messageArrived)) != 0)
        printf("From MQTT subscribe: %d\n\r", rc);

    MQTT::Message message;

    for (int i=0;i<5;i++){
        // QoS 0
        char buf[100];
        sprintf(buf, "%d!  QoS 0 message \n", i);
        message.qos = MQTT::QOS0;
        message.retained = false;
        message.dup = false;
        message.payload = (void*)buf;
        message.payloadlen = strlen(buf)+1;
        rc = client.publish(topic, message);
        while (arrivedcount < 1)
            client.yield(100);
        wait(2);
    }

    client.unsubscribe(topic);
    client.disconnect();
    ipstack.disconnect();

    return 0;
}
*****************************************************************
```

Exercise 12.3

Using the above program as an example, modify the example code 12.1 so that it sends the temperature reading out at MQTT messages.

You can view the standard MQTT clients, such as IBM's WMQTT IA92 Java utility:

https://github.com/mqtt/mqtt.github.io/wiki/ia92

Just download the software and follow the instructions to install and run it. If you have not installed Java before, you do need to install Java first before you can run the program. Figure 12.3 shows the screenshot of the program and the messages it receives.

Further Information about Java:

https://www.java.com/en/
https://www.java.com/en/download/help/index_installing.xml
http://www.oracle.com/technetwork/topics/newtojava/learn-141096.html

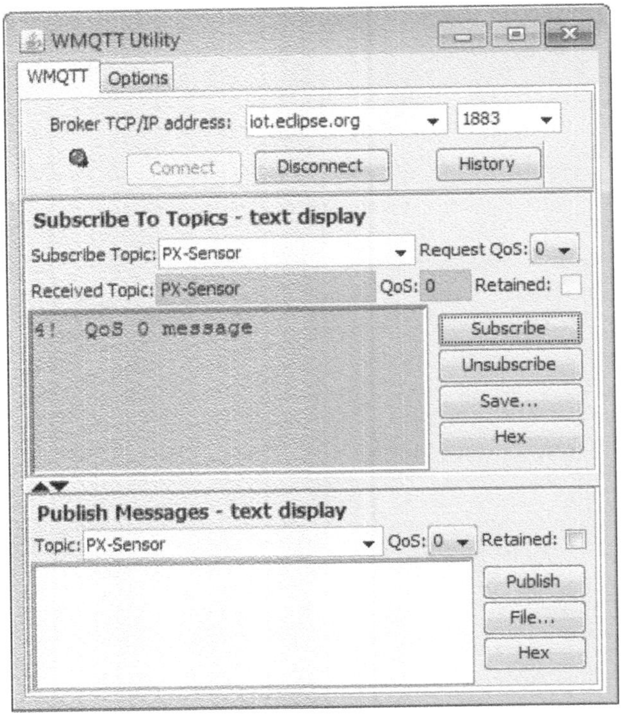

Figure 12.3 The WMQTT IA92 Java utility.

12.2 Smart Lighting

Lighting and heating are the two most significant parts of utility bill. Smart lighting can help to reduce costs. In this project, we use a LED to represent the light of a room, a PIR (passive infrared) sensor to detect if a person is present in the room, a LDR (light-dependent resister) to detect the ambient light, i.e., whether it is daylight, or night. Figure 12.4 shows the schematic circuit diagram of the project.

Hardware Required
- Arm® Mbed™ FRDM-K64F development board
- LED
- LDR + 10kΩ resister
- PIR sensor (HiLetgo HC-SR501)
- Mini USB cable and Ethernet cable

Software Required
- An Internet browser
- Java compiler

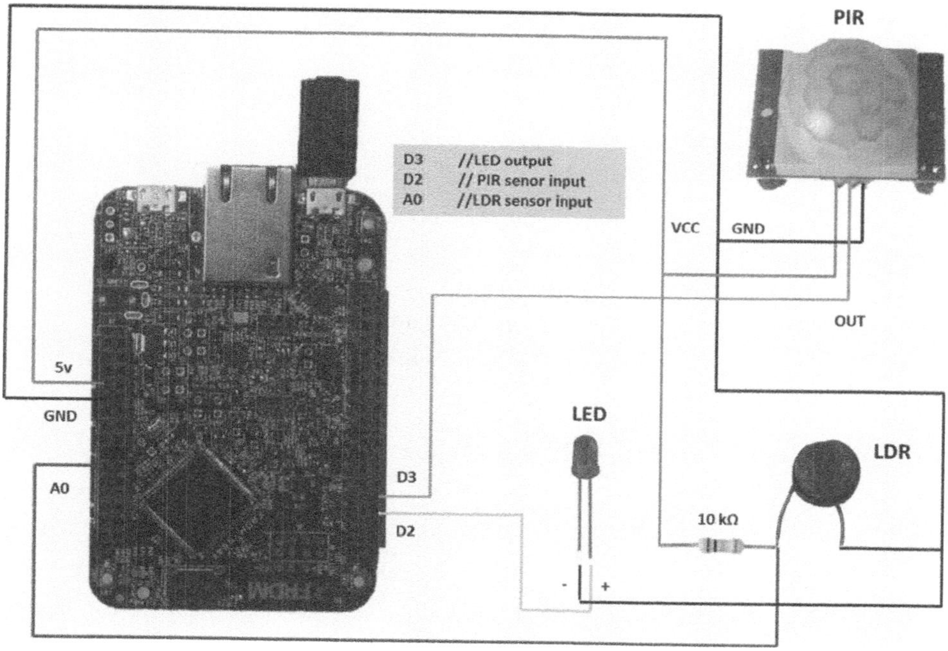

Figure 12.4 The schematic circuit diagram of the smart lighting project.

Procedure

Following are the schematic diagram of the circuit and corresponding software code. You will need to import the following libraries.

- *"EthernetInterface"* library—for Ethernet connection https://os.mbed.com/users/mbed_official/code/EthernetInterface/
- *"mbed-rtos"* library—for EthernetInterface and multithreading https://os.mbed.com/users/mbed_official/code/mbed-rtos/

The program runs in two separate threads, one for receiving commands from the Internet using TCP Socket and one for controlling the light. There are three modes:

- **Mode 0:** default off mode. In this mode, the light is always off.
- **Mode 1:** on mode. In this mode, the light is always on.
- **Mode 2:** automatic mode. In this mode, if there is someone present in the room and ambient light is dark, switch the light on; otherwise, keep the light off.

```
*****************************************************************
// Example 12.6

#include "mbed.h"
#include "EthernetInterface.h"
#include "rtos.h"
#define SERVER_PORT   9999

EthernetInterface eth;
DigitalOut light(D3);   //LED output
DigitalIn pir(D2);      // PIR senor input
AnalogIn ldr(A0);      //LDR sensor input

int val=0;
int mode = 0;      //0: off;  1: on;  2:  auto

void socket_thread(void const *args) {

    TCPSocketServer server;
    server.bind(SERVER_PORT);
    server.listen();

    while (true) {
        TCPSocketConnection client;
        server.accept(client);
        client.set_blocking(false, 1500); // Timeout after (1.5)s
        printf("Connection from: %s\n", client.get_address());
        char buffer[256];
        while (true) {
            int n = client.receive(buffer, sizeof(buffer));
            if (n <= 0) break;

            // print received message to terminal
            buffer[n] = '\0';
            printf("Received message from Client :'%s'\n",buffer);

            if(strcmp(buffer,"off")==0)
            {
                mode=0;
            }
            else if(strcmp(buffer,"on")==0)
            {
                mode=1;
            }
            else if(strcmp(buffer,"auto")==0)
            {
                mode=2;
            }   //0: auto;  1: on;  3:  off
```

```
            }
        client.close();
      }
  }

 void light_thread(void const *args) {
     while (true) {
        if (mode ==0)                                 //default off mode
        {
             light=0;
        }
        else if (mode ==1)                            // on mode
         {
             light=1;
         }
         else                                     // automatic mode
         {
             val = pir.read();
             if (val==0) {
                if (ldr.read()>0.7)               //LDR 1k ohm: full light
                {                                 //   40k omh: dark
                   light=1;
                }
             }
             else{
                light=0;
             }
         }
        Thread::wait(500);
     }
}
int main()
{
    eth.init();
    eth.connect();
    printf(" IP address: %s \r\n",eth.getIPAddress());

    Thread thread(socket_thread, NULL, osPriorityNormal,
DEFAULT_STACK_SIZE);
    Thread thread_2(light_thread, NULL, osPriorityNormal,
DEFAULT_STACK_SIZE);
    while(1){}
}
******************************************************************
```

Exercise 12.4

Modify the above example so that it uses UDP server to receive the messages.

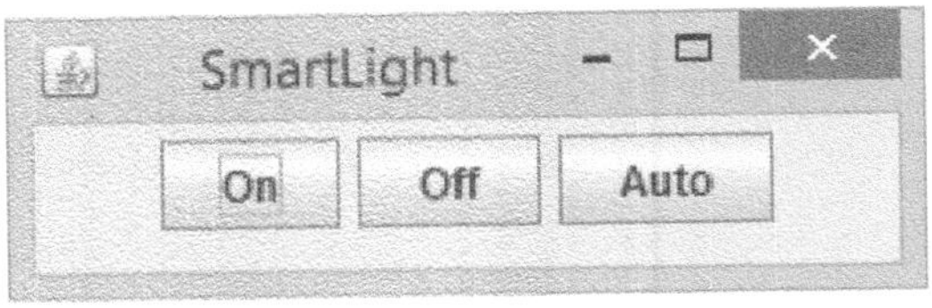

Figure 12.5 The Java Socket client program

Following is a Java Socket client example that can send *"on," "off,"* and *"auto"* command to the Arm® Mbed™ development board. Figure 12.5 shows its graphical user interface.

```
************************************************************
//  Example 12.7

import java.awt.*;
import java.awt.event.*;
import javax.swing.*;
import java.io.*;  //Imports java input-output libraries (for
StreamReaders)
import java.net.*; //Imports java network libraries (for sockets)

public class SmartLight {

    static String SERVER="192.168.137.1";
      static int PORT = 9999;
    /**
     * Create the GUI and show it.
     */
    private static void createAndShowGUI() {
        //Create and set up the window.
        JFrame frame = new JFrame("SmartLight");
        frame.setDefaultCloseOperation(JFrame.EXIT_ON_CLOSE);

        JButton openButton = new JButton("On");
        JButton closeButton = new JButton("Off");
        JButton autoButton = new JButton("Auto");
             frame.getContentPane().setLayout(new FlowLayout ());
        frame.getContentPane().add(openButton);
        frame.getContentPane().add(closeButton);
        frame.getContentPane().add(autoButton);

             openButton.addActionListener(new ActionListener() {
                 public void actionPerformed(ActionEvent e) {
                       sendcmd(SERVER, PORT,"on");
                 }
```

```
            });
            closeButton.addActionListener(new ActionListener() {
                  public void actionPerformed(ActionEvent e) {
                        sendcmd(SERVER, PORT,"off");
                  }
            });
            autoButton.addActionListener(new ActionListener() {
                  public void actionPerformed(ActionEvent e) {
                        sendcmd(SERVER, PORT,"auto");
                  }
            });
        //Display the window.
        frame.pack();
        frame.setVisible(true);
    }
       private static void sendcmd(String server, int port,String cmd)
       {

          Socket echoSocket;     //Declares a socket
        PrintWriter out;   //Declares a PrintWriter object to write
to the socket
        BufferedReader in; //Declares a Buffered reader to read
from the socket
          try
            {

          //Instantiates a new socket with the server IP address
and port number
          echoSocket = new Socket(server, port);

         //Creates a new output stream in order to write to the socket
          out = new PrintWriter(echoSocket.getOutputStream(),true);

         //Input from the socket with a bufferedreader
         in = new BufferedReader(new InputStreamReader(
            echoSocket.getInputStream()));

         //Writes the user input into the socket for transmission
         out.println(cmd);

         //Writes the received "echo" line to the screen
         JOptionPane.showMessageDialog(null, in.readLine(), "",
JOptionPane.INFORMATION_MESSAGE);

         //Close all the input and output streams
          out.close();
          in.close();
```

```
            echoSocket.close();
        }
        catch(Exception e)
        {
              System.out.println("Error: "+e.toString());
              System.exit(-1);
        }
      }//end sendcmd
   public static void main(String[] args) {
       //creating and showing this application's GUI.
       javax.swing.SwingUtilities.invokeLater(new Runnable() {
           public void run() {
               createAndShowGUI();
           }
       });
   }
}
****************************************************************
```

Further Information about Java Sockets:

http://docs.oracle.com/javase/tutorial/networking/sockets/
https://www.tutorialspoint.com/java/java_networking.htm

12.3 Voice-Controlled Door Access

Modern mobile phones come with many useful features and functions that can enrich your IoT projects. In this project, we will use the speech recognition feature of Android phones, to make a voice controller door access. We will develop the phone app using MIT App Inventor 2 (AI2), which is an excellent web-based, graphical programming tool, developed by the Massachusetts Institute of Technology in the United States. The phone app will use speech recognition to take commands. In this case, when the phrase "open sesame" is detected, it will send a command to FRDM-K64F development board to open the door through a servo motor. Figure 12.6 shows a schematic circuit diagram of the project.

Hardware Required
- Arm® Mbed™ Ethernet IoT Starter Kit (FRDM-K64F + mbed application shield)
- A servo motor
- An Android phone
- Mini USB cable and Ethernet cable

Software Required
- An Internet browser
- MIT AI2 online compiler

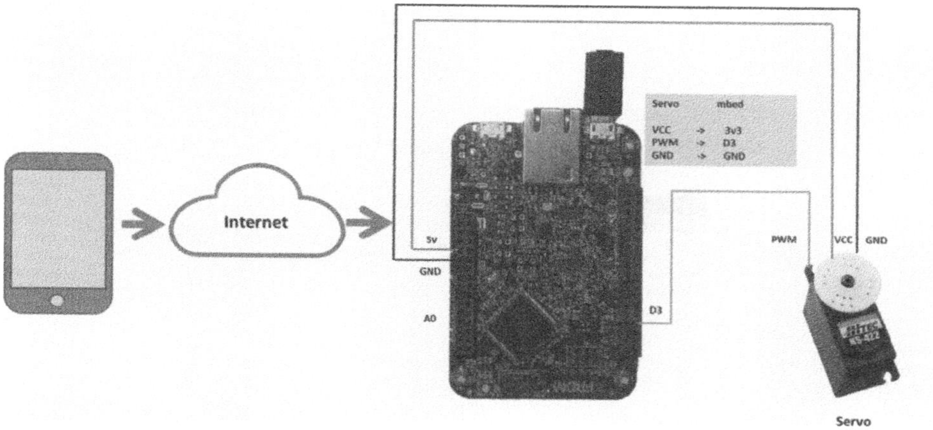

Figure 12.6 The schematic circuit diagram of the voice-controlled door access project.

Servo motors (or servos, RC servos etc.) are small, cheap, mass-produced motors typically having a drive wheel that is controlled by a PWM coded signal. A typical radio control (RC) servo is shown in Figure 12.6. The wheel moves around 0 to 180 degrees. Servo motors are ideal for hobbyist and student robotics applications. You can easily get servo motors from Amazon, Sparkfun, eBay, Cool Components, etc. Hitec and Futaba are the leading servo manufacturers.

Procedure

To use MIT AI2, just log in to the MIT AI2 website:

http://ai2.appinventor.mit.edu

Then follow the instructions to register and log in. You can also use your Google account to log in. After logging in, you can create a new project by clicking "Projects -> Start new project". You need to give a name to your project—in this example, we called it "IoTProject" (Figure 12.7). The "Viewer" window in the middle shows the front end of your phone app, i.e., how it looks when running.

From the left side "Pallette," under the section of "User Interface," drag a "Button," a "Textbox," a "Label" into the screen. This will be the graphic interface of your phone app. Then, from the section of "Media," drag a "SpeechRecognizer" component into the screen. Please note, this is an invisible component of your phone app. From the section of "Connectivity," drag a "Web" component into the screen. This is also an invisible component.

Next click the "Blocks" button on the top right corner, this will bring the backend of your phone app (Figure 12.8). You can switch between the frontend view and backend view of your phone app by clicking the "Designer" button and "Blocks" button.

From the backend view, create your program using blocks as illustrated in Figure 12.8. To compile your program, select *"Build -> App (provide QR code for .apk)"* as shown in Figure 12.9. After successful compilation, a 2D QR code will pop up, as shown in Figure 12.10. Use your mobile phone to scan the QR code to install the phone app.

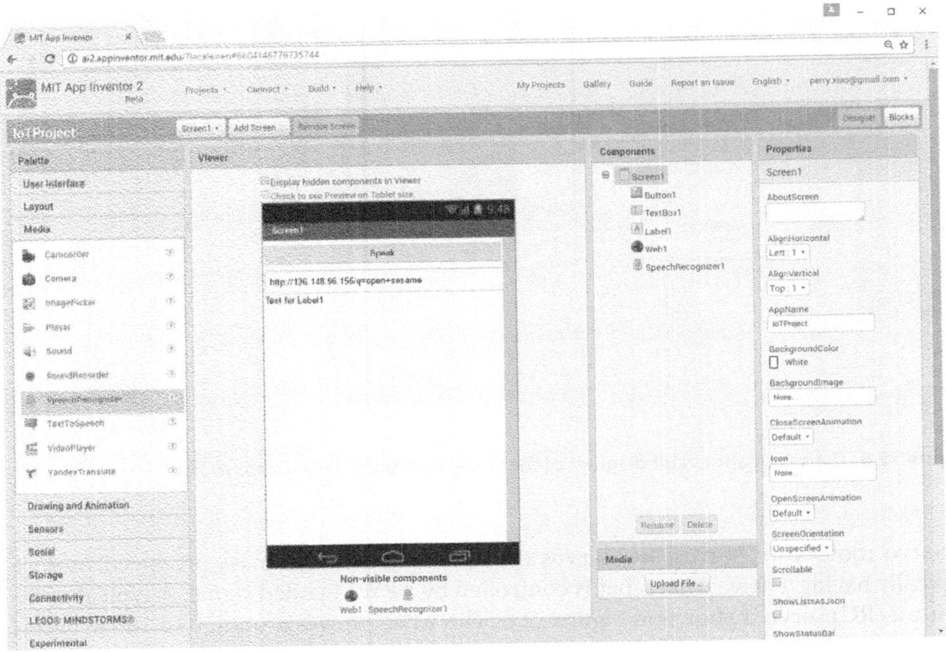

Figure 12.7 The MIT AI2 project development web page, Designer view (frontend).

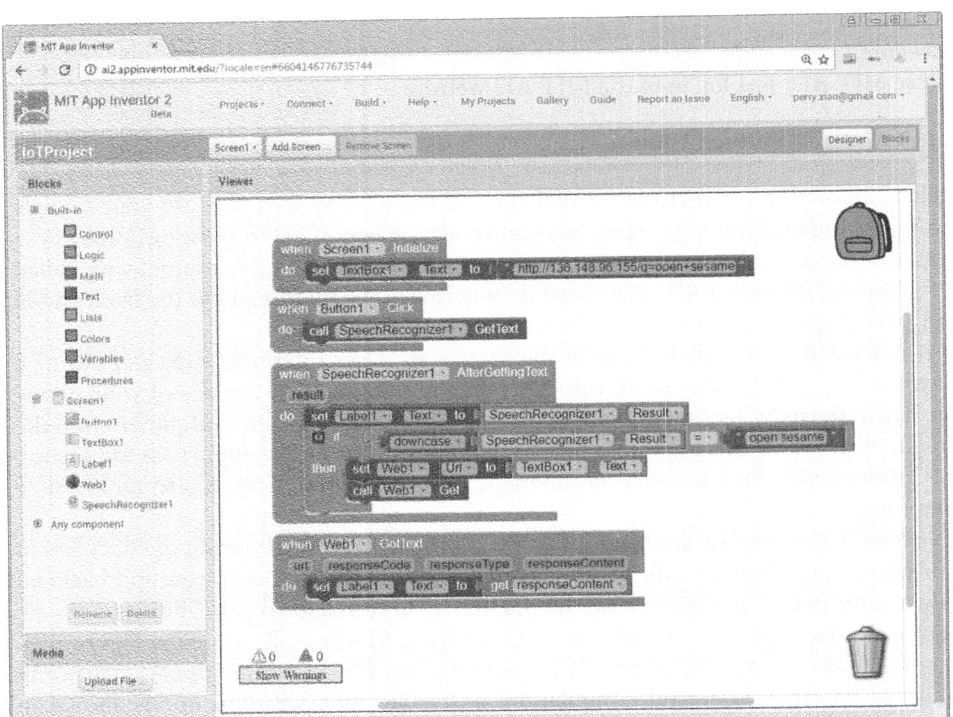

Figure 12.8 The MIT AI2 project development web page, Blocks view (backend).

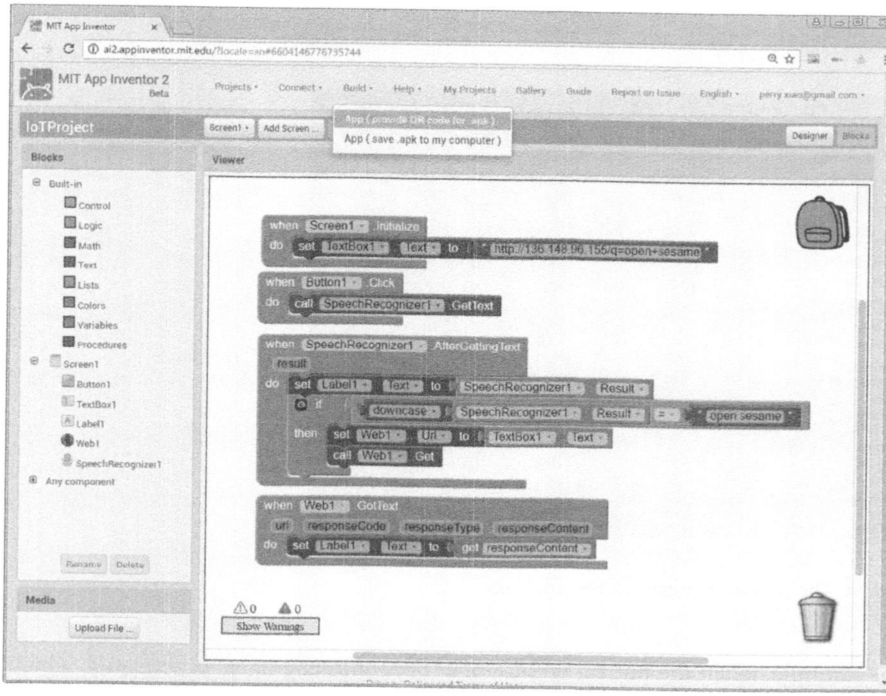

Figure 12.9 The MIT AI2 project compilation.

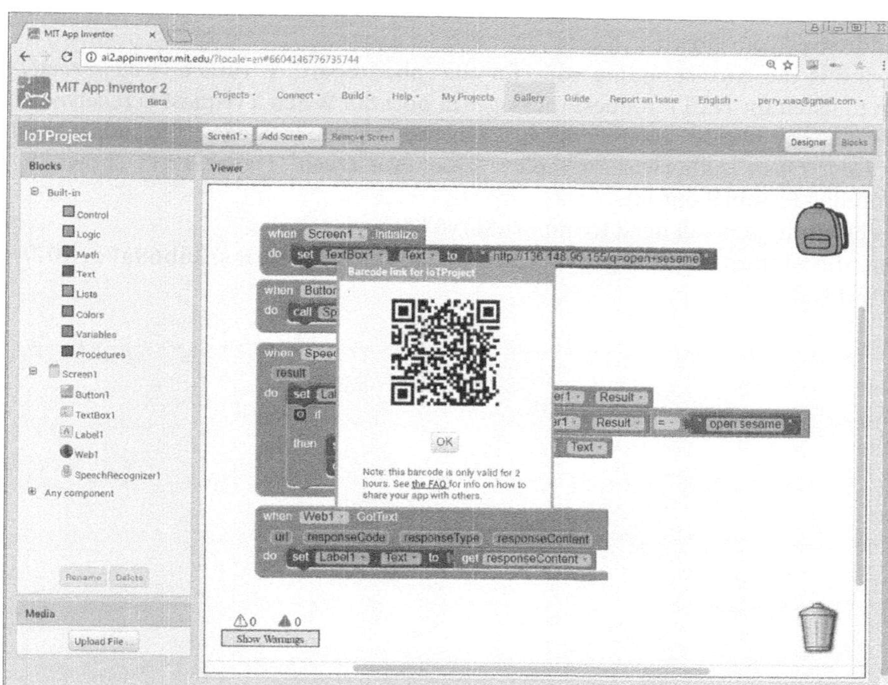

Figure 12.10 The QR code of MIT AI2 project.

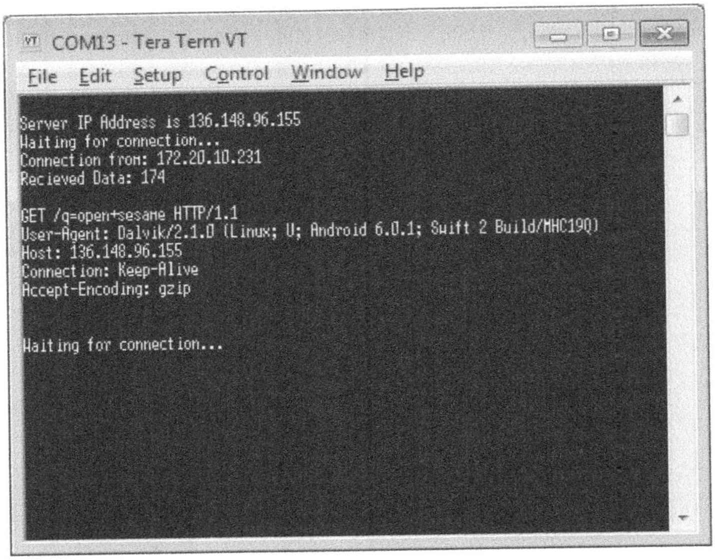

Figure 12.11 The Tera Term output of the program.

In this example, when the button is clicked, a "SpeechRecognizer" is activated, which will listen to what you speak and convert it to text, shown in the label. If what you said is a secret phrase, e.g., "open sesame", it will send a web request "http://x.x.x.x/q=open+sesame" to you FRDM-K64F development board, where "x.x.x.x" should be the IP address of your board.

Following is the corresponding code for the mbed board. It runs a *"web_server()"* function to listen for HTTP request messages at port 80. When a request is received, it looks for the key phrase *"q=open+sesame"* in the request message. If found, it then replies *"Door Open"*; otherwise, it replies *"Door Not Open."* Figure 12.11 shows the corresponding terminal outputs.

In this example, you will need to import Servo library:

https://developer.mbed.org/users/simon/code/Servo/docs/36b69a7ced07//classServo.html

```
******************************************************************
//  Example 12.8

#include "mbed.h"
#include "EthernetInterface.h"
#include <stdio.h>
#include <string>
#include "rtos.h"
#include "Servo.h"

#if defined(TARGET_K64F)
    Servo myservo(D9);
```

```
#elif defined(TARGET_LPC1768)
    Servo myservo(p21);
#endif

#define PORT   80

void web_server(void const *args)
{
    TCPSocketServer server;
    TCPSocketConnection client;

    server.bind(PORT);
    server.listen();

    while(true){
        printf("Waiting for connection...\r\n");
        int32_t status = server.accept(client);
        printf("Connection from: %s\r\n", client.get_address());

        if (status>=0)
        {
          char buffer[1024] = {};
          int n= client.receive(buffer, 1023);
          printf("Received Data:
%d\n\r\n\r%.*s\n\r",strlen(buffer),strlen(buffer),buffer);

          //GET /q=open+sesame HTTP/1.1
          char item[13];
          for(int k=0; k<13; k++){
              item[k]= buffer[k+5];
          }

          char Body[1024] = {};

          if (strcmp(item,"q=open+sesame")==0){
              sprintf(Body,"<html><title></title><body><h1>Door
Open</h1></body></html>\n\r\n\r");
              //move the servo to open the door,
              myservo = 1;
              // wait for 5 seconds, close the door
              wait(5);
              //close the door
              myservo = 0;
          }
          else{
             sprintf(Body,"<html><title></title><body><h1>Door Not
Open</h1></body></html>\n\r\n\r");
```

```
                //do nothing with the door
            }


            char Header[256] = {};
        sprintf(Header,"HTTP/1.1 200 OK\n\rContent-Length: %d\n\rContent-
Type: text/html\n\rConnection: Keep-Alive\n\r\n\r",strlen(Body));
            client.send(Header,strlen(Header));
            client.send(Body,strlen(Body));

            client.close();
        }
    }
}
int main() {
    EthernetInterface eth;
    eth.init();
    eth.connect();
    printf("\r\nServer IP Address is %s\r\n", eth.getIPAddress());

    //close the door
    myservo = 0;
    //wait for instructions
    web_server("");
    while(1){}
}
*************************************************************
```

You can also test your program using a web browser; just type in "http://x.x.x.x/q=open+sesame" as the URL to connect to your FRDM-K64F development board, where "x.x.x.x" should be the IP address of your board. You should get the same results.

Exercise 12.5

Modify the above example so that it also checks the client's IP address, and only messages from allowed client addresses can be accepted.

Further Information about MIT App Inventor 2:

http://appinventor.mit.edu/explore/
http://appinventor.mit.edu/explore/ai2/tutorials.html

Further Information about mbed Servo Motor:

https://os.mbed.com/users/4180_1/notebook/an-introduction-to-servos/
https://os.mbed.com/cookbook/Servo
https://os.mbed.com/users/simon/code/Servo/docs/36b69a7ced07/classServo.html

12.4 RFID Reader

RFID is promising technology that has been increasingly used for tracking and identification. In this project, we will illustrate how to use FRDM-K64F development board and SunFounder 13.56 MHz RC522 RFID reader and tags kit (Figure 12.12) to make an RFID reader.

Hardware Required

- Arm® Mbed™ FRDM-K64F development board
- RC522 RFID reader and tags
- Mini USB cable and Ethernet cable

Software Required

- An Internet browser

Figure 12.13 shows the wiring of FRDM-K64F development board and RFID-RC522 reader.

Procedure

Following is an example code for the RFID reader. It first initializes the RFID reader using the "*RFID.PCD_Init()*" function, then uses the "*RFID.PICC_IsNewCardPresent()*" function to check if a new RFID tag is present. It uses the "*RFID.PICC_ReadCardSerial()*" to read the information out of the tag, and finally, prints out the details (Figure 12.14).

In this program, you will need the RFID MFRC522 Library:

https://os.mbed.com/users/AtomX/code/MFRC522/

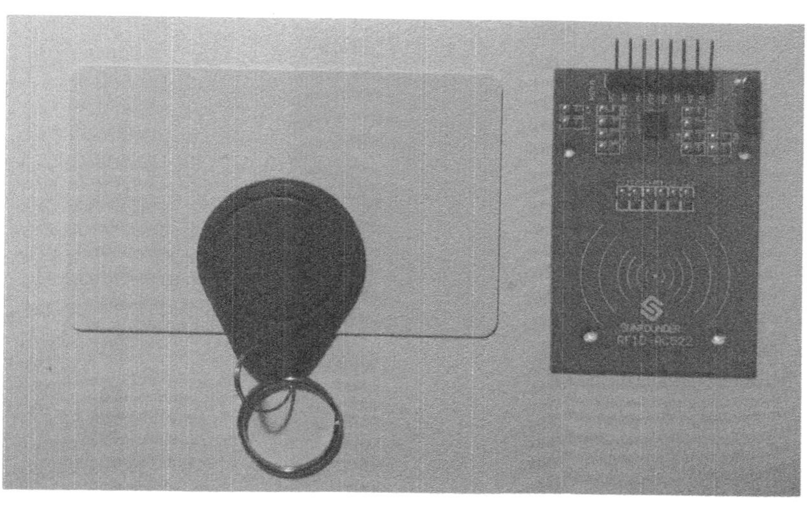

Figure 12.12 The SunFounder 13.56 MHz RFID-RC522 reader and RFID tags kit.

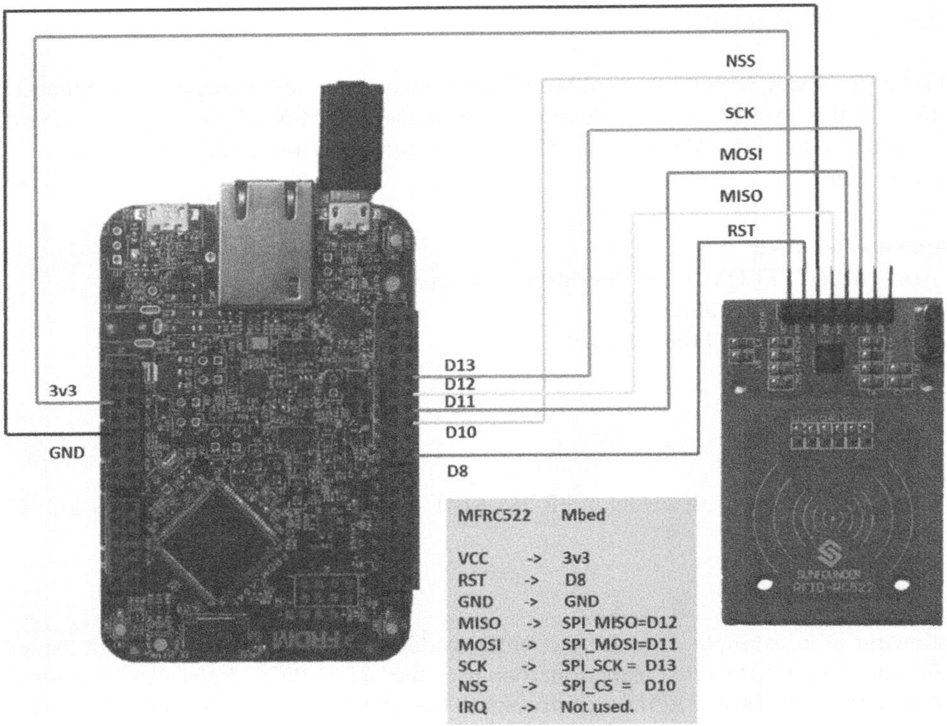

Figure 12.13 The wiring of FRDM-K64F development board and RFID-RC522 reader.

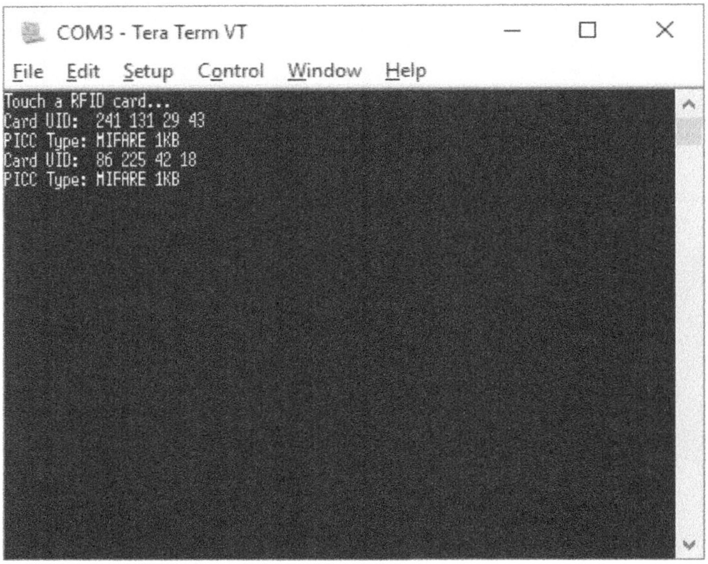

Figure 12.14 The output of RFID-RC522 reader.

```
*****************************************************************
// Example 12.9

#include "mbed.h"
#include "MFRC522.h"

//MFRC522  RfChip  (SPI_MOSI, SPI_MISO, SPI_SCK, SPI_CS, MF_RESET);
MFRC522    RFID   (D11, D12, D13, D10, D8);

int main(void) {
  printf("Touch a RFID card...\r\n");

  // Init. RC522 Chip
  RFID.PCD_Init();

  while (true) {

    // Look for new cards
    if ( ! RFID.PICC_IsNewCardPresent())
    {
      wait_ms(500);
      continue;
    }

    // Select one of the cards
    if ( ! RFID.PICC_ReadCardSerial())
    {
      wait_ms(500);
      continue;
    }

    // Print Card UID
    printf("Card UID: ");
    for (uint8_t i = 0; i < RFID.uid.size; i++)
    {
       printf(" %d", RFID.uid.uidByte[i]);
    }
    printf("\n\r");
    // Print Card type
    uint8_t piccType = RFID.PICC_GetType(RFID.uid.sak);
    printf("PICC Type: %s \n\r", RFID.PICC_GetTypeName(piccType));

    wait_ms(500);
  }
}
*****************************************************************
```

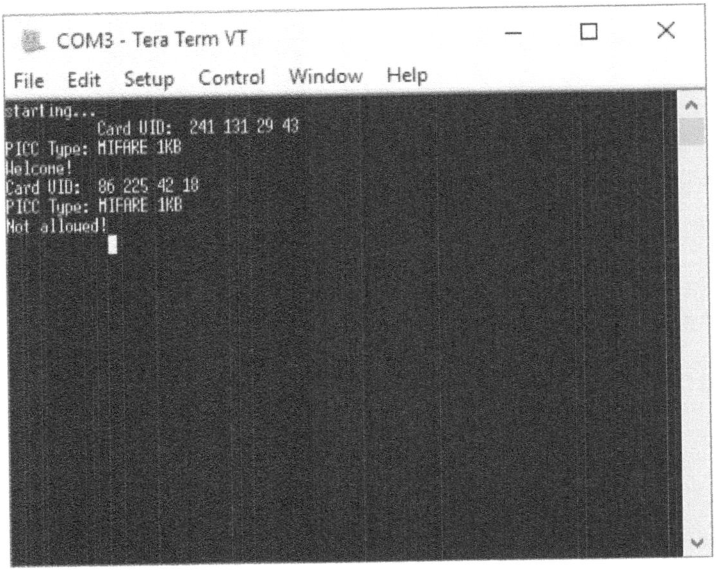

Figure 12.15 The output of RFID-RC522 reader.

With the RFID reader you can also create an access control, such as door access control, depending on the tags, and door access can be granted or refused. In the following example, a list of tag IDs that can be granted for access is created in an array called "*int cards[][4]*". If the tag's ID matches one of the IDs in the list, it will display "Welcome!" and "*LedGreen*" will light up; otherwise, it will display "Not Allowed!" and "*LedRed*" will light up. Figure 12.15 shows the corresponding terminal output.

```
*****************************************************
// Example 12.10

#include "mbed.h"
#include "MFRC522.h"  //https://developer.mbed.org/users/AtomX/code/
MFRC522/

DigitalOut LedRed(LED1);
DigitalOut LedGreen(LED2);

//MFRC522  RfChip  (SPI_MOSI, SPI_MISO, SPI_SCK, SPI_CS, MF_RESET);
MFRC522    RFID   (D11, D12, D13, D10, D8);

int cards[][4] = {
  {241,131,29,43}, // card 1
  {98,225,42,38}   // card 2
};
bool access = false;
```

```
int main(void) {
  printf("starting...\n");

  // Init. RC522 Chip
  RFID.PCD_Init();

  while (true) {
    LedRed = 1;
    LedGreen = 1;

    // Look for new cards
    if ( ! RFID.PICC_IsNewCardPresent())
    {
      wait_ms(500);
      continue;
    }

    // Select one of the cards
    if ( ! RFID.PICC_ReadCardSerial())
    {
      wait_ms(500);
      continue;
    }



    // Print Card UID
    printf("Card UID: ");
    for (uint8_t i = 0; i < RFID.uid.size; i++)
    {
       printf(" %d", RFID.uid.uidByte[i]);
    }
    printf("\n\r");
    // Print Card type
    uint8_t piccType = RFID.PICC_GetType(RFID.uid.sak);
    printf("PICC Type: %s \n\r", RFID.PICC_GetTypeName(piccType));

    for(int x = 0; x < sizeof(cards); x++){
        for(int i = 0; i < sizeof(RFID.uid.size); i++ ){
            if(RFID.uid.uidByte[i] != cards[x][i]) {
                access = false;
                break;
            } else {
                access = true;
            }
        }
        if(access) break;
```

```
    }
    if(access){
          printf("Welcome!\n\r");
          LedGreen = 0;
    } else {
           printf("Not allowed!\n\r");
           LedRed = 0;
    }

    wait_ms(500);
  }
}
************************************************************
```

Exercise 12.6

Modify the above example so that it only allows the access of a certain RFID card between 9 a.m. and 5 p.m.

Further Information about RFID RC522:

https://os.mbed.com/users/kirchnet/code/RFID-RC522/
https://os.mbed.com/users/nivmukka/code/Personal-Alert-System-using-RFID-with-FR/
https://www.sunfounder.com/wiki/index.php?title=Mifare_RC522_Module_RFID_Reader

12.5 Cloud Example with IBM Watson Bluemix

IBM Watson IoT Platform is a powerful cloud-based platform that allows you rapidly create analytics applications, visualization dashboards, and mobile IoT apps. With Arm® Mbed™ IBM Ethernet IoT Starter Kit, you can easily connect to IBM Watson IoT platform.

Hardware Required
- Arm® Mbed™ Ethernet IoT Start Kit (FRDM-K64F + mbed application shield)
- Mini USB cable and Ethernet cable

Software Required
- An Internet browser

Procedure
Just connect your IoT Starter Kit to your computer through USB, and to the Internet through an Ethernet cable. From the following link (also see Figure 12.16), import the *"IBMIoTClientEthernetExample"* program into your online compiler, compile it, and load it up to your board. This program will send an accelerometer, temperature sensor,

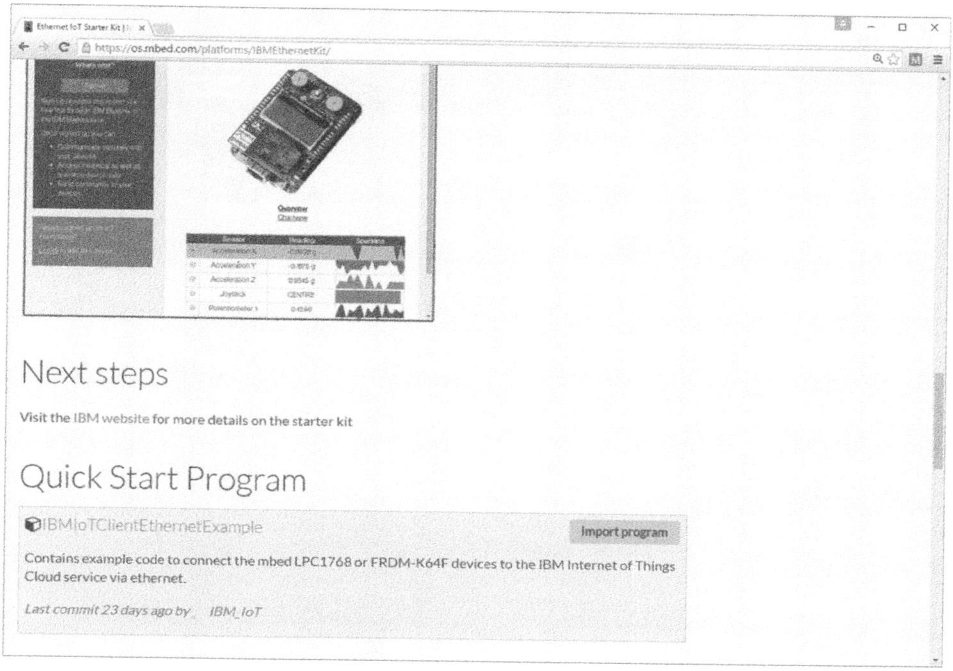

Figure 12.16 The IBM IoT Client Ethernet Example program page.

joystick, potentiometer 1 and 2 data to IBM Water IoT platform by MQTT. It also displays an information menu on the LCD. You can use a joystick to roll up and down the menu to get information on device ID, MQTT status, Ethernet status, socket status, and IP address, for example.

https://os.mbed.com/platforms/IBMEthernetKit/

12.5.1 IBM Quickstart Service

By default, the example program uses IBM Quickstart Service to send the data, you can view the by going to the following website (also see Figure 12.17). No registration is needed.

https://ibm.biz/iotqstart/

Just type in the correct device ID, and click the "**Go**" button. Your device ID is basically the MAC address, you can get your device ID from the LCD menu display by rolling up or down the joystick. Now you should see all the sensor data, and the corresponding chart display. You can select which sensor data to display in the chart, as shown in Figure 12.18.

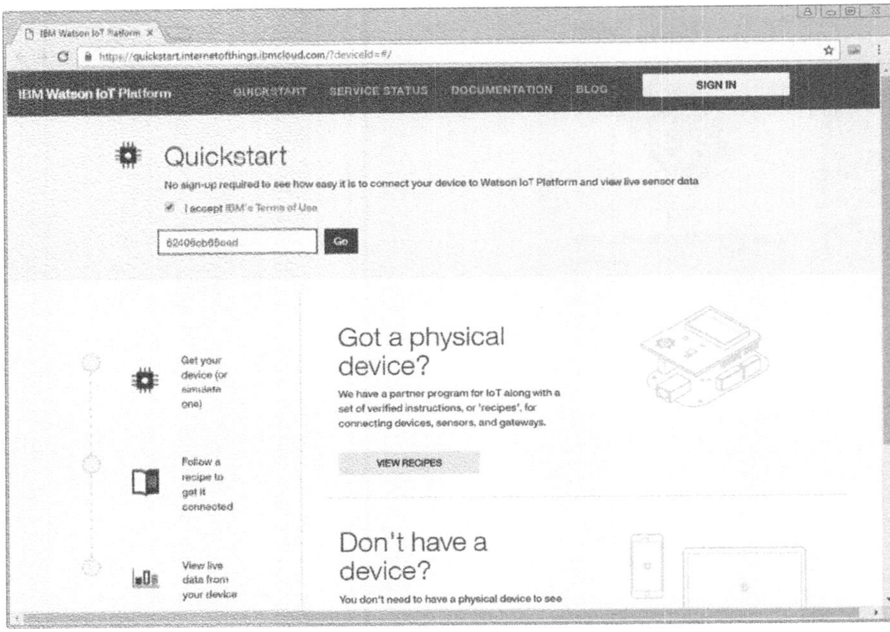

Figure 12.17 The IBM Quickstart Service page.

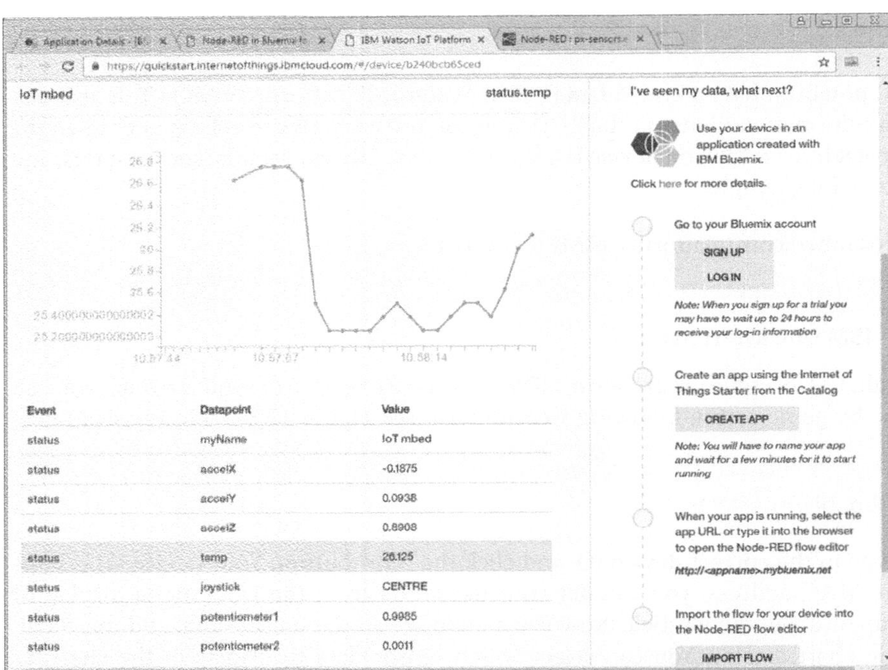

Figure 12.18 The IBM Quickstart Service chart and data display.

12.5.2 IBM Registered Service (Bluemix)

To create your own application for viewing and processing your data you will need to register on IBM Watson Bluemix website:

https://console.ng.bluemix.net/

You can register for a 30-day free trial account (Figure 12.19). After 30 days, you will need to continue with credit card details, but as long as the number of devices connected and the data metric are less than certain levels, it is still free. See IBM pricing website for details:

https://www.ibm.com/internet-of-things/platform/pricing/

Just follow the instruction and register and log in. After logging in, you will be asked to choose your country and create an organization and project name, as shown in Figure 12.20. In this case, the country is "UK," organization is "London South Bank University," and project name is "IoT Projects." Then click the blue "*Create App*" button to create your application program.

This will bring you to the IBM Watson catalog page. Select the "*Internet of Things Platform Starter*" within the Boilerplates (Figure 12.21). Boilerplates are simply the ready-made software modules that can be reused in your applications.

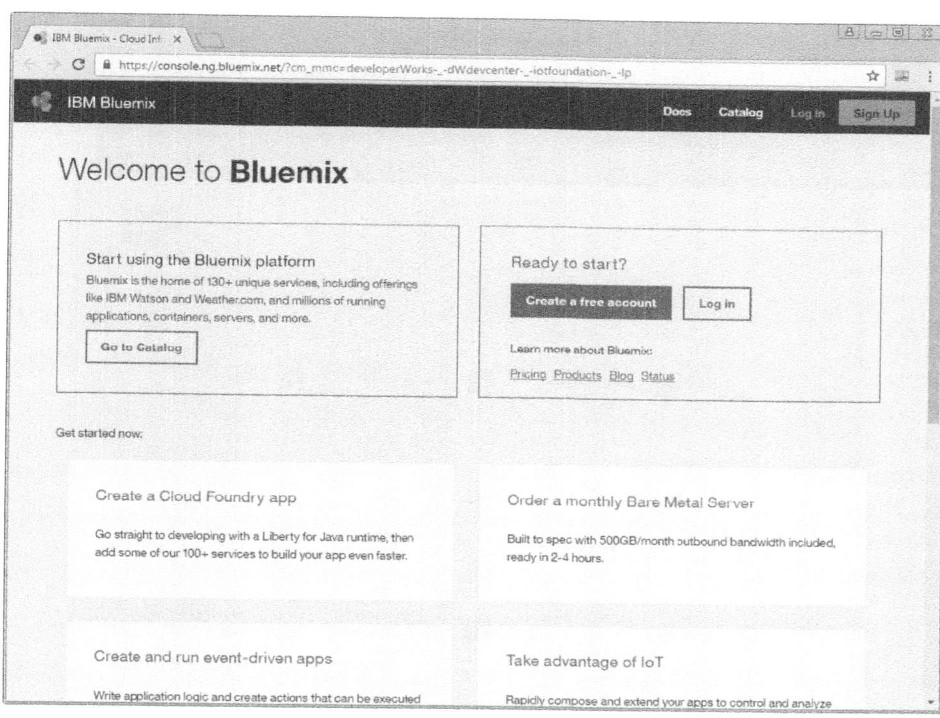

Figure 12.19 The IBM Bluemix registration and login page.

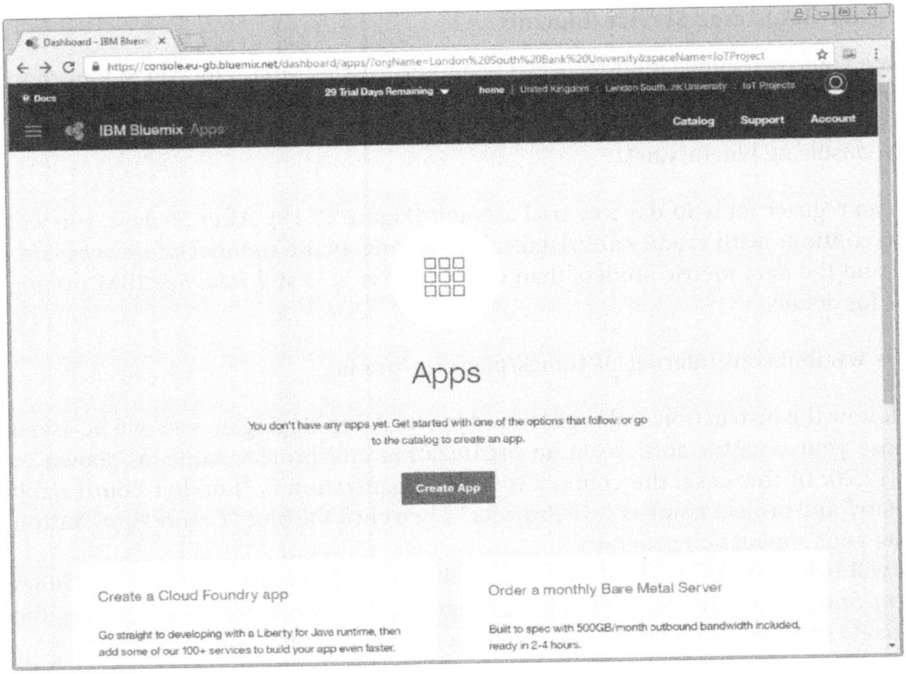

Figure 12.20 You Bluemix homepage after login.

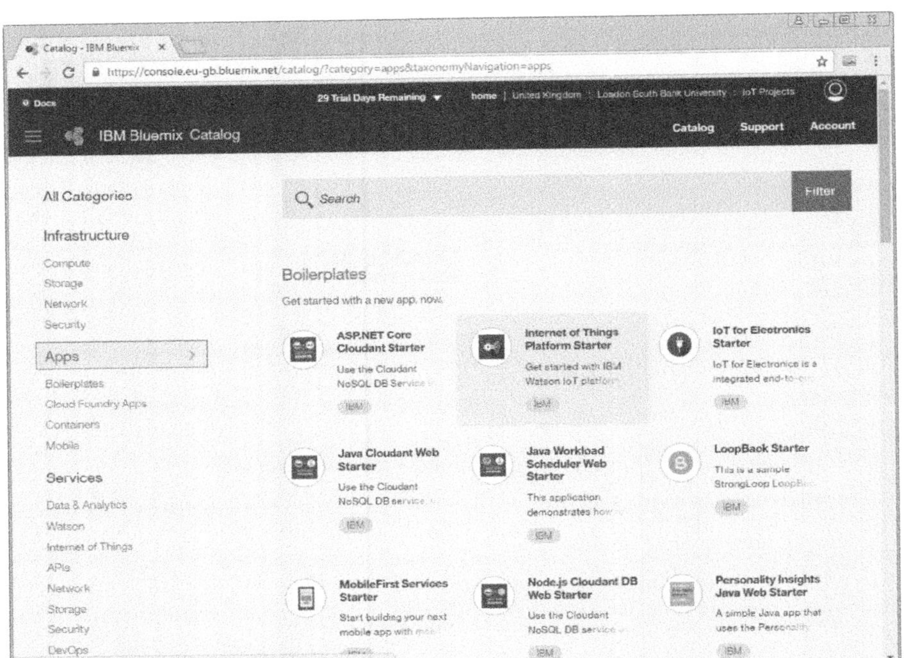

Figure 12.21 The available Boilerplates in IBM Bluemix catalog page.

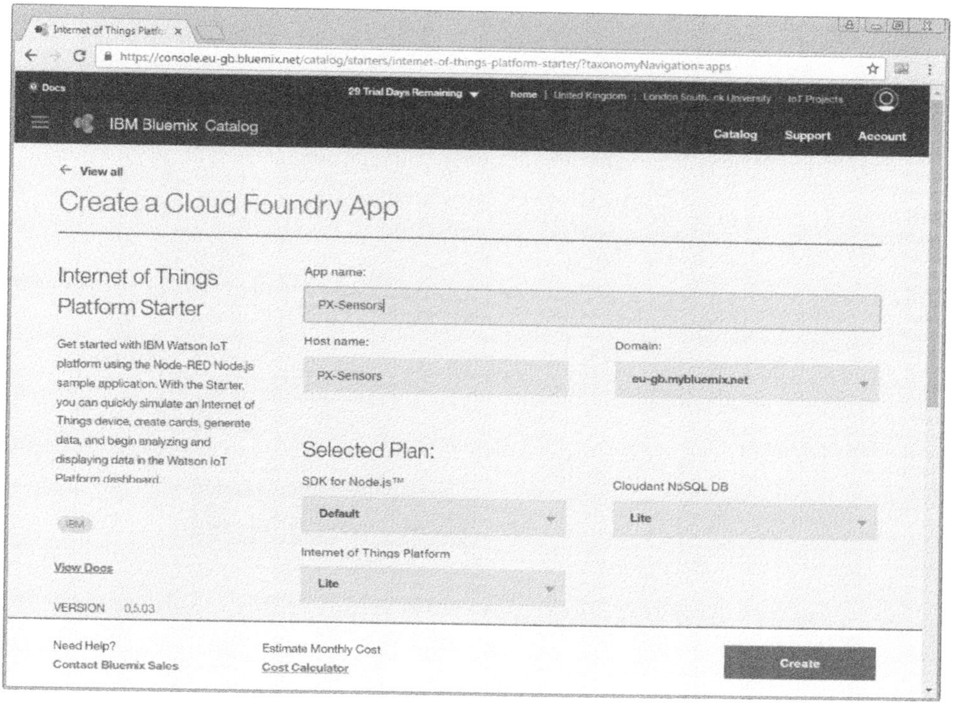

Figure 12.22 Application creation page.

This will bring you to a new page. Provide names to your Application and Host, then click on the "*Create*" button (Figure 12.22). This will take a few minutes to create your application. When it is ready, click the application URL to open the Node-RED Internet of Things landing page (Figure 12.23).

Click "*Go to you Node-RED flow editor*" button, a default IBM IoT QuickStarter Node-RED program page will appear (Figure 12.24). There are two parts, or two flows in the program. The top flow allows you to send data to the IBM Ethernet IoT Starter Kit, the bottom flow allows you to receive temperature data from the Starter Kit. To make the program work, click the blue "*IBM IoT App In*" block in the bottom flow, a configuration panel will appear (Figure 12.25), just type in the correct device ID. Please ensure the "*Authentication*" is "*Quickstart*".

When the device ID is corrected configured, select the green "*device data*" block, then click the "debug" tab on the right-hand side of the page, you should be able to see the temperature readings starting coming in, as shown in Figure 12.26.

Click the "*temp thresh*" block, this allows you to set a temperature threshold, if the temperature goes above this threshold it will display a warning (Figure 12.27).

You can also read potentiometer from the Starter Kit. Just drag a function block from the left "*function*" panel, put it under the bottom flow, and wired it up with the blue "*IBM IoT App In*" block, as shown in Figure 12.28. Click the function block and enter the configuration information as shown in Figure 12.29.

From the left "*output*" panel, drag a debug block into the bottom flow, change it to "*msg. payload*", and connect it to the "*Potentiometer 1*" function block, as shown in Figure 12.30.

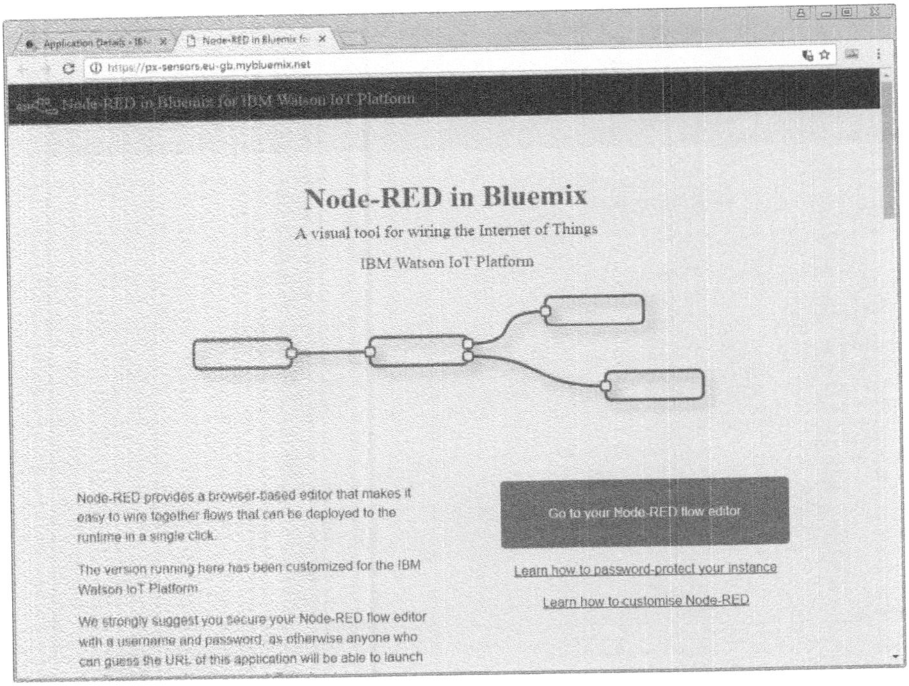

Figure 12.23 The Bluemix Node-RED homepage.

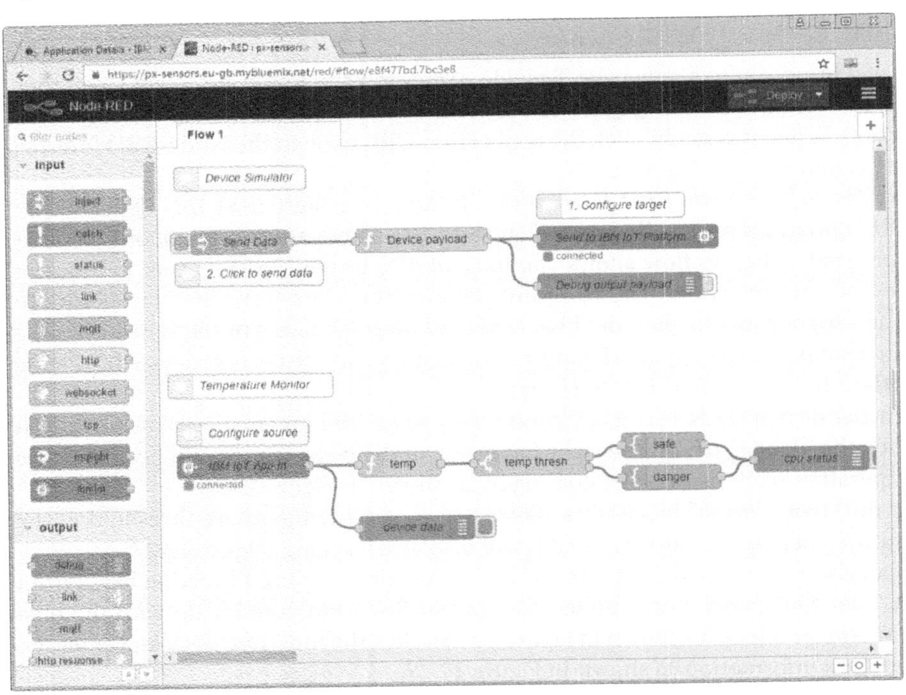

Figure 12.24 The default IBM IoT Node-RED application page.

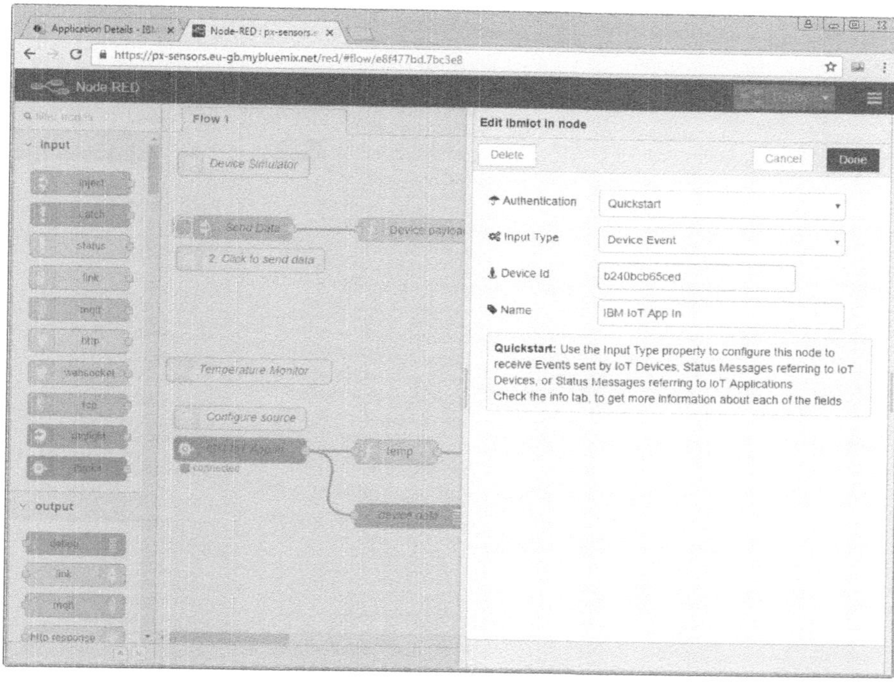

Figure 12.25 The configuration of Device Id of the blue "*IBM IoT App In*" block.

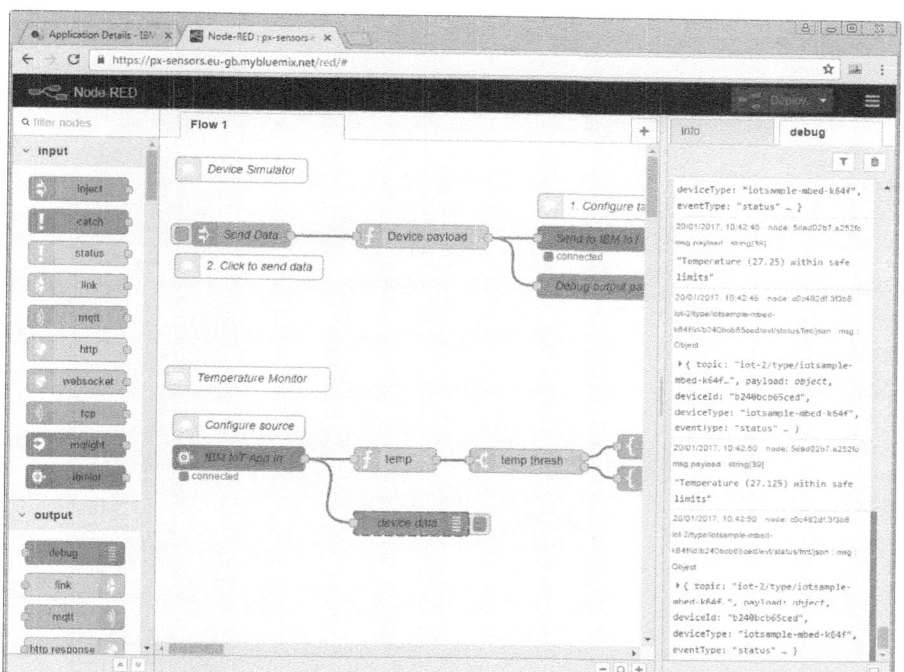

Figure 12.26 The configuration of Device Id of the blue "*IBM IoT App In*" block.

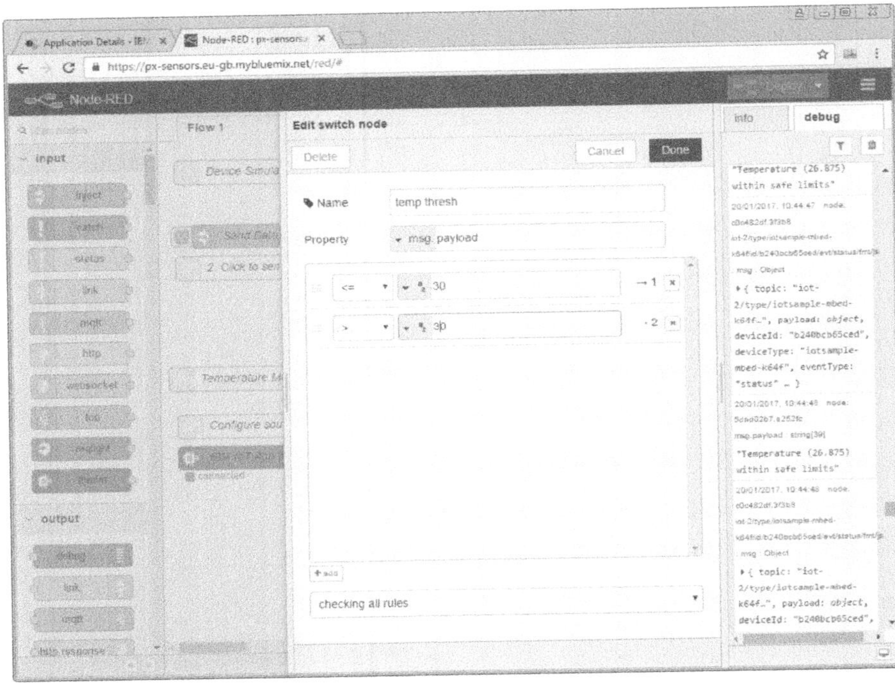

Figure 12.27 The temperature threshold configuration of the "temp thresh" block.

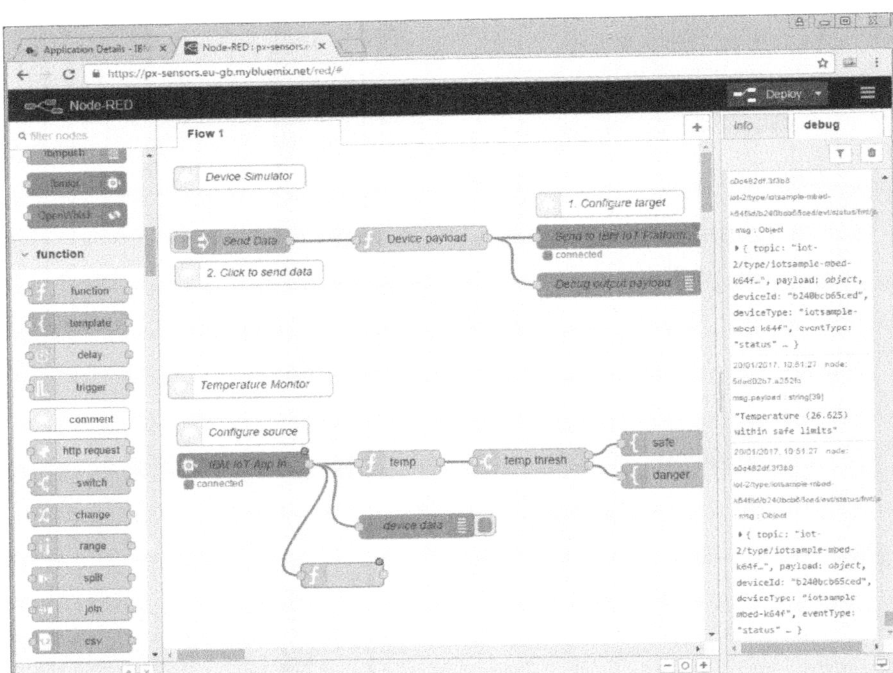

Figure 12.28 Add a function block to the bottom flow.

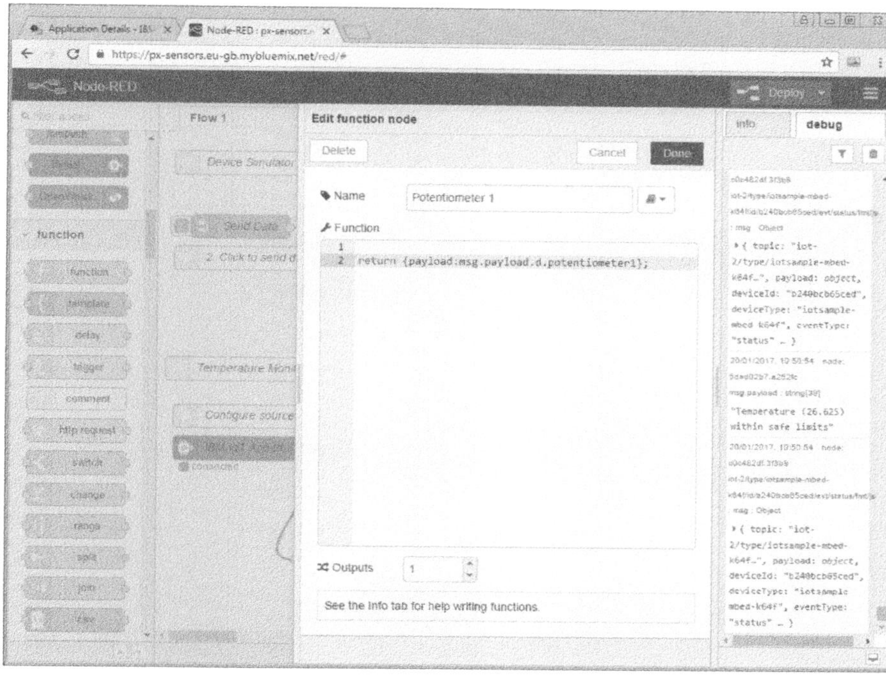

Figure 12.29 The configuration of the *"Potentiometer 1"* function block.

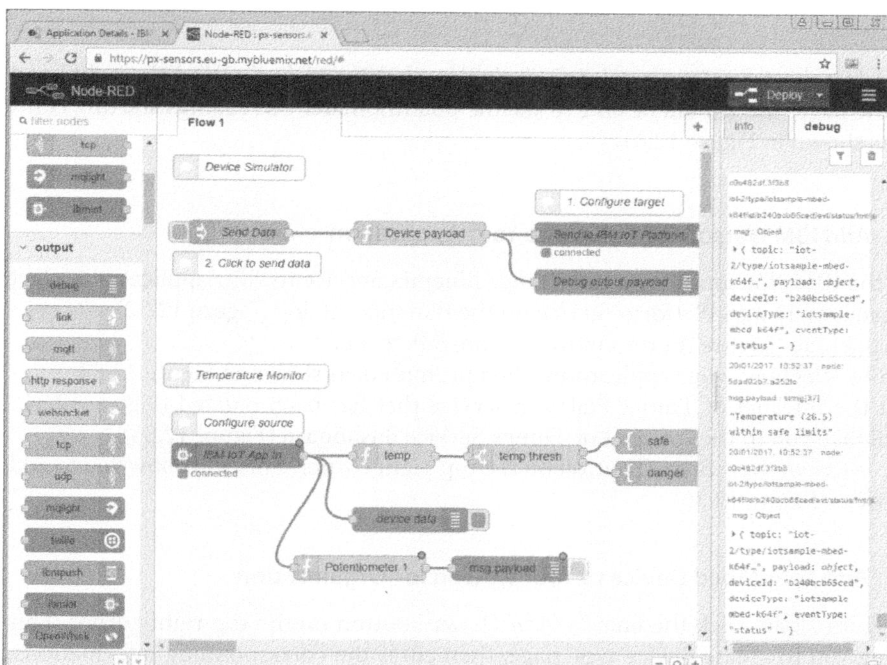

Figure 12.30 Add a debug block to *"Potentiometer 1"* function block.

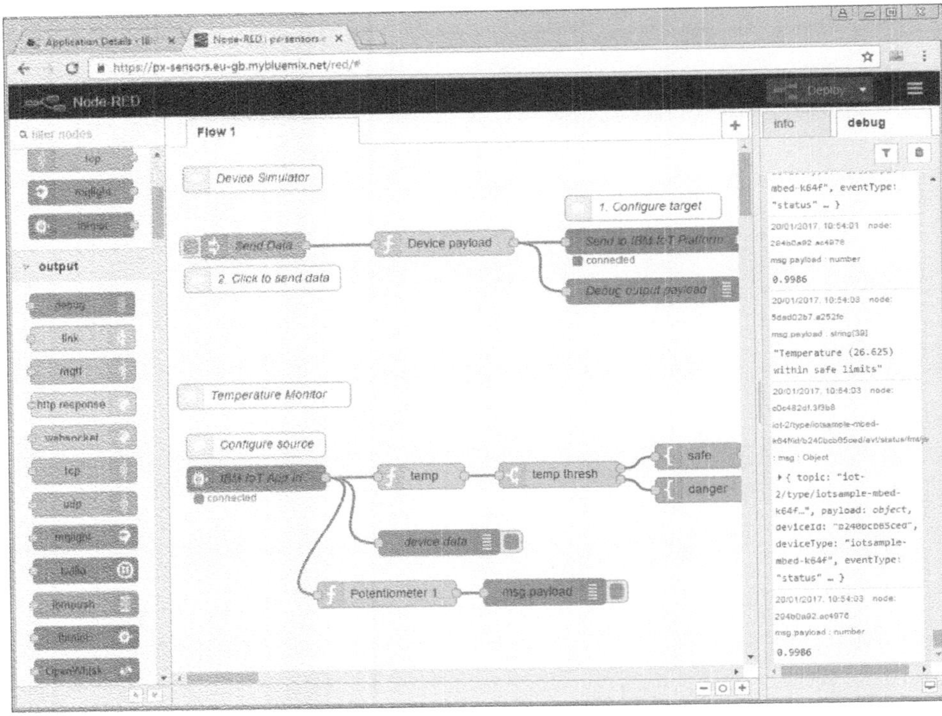

Figure 12.31 The debug outputs from the *"msg.payload"* debug block.

Now select the green *"msg.payload" block, then click the "debug"* tab on the right hand side of the page, you should be able to see the potentiometer 1's readings starting coming in, as shown in Figure 12.31.

12.5.3 Add IBM Watson IoT Service to Your Application

To add the IBM Watson Internet of Things Bluemix Service to your application, select the Internet of Things Platform service on the Bluemix catalog (Figure 12.32).

Click the blue "*Create*" button in the bottom left corner.

Select to "***Restage***" your application when prompted, as shown in Figure 12.33.

Select the Internet of Things Platform service that has been created (Figure 12.34). This will take you to the Internet of Things Service dashboard (Figure 12.35).

Click "+ *Create New Board*" button on the top-right corner. This will allow you to add new devices (Figure 12.36).

12.5.4 Add Your Mbed Device to Your Watson IoT Organization

From Figure 12.36, click the blue "+ *Add Device*" button on the top right corner. This will bring you to create device type page, then enter the corresponding informations, such as device type, device ID etc., click "*Next*" button after finishing each page (Figure 12.37–12.40).

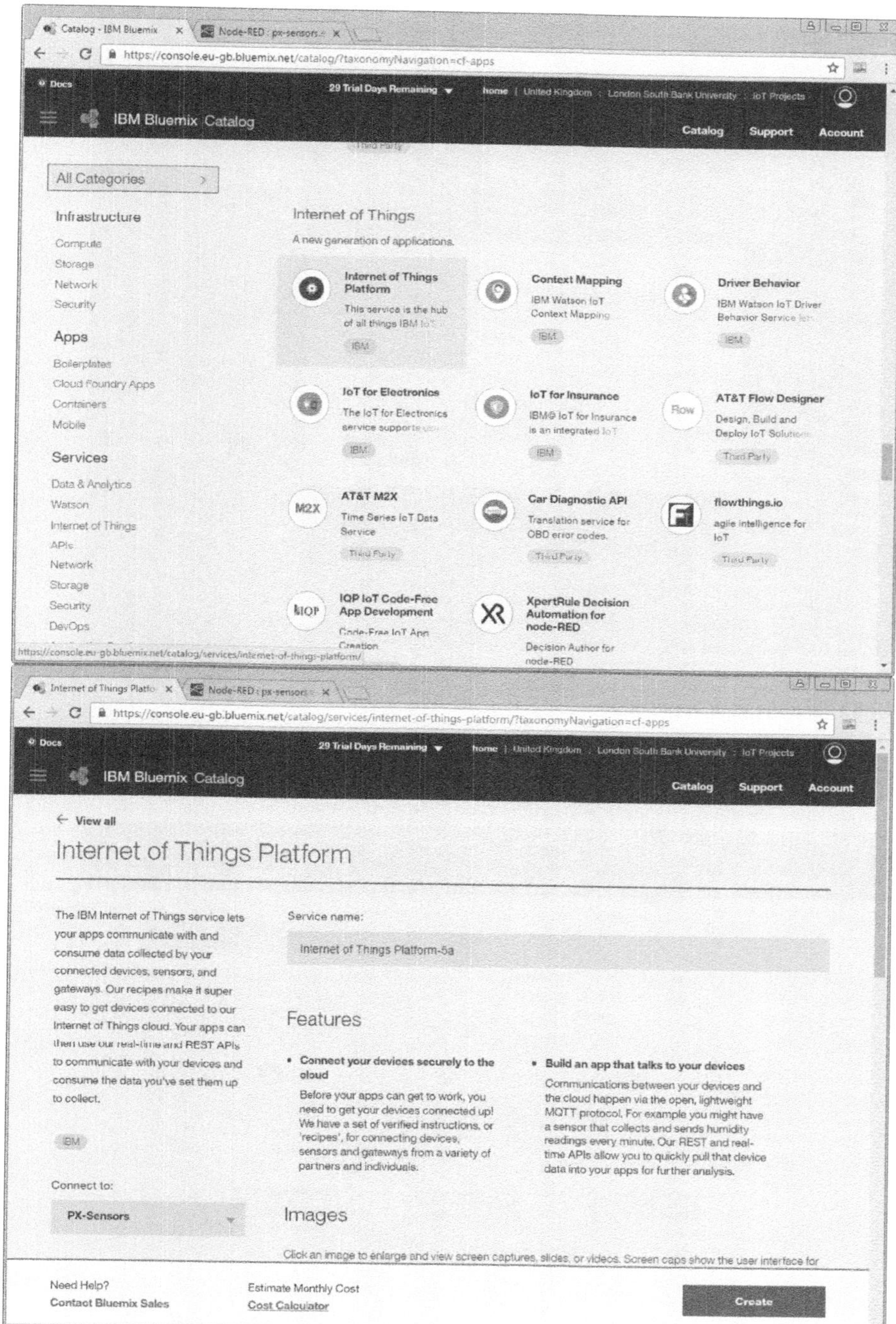

Figure 12.32 The IBM Bluemix service catalog (top) and the Internet of Things Platform service (bottom).

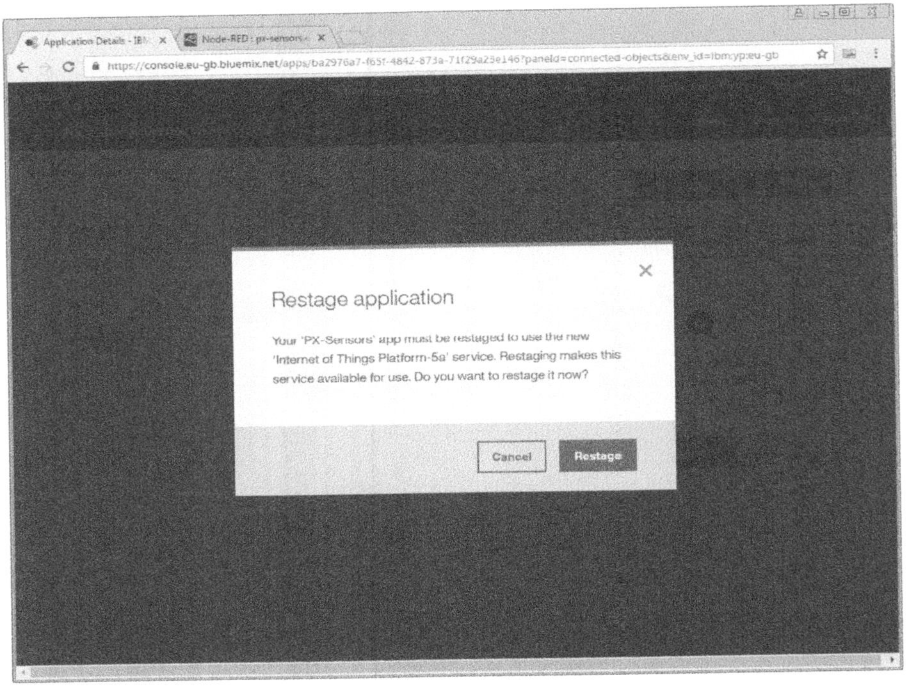

Figure 12.33 Restage your application.

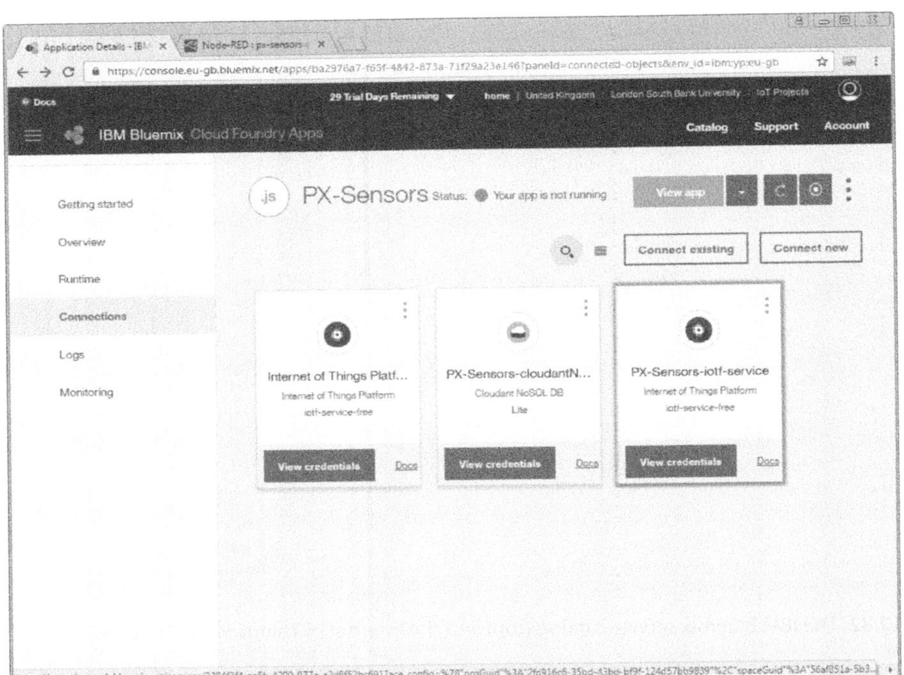

Figure 12.34 The Internet of Things Service platform.

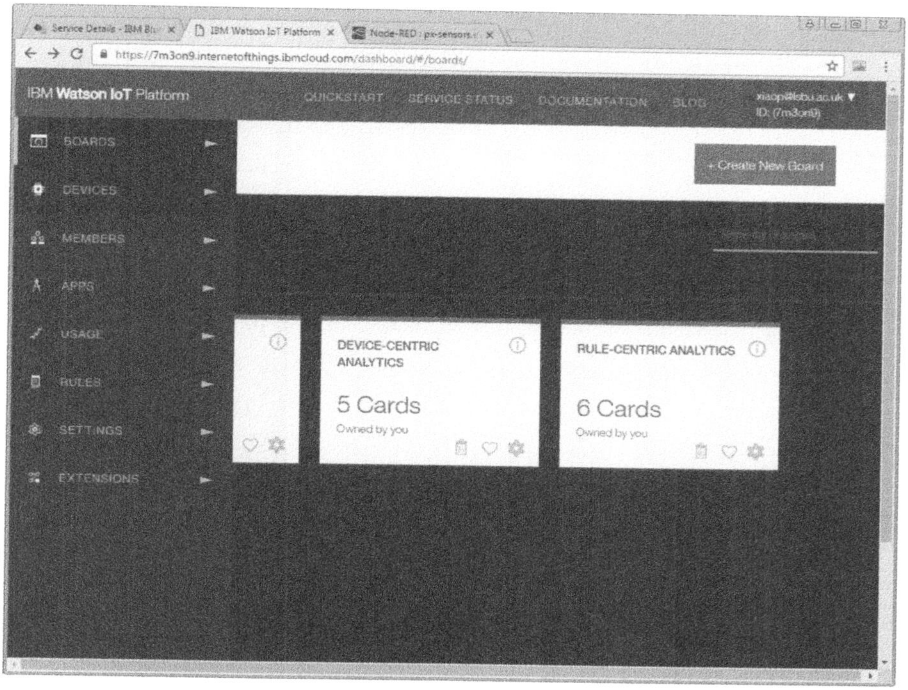

Figure 12.35 The Internet of Things Service dashboard.

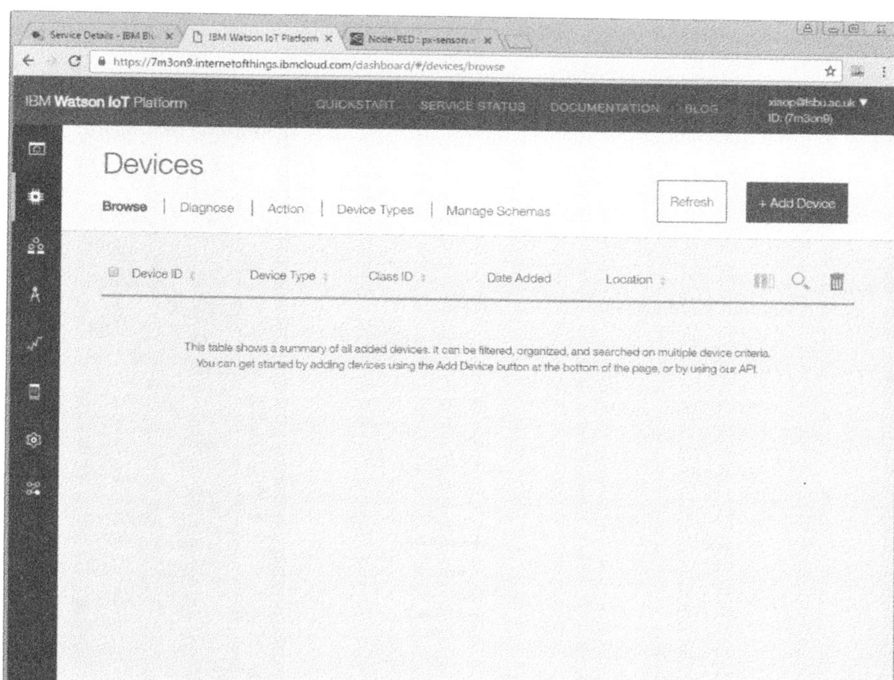

Figure 12.36 Adding devices to your IoT application.

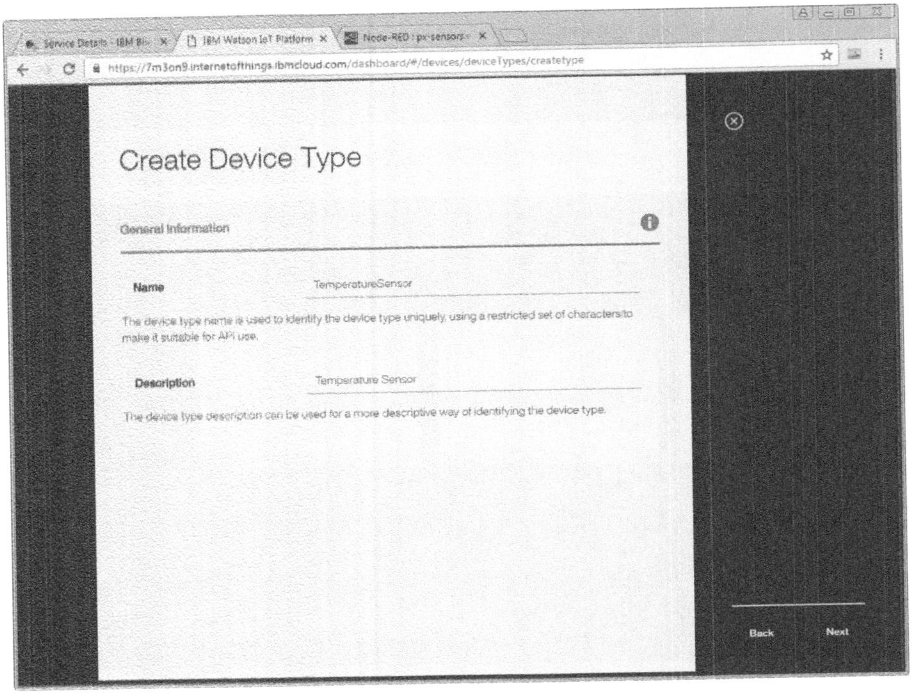

Figure 12.37 Create Device Type—General Information.

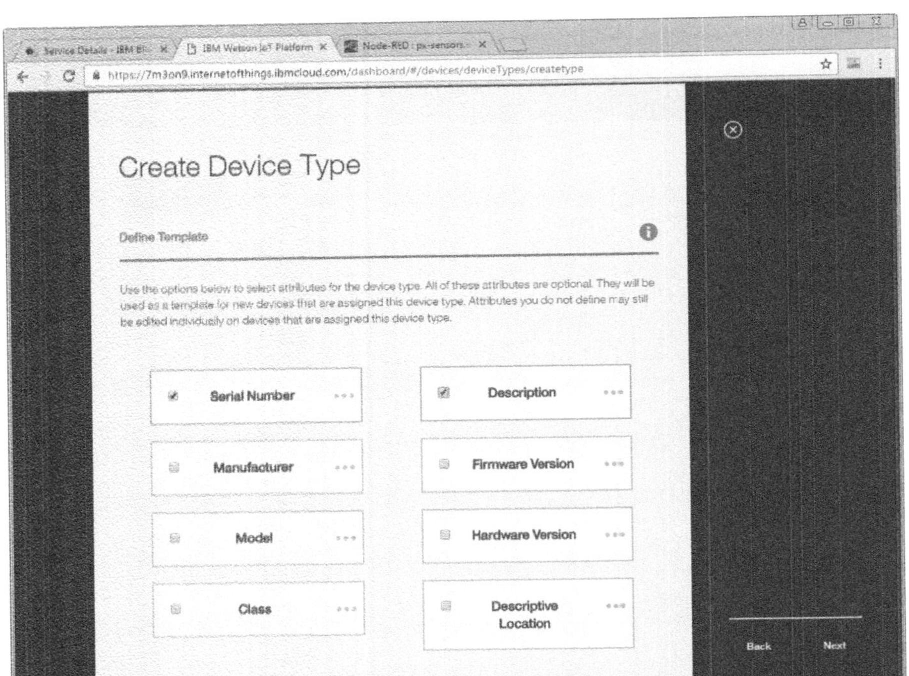

Figure 12.38 Create Device Type—Define Template.

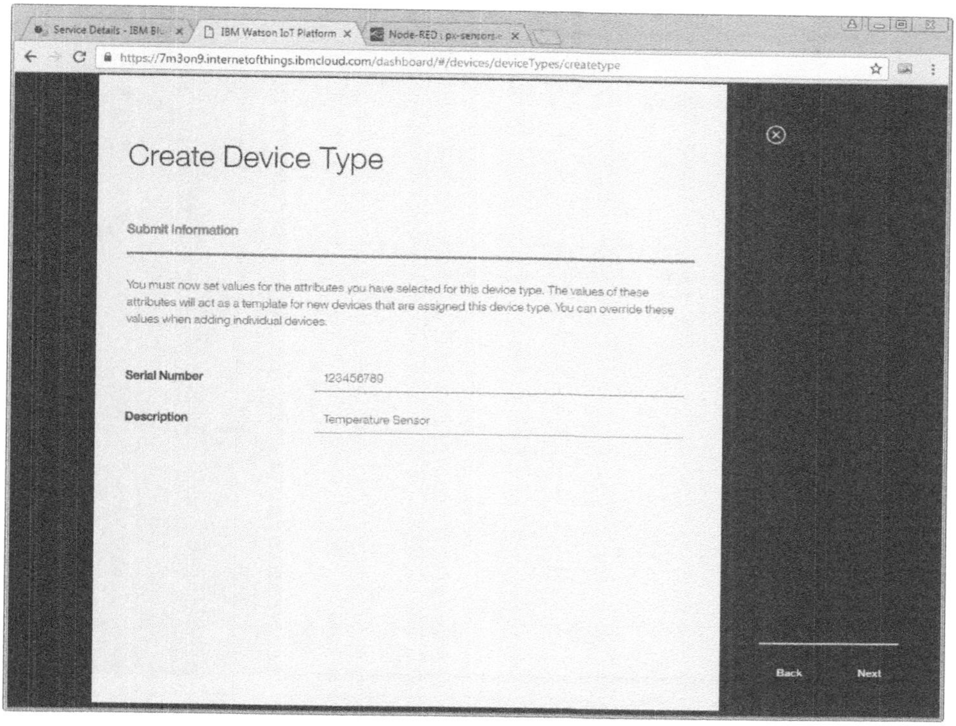

Figure 12.39 Create Device Type—Submit Information.

12.5.5 Adding Credentials onto Your Mbed Device

You can now get your device credentials, as shown in Figure 12.41. Copy them and add them to your IBM IoT Client Ethernet Example program code, as shown in Figure 12.42. Compile and run the program on your mbed device. The device will now run in registered mode.

Now you should be able to see all your sensor information of the mbed board from your IBM Bluemix application, as shown in Figure 12.43.

12.5.6 Link Your IBM IoT Watson Application to Your Mbed Device

From the IBM Bluemix dashboard, as shown Figure 12.44, select your application and click the application URL to open the Node-RED landing page. Select "*Go to your Node-RED flow editor*" button to view your application. Double-click the "*IBM IoT App In*" node in the flow editor and configure the node with correct device type and device ID, as shown in Figure 12.45. Please ensure the "*Authentication*" is "*Bluemix Service.*"

Click the red "*Deploy*" button to run your application.

Now you should be able to see your mbed device and the corresponding sensor readings from Device-Centric Analytics page, as shown in Figure 12.46.

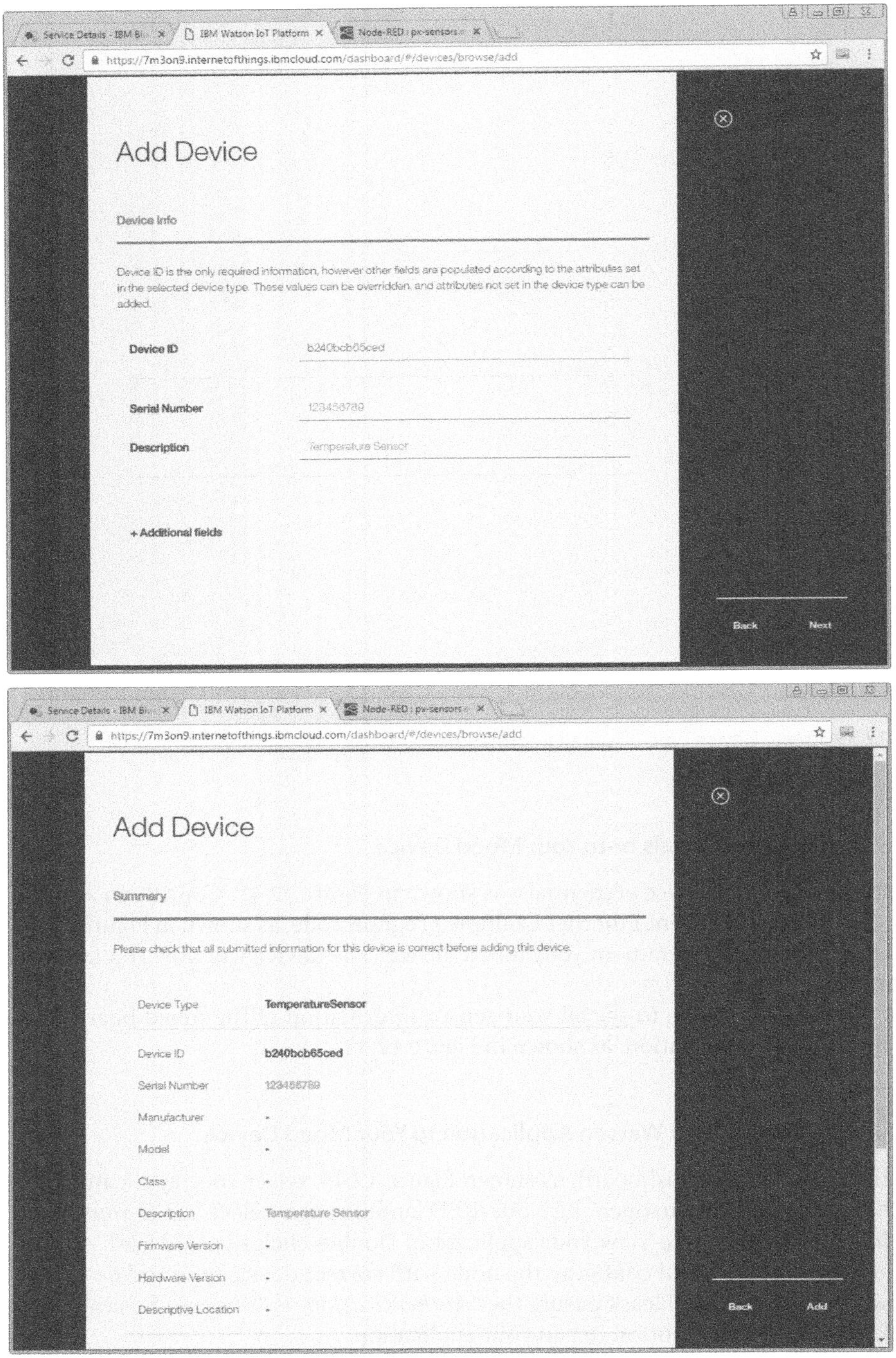

Figure 12.40 Add Device—Device Info (top) and Add Device—Summary (bottom).

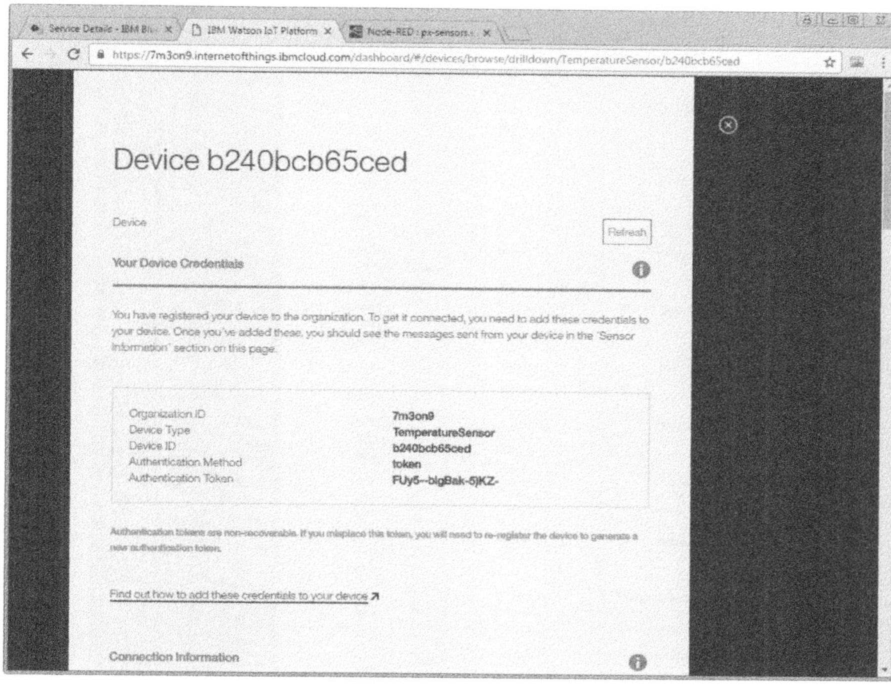

Figure 12.41 Your Device Registration Credentials.

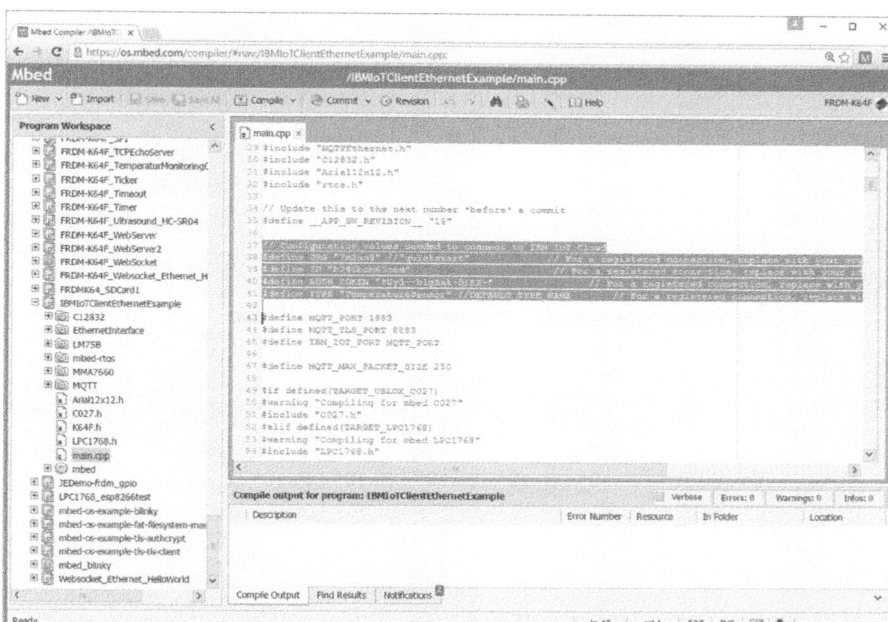

Figure 12.42 Using your device credentials in your mbed program.

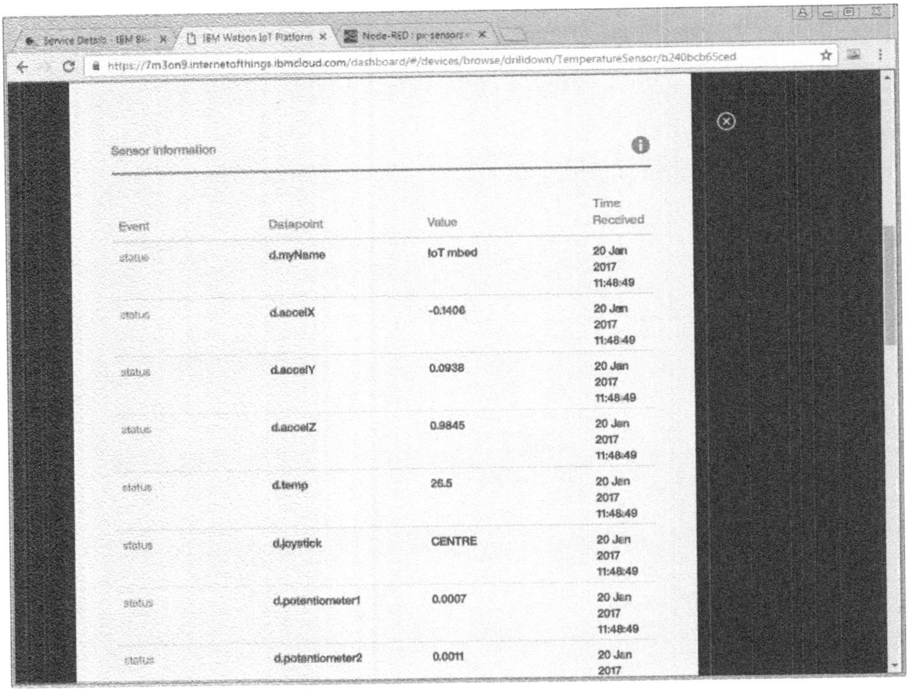

Figure 12.43 The sensor information of your mbed device on your IBM Bluemix application.

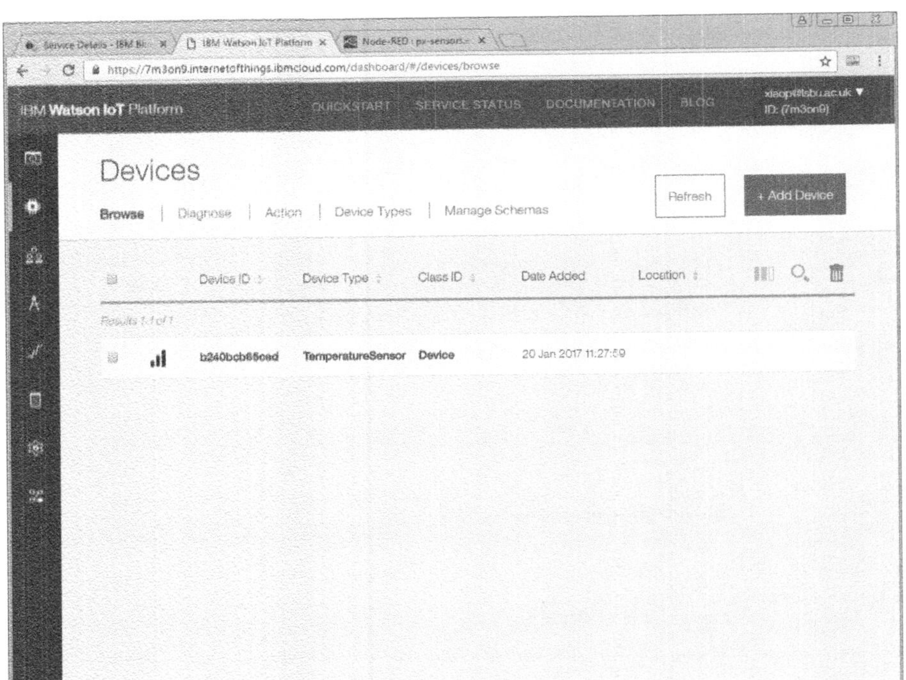

Figure 12.44 Your IBM Bluemix dashboard.

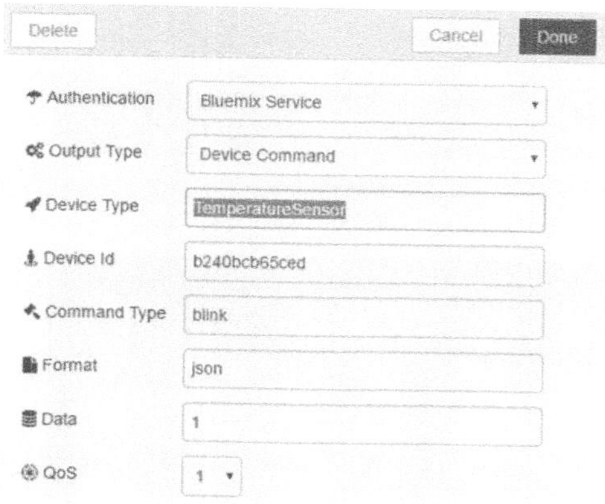

Figure 12.45 Your IBM Watson IoT Application Node-RED page.

12.5.7 Sending Commands from Your IBM IoT Watson Application to Your Mbed Board

From your Node-RED editor, navigate to the menu at the top right of the screen and select "*Import*" → "*Clipboard*," as shown in Figure 12.47.

Copy the JSON string from the link below and paste it into the dialog box in Node-RED and select "*Import*," and the imported JSON code should generate a new sub-flow, as illustrated in Figure 12.48.

https://raw.githubusercontent.com/ibm-messaging/iot-device-samples/master/mbed/ARM-mbed-Blink-LED.json?cm_mc_uid=14431715020414847496313&cm_mc_sid_50200000=1484906035

Connect "*blick rate*" node with "*Potentiometer 1*", as shown in Figure 12.49.

Double-click the "*IBM IoT Out*" node, and enter corresponding configuration information, as shown in Figure 12.50. When finished, click "*Deploy*" button to activate the program.

Now, by twisting the potentiometer 1, you should be able to change the flashing rate of the blue LED. Figure 12.51 shows the corresponding terminal outputs.

12.5.8 More with Node-RED

From your Node-RED editor, navigate to the menu at the top right of the screen and select "*Manage palette*," as shown in Figure 12.52.

An "*Install*" tab will appear on the left hand side of Node-RED editor, as shown in Figure 12.53. Search for "*dashboard*" and select "node-red-dashboard," and click the red button "*Done*."

The Node-RED dashboard module should now be installed and appear at the left side of the Node-RED editor, as shown in Figure 12.54. As you can see, there is a whole range of gadgets available, e.g., buttons, switches, sliders, gauges, forms, charts etc.

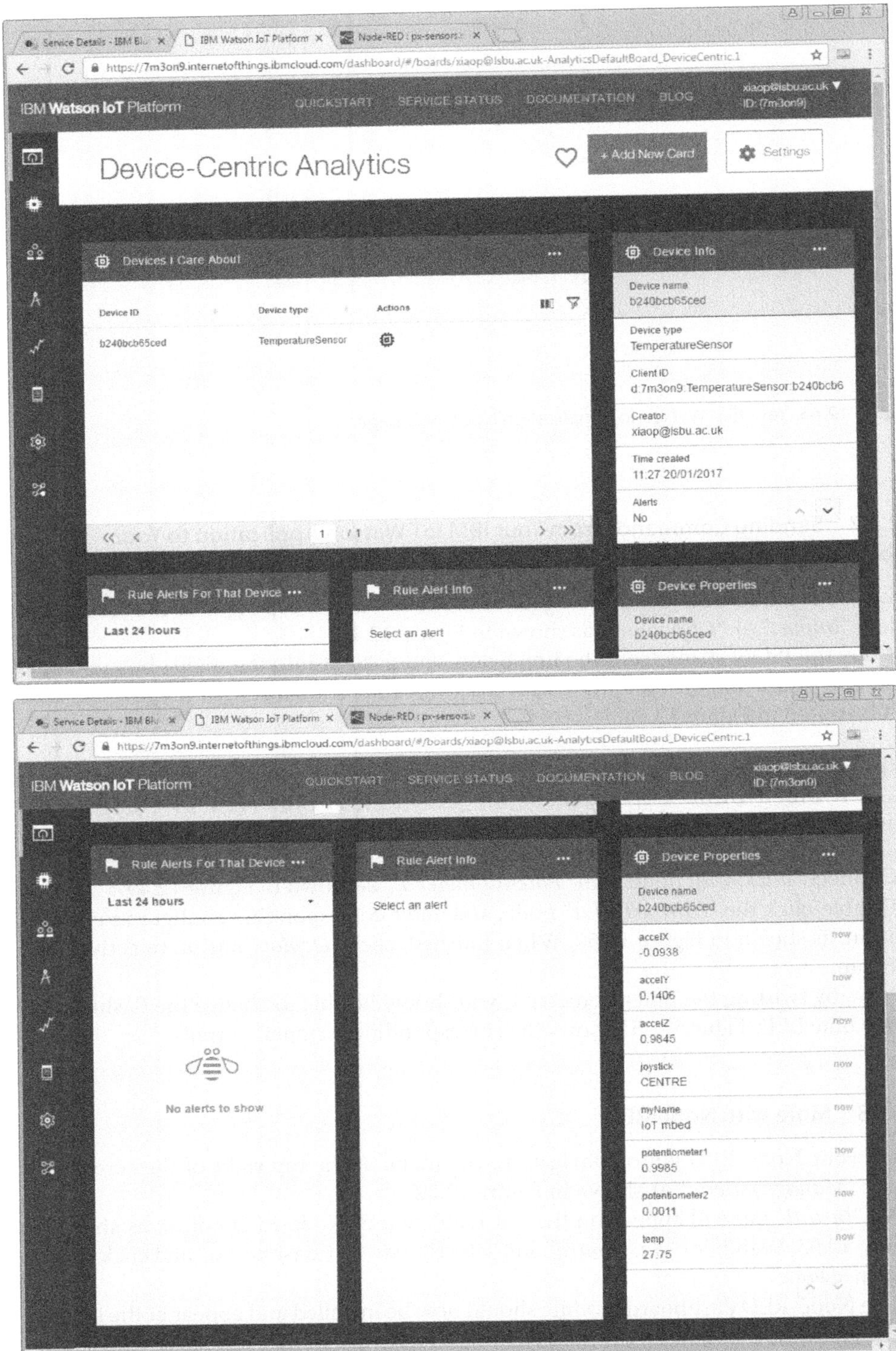

Figure 12.46 Device-Centric Analytics page, with device info (top) and device properties (bottom).

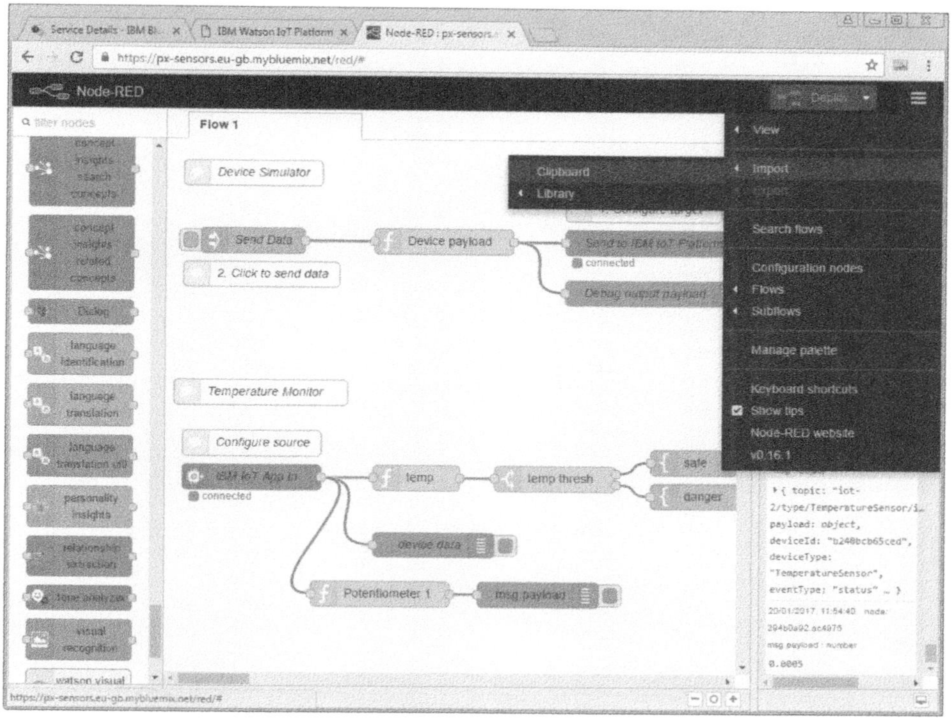

Figure 12.47 Import menu in Node-RED.

Figure 12.55 shows adding a chart to the program, connected to the "temp" node, and Figure 12.56 shows the corresponding display of the chart from the application web page.

You can also send emails or Twitter messages using Node-RED. Figure 12.57 shows how to add an "*email*" node (available under "*social*" category) to your program. In this case, when the temperature is exceeding the threshold value, it will send you an email.

Figure 12.58 shows how to add a "*twitter*" node (available under "*social*" category) to your program and the corresponding configuration. In this case, when the temperature exceeds the threshold value, it will send you a Twitter message.

Further Information about IBM Watson and IoT Starter Kit:

https://console.ng.bluemix.net/docs/starters/IoT/iot500.html
http://www.instructables.com/id/Making-a-IoT-cloud-service-with-ARM-mbed-platform-/?ALLSTEPS
https://developer.ibm.com/recipes/tutorials/arm-mbed-iot-starter-kit-part-1/
https://developer.ibm.com/recipes/tutorials/arm-mbed-iot-starter-kit-part-2/

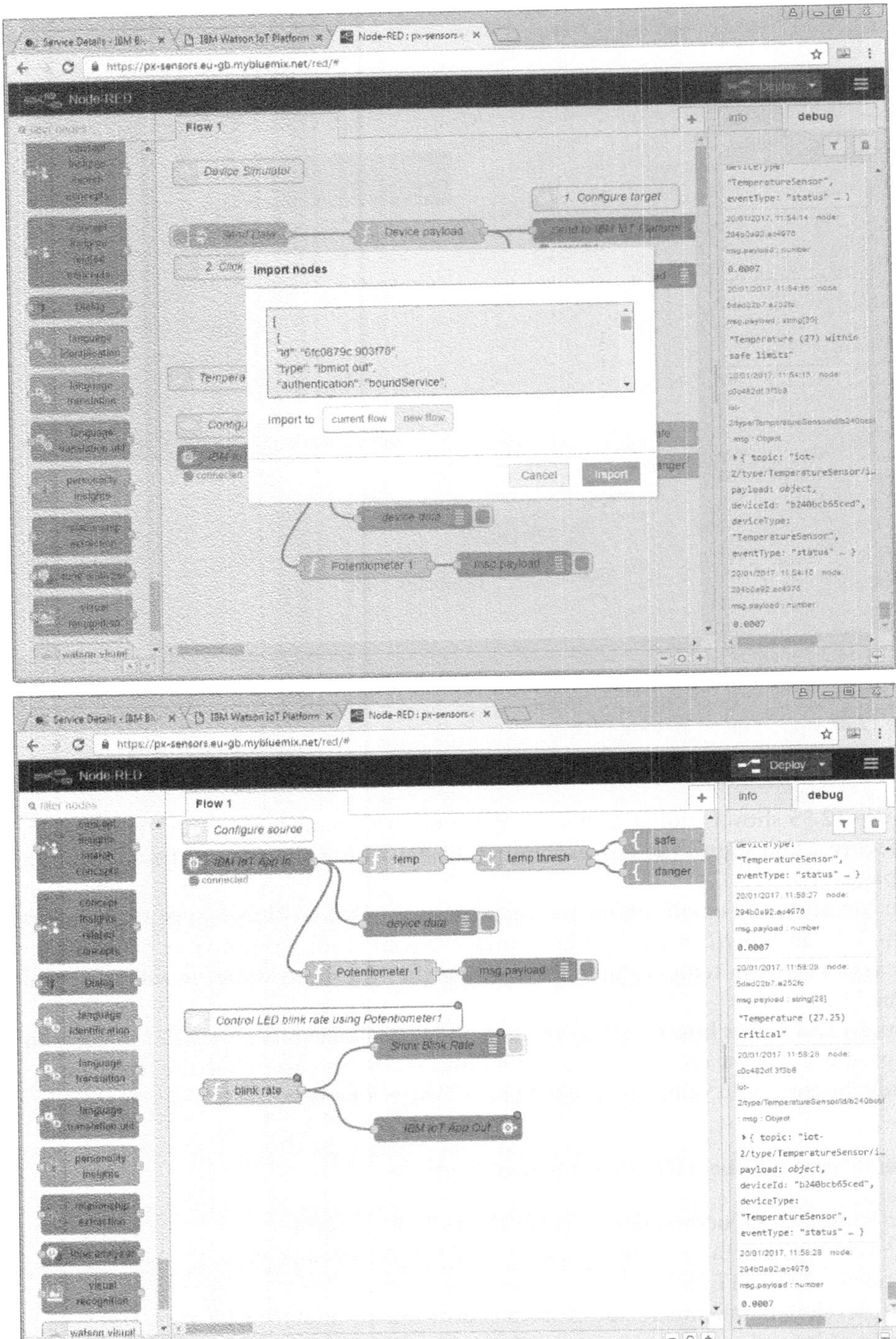

Figure 12.48 Import JSON code using the clipboard in Node-RED (top) and the new imported flow in Node-RED.

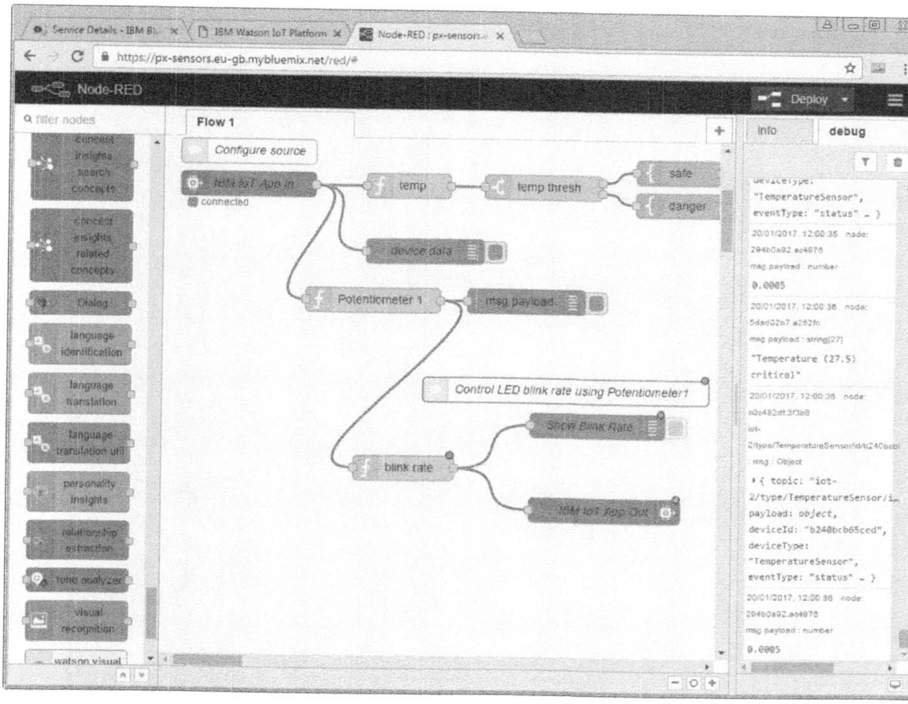

Figure 12.49 Connect "*blick rate*" node with "*Potentiometer 1.*"

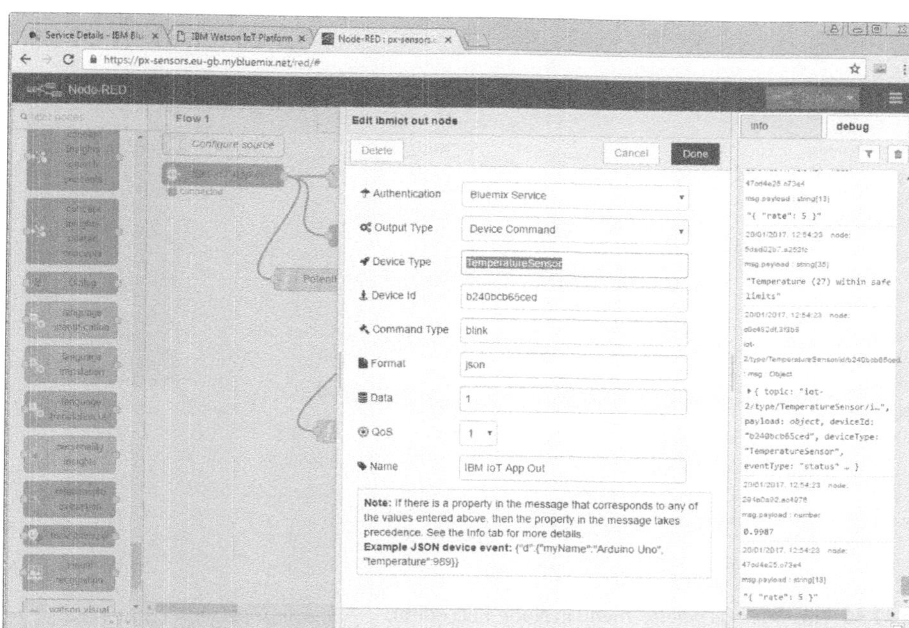

Figure 12.50 The configuration information of "*IBM IoT Out*" node.

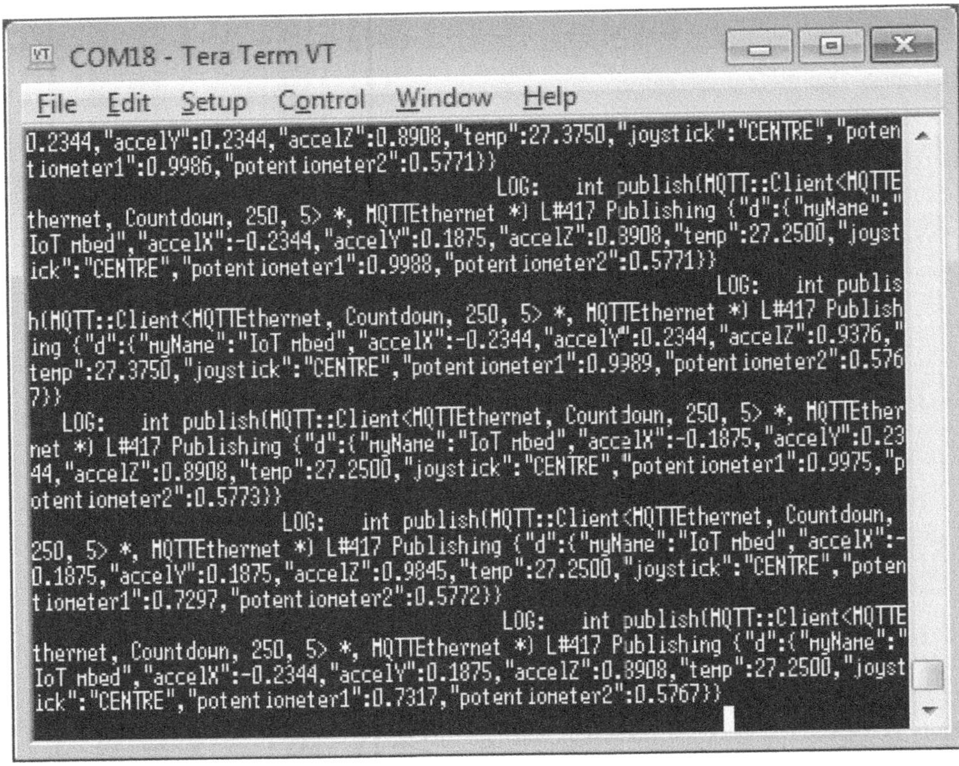

Figure 12.51 The corresponding Tera Term outputs.

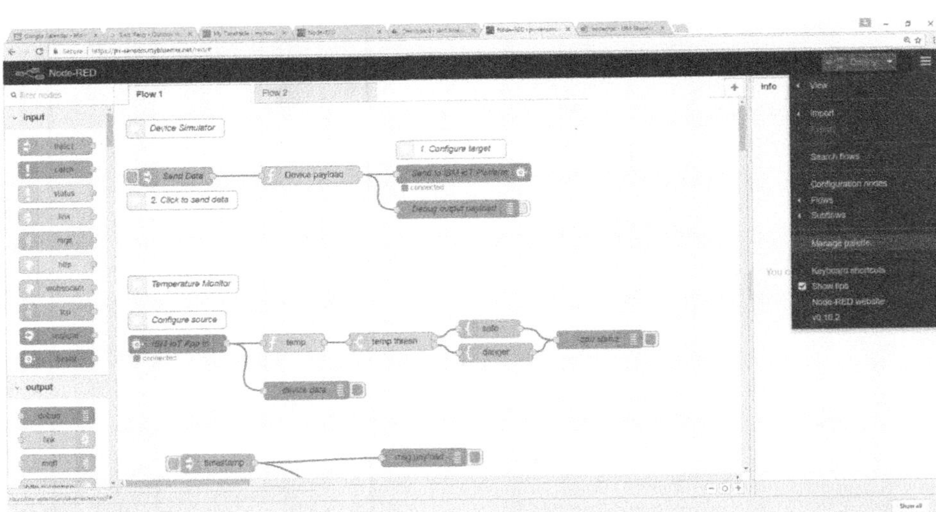

Figure 12.52 The *"Manage palette"* menu in Node-RED editor.

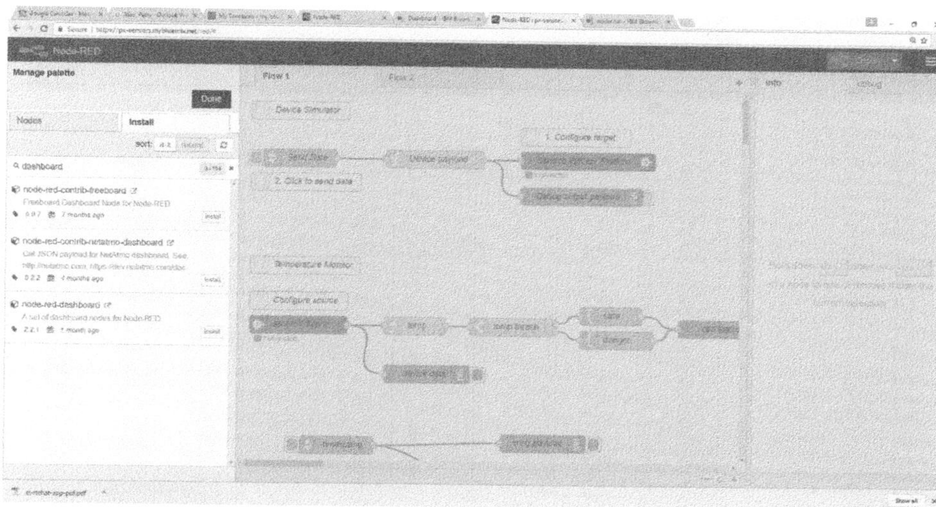

Figure 12.53 The "*Install*" tab on the left in Node-RED editor.

Figure 12.54 The "*dashboard*" module in Node-RED editor.

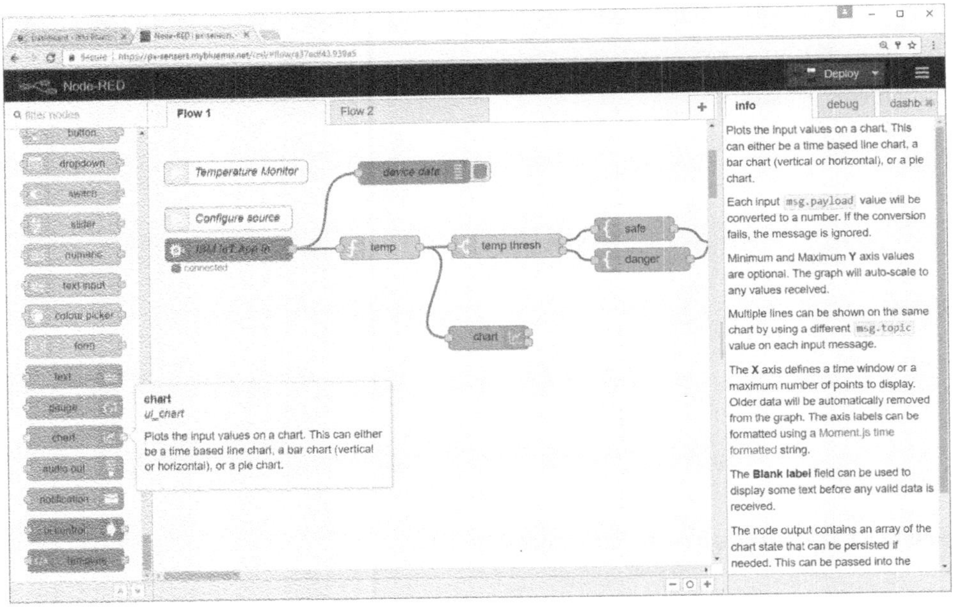

Figure 12.55 Adding a "*chart*" node to the program in Node-RED editor.

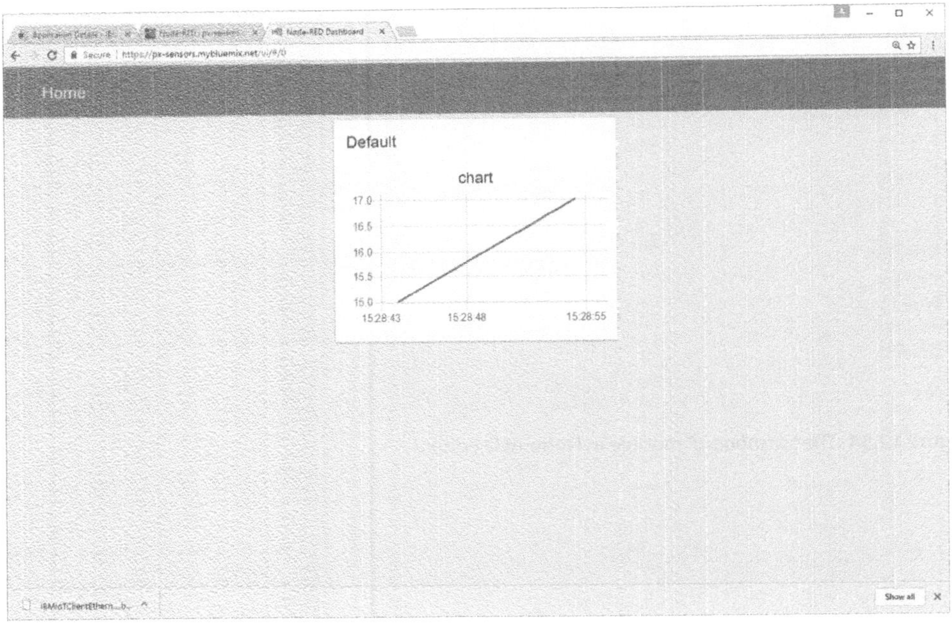

Figure 12.56 The "*chart*" display.

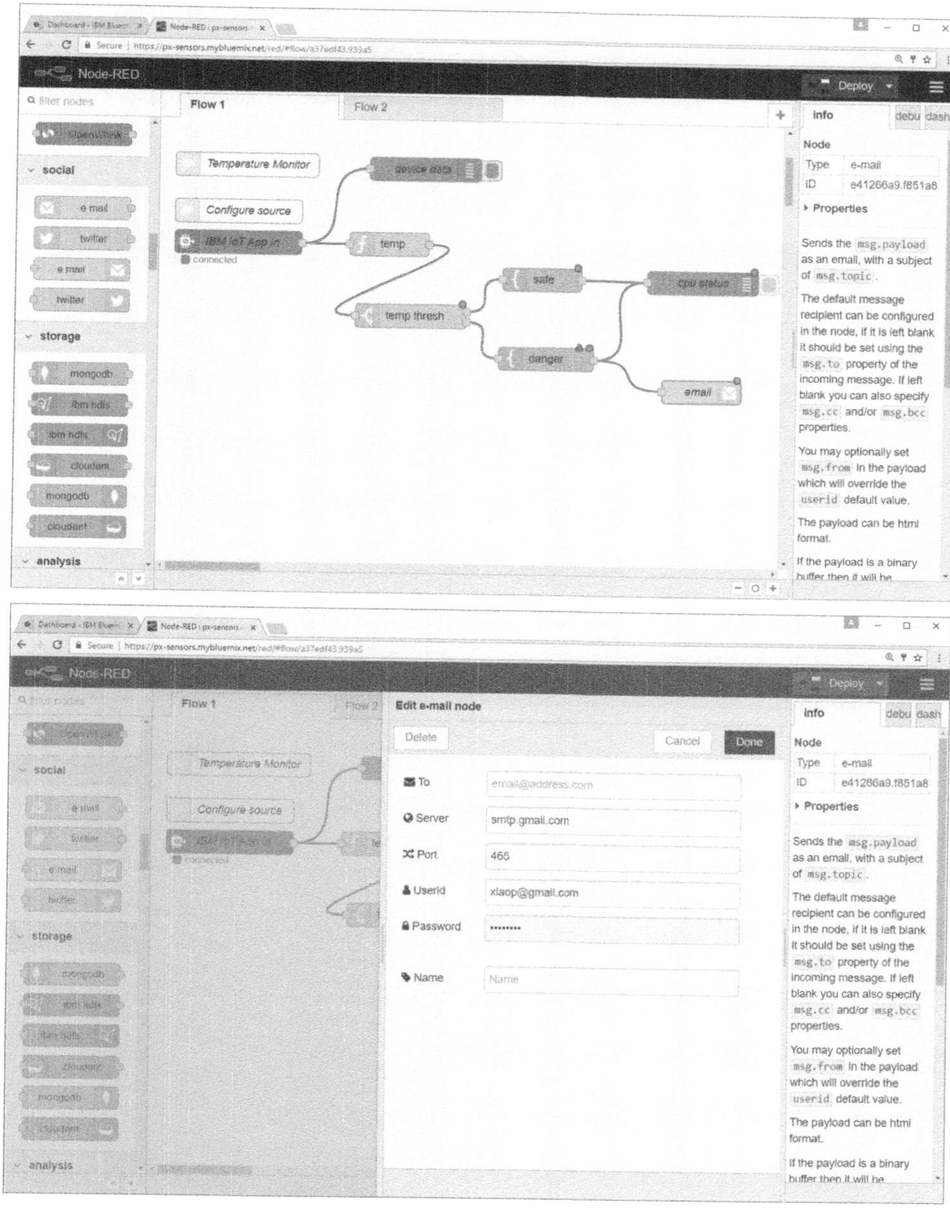

Figure 12.57 Adding "*email*" node to your program (top) and the corresponding email configurations (bottom).

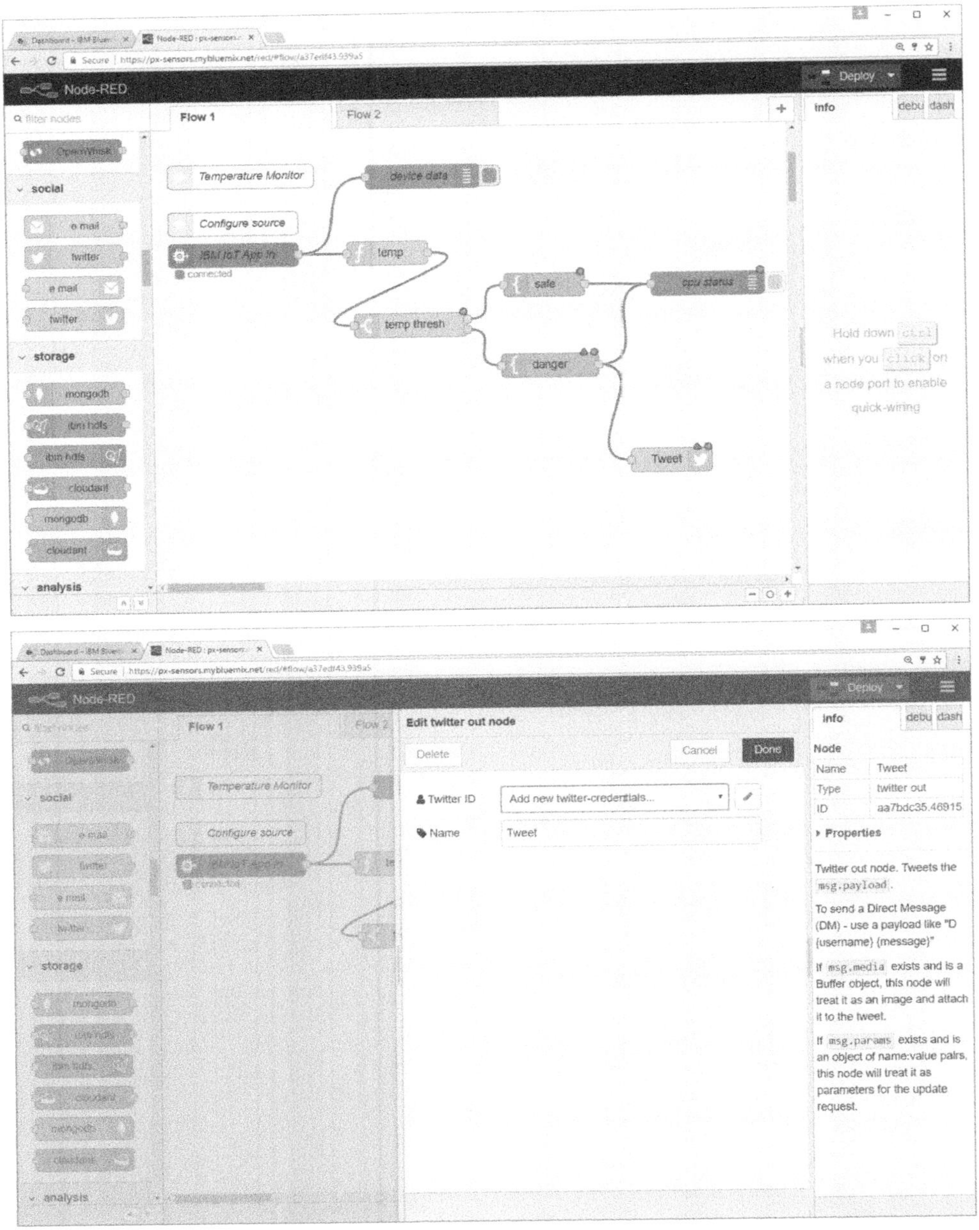

Figure 12.58 Adding "***twitter***" node to your program (top) and the corresponding Twitter configurations (bottom).

12.6 Real-Time Signal Processing

In many applications, such as real-time voice and signal processing, you will need to read the analog input and process it as fast as possible. In this project, we will demonstrate how to develop a real-time signal processing application by using fast analog inputs and fast analog output.

Hardware Required

- Arm® Mbed™ FRDM-K64F (or LPC1768) development board
- Mini USB cable and Ethernet cable
- PicoScope (https://www.picotech.com/oscilloscope/2000/picoscope-2000-overview)

Software Required

- An Internet browser

Procedure

The following example uses A0 as the analog input (for LPC1768 is P17) and uses the Timer to record the time. It first use a *while* loop to read 4096 data points from the analog input, it also use Timer's *"read_us()"* function to get the time in microseconds, then it uses a *for* loop to print the time (seconds) and data to computer through virtual COM port.

```
*************************************************************
// Example 12.11

#include "mbed.h"

#if defined(TARGET_K64F)
   AnalogIn   ain(A0);
#elif defined(TARGET_LPC1768)
   AnalogIn   ain(p17);
#endif

Timer t;
double t1=0,t0=0,tp=0;

int main(void)
{
    t.start();
    t0=t.read_us();
    int i=0;
    double dt[4096];
    double val[4096];
    printf("Recording.................\r\n");
    while (i<4096) {
        val[i]=ain.read();
        t1=t.read_us()-t0;
```

```
            dt[i]=((t1)*0.000001);
            tp=t1;
            i++;
        }
        printf("Printing.................\r\n");
        for(i=0;i<4096;i++){
            printf("%f\t%f\r\n",dt[i],val[i]);
        }
        printf("Done......................\r\n");
}
************************************************************
```

Figure 12.59 shows the corresponding Arduino Serial Monitor outputs. The first column shows the time in seconds, and the second column shows the analog input values. As we can see, it can read the input as fast as about 0.00002 s, or 20 μs—that is, about 50 KHz!

Exercise 12.7

Modify the above program so that it saves the time and data values to a text file on an SD card (for LPC1768, local file system).

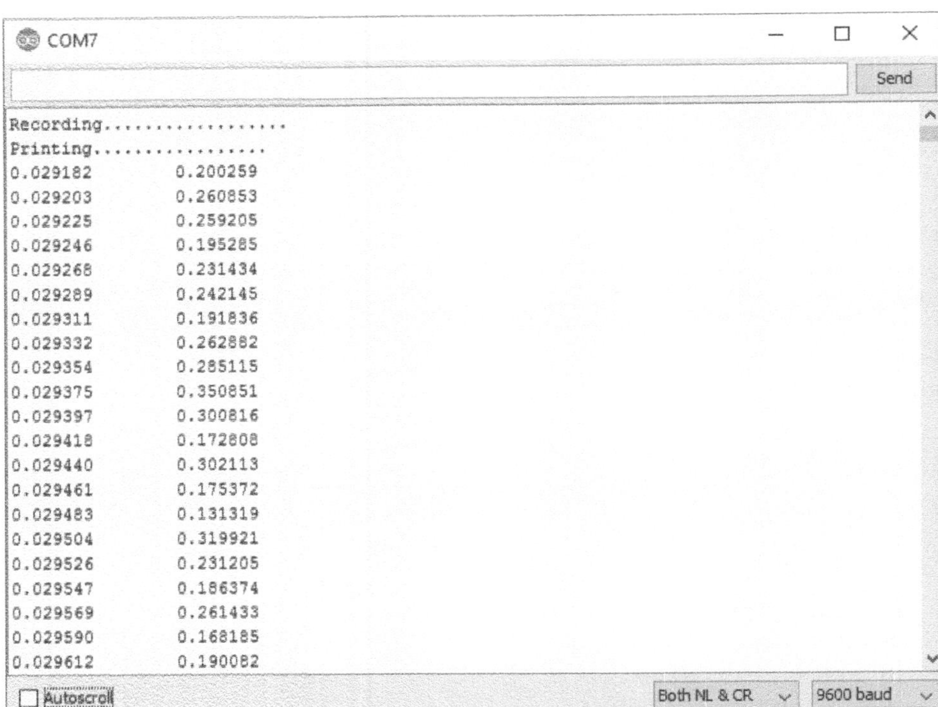

Figure 12.59 The Arduino Serial Monitor outputs.

For LPC1768 development board, there is a FastAnalogIn library; see the following link. It uses the burst feature to read analog input at the background.

https://os.mbed.com/users/Sissors/code/FastAnalogIn/

The following example uses FastAnalogIn library to read analog input from P15 pin. It also uses Timer to get the time information. It first start the time, then uses a *"for"* loop to read in 4096 data points (as 16-bit integers), and then uses another *"for"* loop to write the time (microseconds) and data to a text file called *"out2.txt"*.

```
************************************************************
// Example 12.12

#include "mbed.h"
#include "FastAnalogIn.h"

FastAnalogIn input1(p15);
Timer t;
LocalFileSystem local("local");
struct packet{
   int times[4096];
   uint16_t samples1[4096];
};
int main() {
    t.start();
    packet sample_data;
    for(int i=0; i<4096; i++) {
        sample_data.times[i] = t.read_us();
        sample_data.samples1[i] = input1.read_u16();
    }
    FILE *fp = fopen("/local/out2.txt", "w");
    for(int i=0;i<4096;i++){
       fprintf(fp, "%d\t%d\r\n",sample_data.times[i],sample_
data.samples1[i]);
    }
    fclose(fp);
}
************************************************************
```

Figure 12.60 shows the content of *"out2.txt."* The first column is the time (microseconds) and the second column is data (as 16-bit integers). As we can see, it can read data as fast as 2 us, or an impressive 500 KHz! That is about 10 times faster than normal analog inputs!

But unfortunately, the FastAnalogIn library only support a few development boards:

- LPC1768
- LPC4088
- LPC11u24

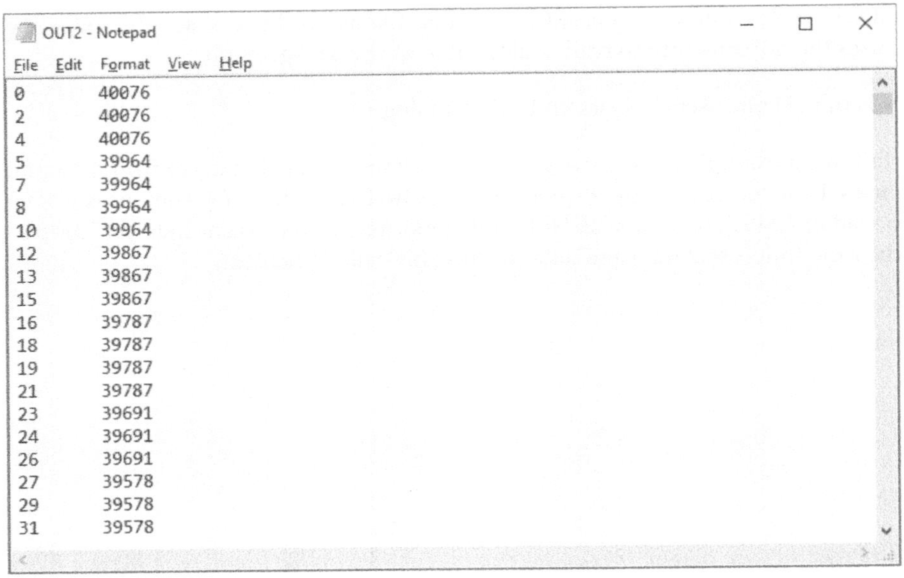

Figure 12.60 The content of out2.txt file.

- KLxx
- K20D50M

and FRDM-K64F is not supported.

The following example demonstrates how to generate fast analog outputs, but setting the analog output pin *DAC0_OUT* (for LPC1768 is P18) to 0.5 (1.65V) and 0 (0V) alternatively without any delay.

```
***************************************************************
// Example 12.13

#include "mbed.h"

#if defined(TARGET_K64F)
   AnalogOut  aout(DAC0_OUT);
#elif defined(TARGET_LPC1768)
   AnalogIn   aout(p18);
#endif

int main(void)
{
    while (1) {
        aout.write(0.5f);           // or  aout = 0.5f;
        aout.write(0.0f);
    }
}
***************************************************************
```

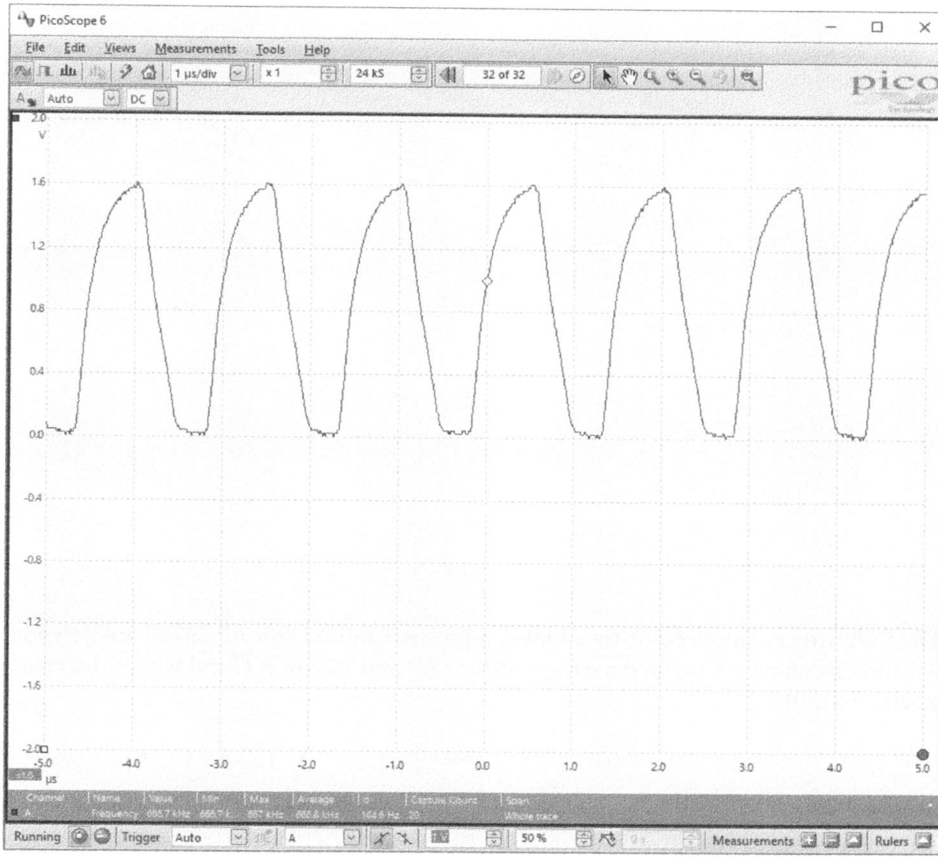

Figure 12.61 The PicoScope output of analog output.

Again, we can observe the analog output by using an oscilliscope. Figure 12.61 shows the corresponding PicoScope output. The results show that we can set analog output as fast as 666.7 Hz.

By combining the fast analog input and fast output, we can make a very useful program that can perform real-time signal processing. The following example reads the analog input (*val*), then updates the (*vals*), which is calculated by using 20% of (*val*) and 80% of previous (*vals*). This is the equivalent of applying a low-pass filter that can smooth out the data and remove high-frequency spikes, and finally, pass the (*vals*) to analog output. In this case, the analog output is the limiting factor, so the program should be able to process voice and signals up to 666.7 Hz.

```
************************************************************
// Example 12.14

#include "mbed.h"

#if defined(TARGET_K64F)
```

```
   AnalogIn    ain(A0);
   AnalogOut   aout(DAC0_OUT);
#elif defined(TARGET_LPC1768)
   AnalogIn    ain(p17);
   AnalogIn    aout(p18);
#endif

int main(void)
{
    double val=0;
    double vals=0;

    while (true) {
      val=ain.read();
      vals = 0.2*val + 0.8*vals;  //low-pass filter smoothing
      aout.write(vals);
    }
}
**************************************************************
```

The following example reads two analog inputs A0 and A1 (for LPC1768 are P16 and P17), and calculates a weighted average, 30% of A0 and 70% of A1, and writes the value to analog output.

```
**************************************************************
// Example 12.15

#include "mbed.h"

#if defined(TARGET_K64F)
   AnalogIn    ain1(A0);
   AnalogIn    ain2(A1);
   AnalogOut   aout(DAC0_OUT);
#elif defined(TARGET_LPC1768)
   AnalogIn    ain1(p16);
   AnalogIn    ain2(p17);
   AnalogIn    aout(p18);
#endif

int main(void)
{
    while (true) {
        aout = 0.3*ain1.read()+ 0.7*ain2.read();
    }
}
**************************************************************
```

The following example reads the analog input A0 (for LPC1768 is P16), delays nine times, and writes the value to analog output.

```
****************************************************************
// Example 12.16

#include "mbed.h"

#if defined(TARGET_K64F)
   AnalogIn    ain(A0);
   AnalogOut   aout(DAC0_OUT);
#elif defined(TARGET_LPC1768)
   AnalogIn    ain(p17);
   AnalogIn    aout(p18);
#endif

int main(void)
{
    double val[10];
    int i=0;
    while (true) {
        val[i%10] = ain.read();
        if(i>=10){
            aout.write(val[i-1]);
        }
        i++;
    }
}
****************************************************************
```

Exercise 12.8

Modify the above program so that it reads analog input A0 and A1, delays A1 nine times, and writes the average value to analog output.

Further Information about FastAnalogIn:

https://os.mbed.com/users/Sissors/code/FastAnalogIn/

12.7 Summary

This chapter provides some IoT project examples using the Arm® Mbed™ Ethernet IoT Starter Kit, such as, temperature monitoring over the Internet, smart lighting, voice-controlled door access, RFID reader, cloud example using IBM Watson Bluemix, and fast analog inputs and outputs.

Part IV

Appendices

In This Part

A

Example Codes

All the example codes used in this book are available on the following website:
http://www.wiley.com\go\xiao\designingembeddedsystemandIoTwitharmmbed

To use these example codes from your online compiler, just click the "***Import!***" button on the top left menu. An "***Import Wizard***" will be displayed in the middle, select the "***Upload***" tab, and an "***Choose File***" button will appear at the bottom (Figure A.1). Please note that the "***Import!***" button on the top right corner of "***Import Wizard***" panel is disabled.

Click the "*Choose File*" button and navigate to the directory that you have downloaded your example code files to. Select the example file you want to import and click "*Open*" button (Figure A.2). All the example codes are compressed in zipped format.

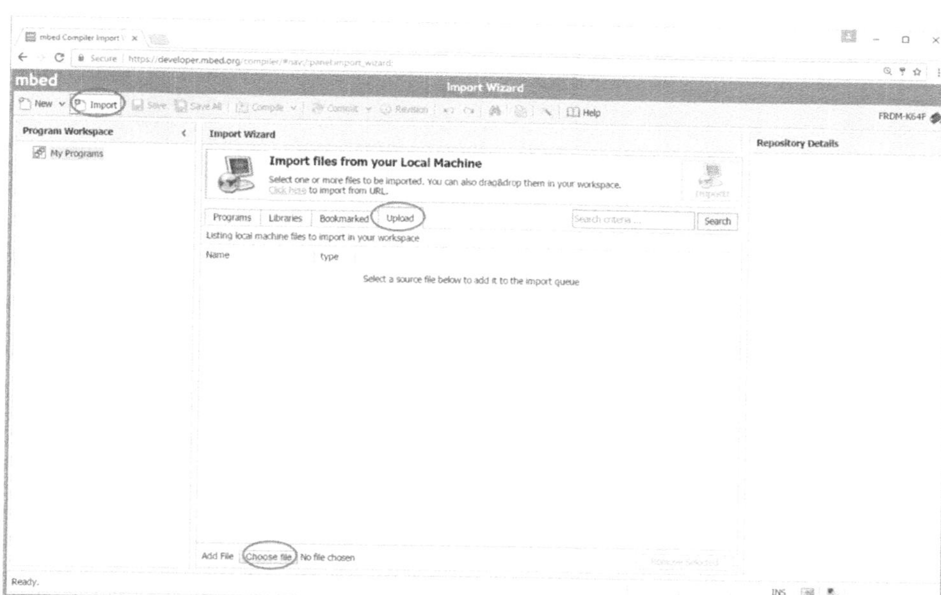

Figure A.1 Import example code from the online compiler.

Designing Embedded Systems and the Internet of Things (IoT) with the ARM® Mbed™, First Edition. Perry Xiao

Companion website: www.wiley.com\go\xiao\designingembeddedsystemandIoTwitharmmbed

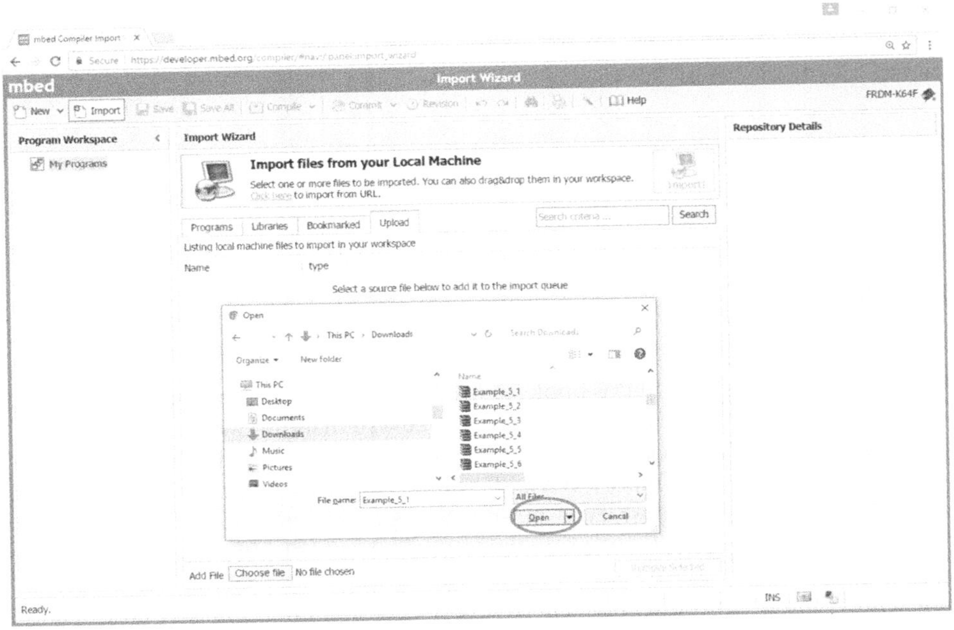

Figure A.2 Choose the example code file to import.

The selected file will then appear in the "*Upload*" tab. Now, the "*Import!*" button on the top right corner of "*Import Wizard*" panel should be enabled. Click the "*Import!*" button to import the file (Figure A.3).

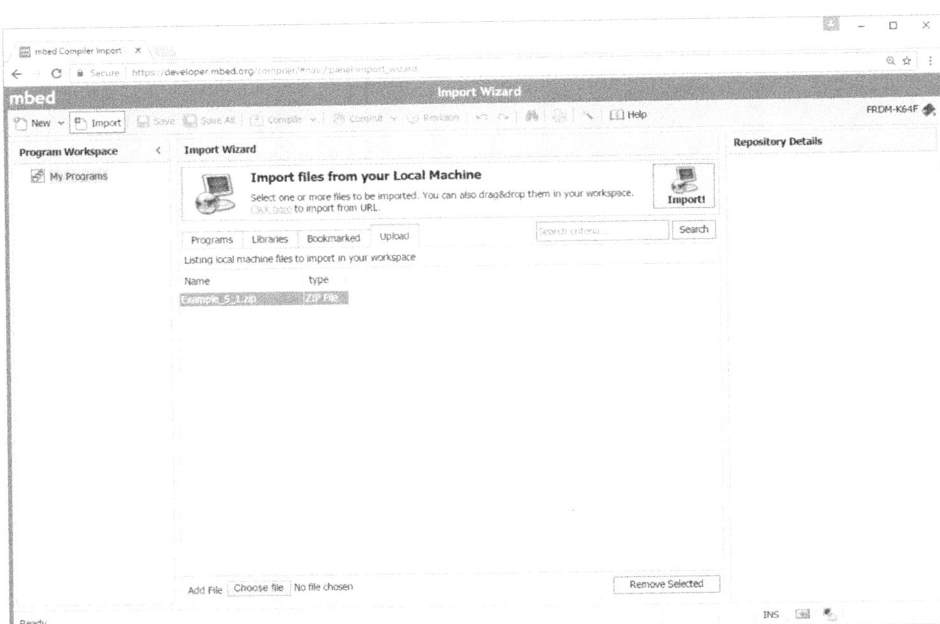

Figure A.3 Import example code by clicking the "*Import!*" button.

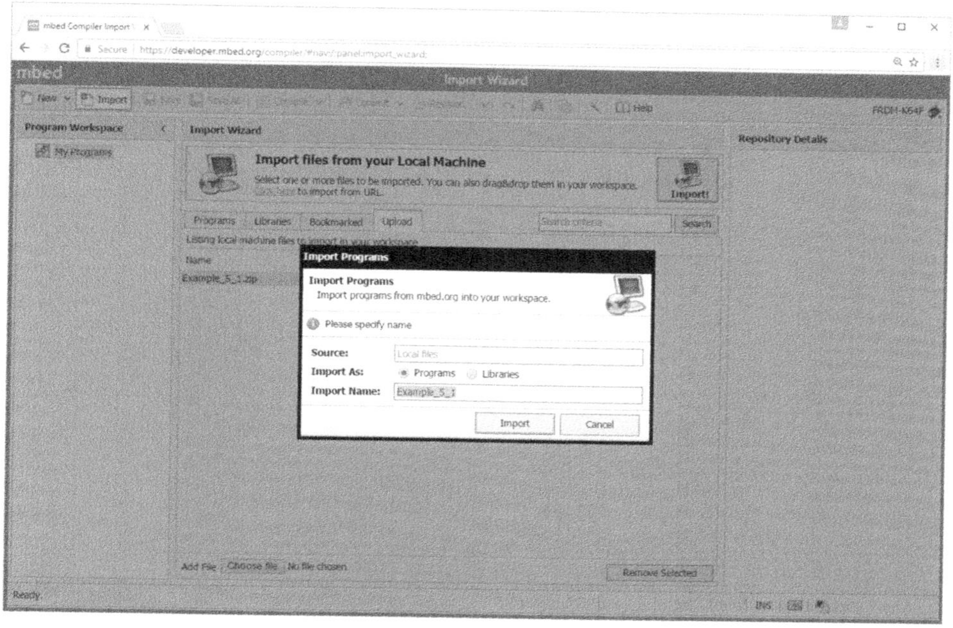

Figure A.4 The "*Import Programs*" pop-up window.

An "*Import Programs*" pop-up window will appear, through which you can change the import name if you want to (Figure A.4).

The example program should now be imported, and now you can compile and run it (Figure A.5).

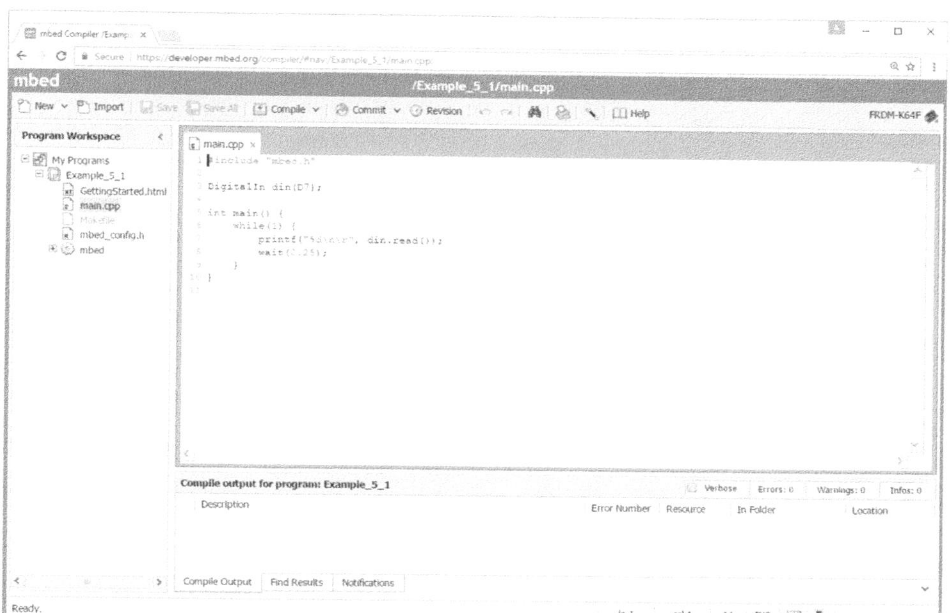

Figure A.5 The imported example program.

B

HiveMQ MQTT Broker

HiveMQ MQTT Broker is a very popular MQTT software that provides Websocket, security, and Socket services. We will use HiveMQ as an example to illustrate how to install and configure a MQTT broker software.

Just go to the HiveMQ website, as shown in Figure B.1, and follow the instructions to download the software and unzipped to any folder, as shown in Figure B.2. In this case, it is downloaded and unzipped to "*This PC > Downloads*" folder. Go to the HiveMQ's "*bin*" directory and double-click "*run*" file to run HiveMQ MQTT broker software. If everything is fine, you will see the software screen outputs like Figure B.3. As we did not purchase any licenses, it is limited to 25 connections.

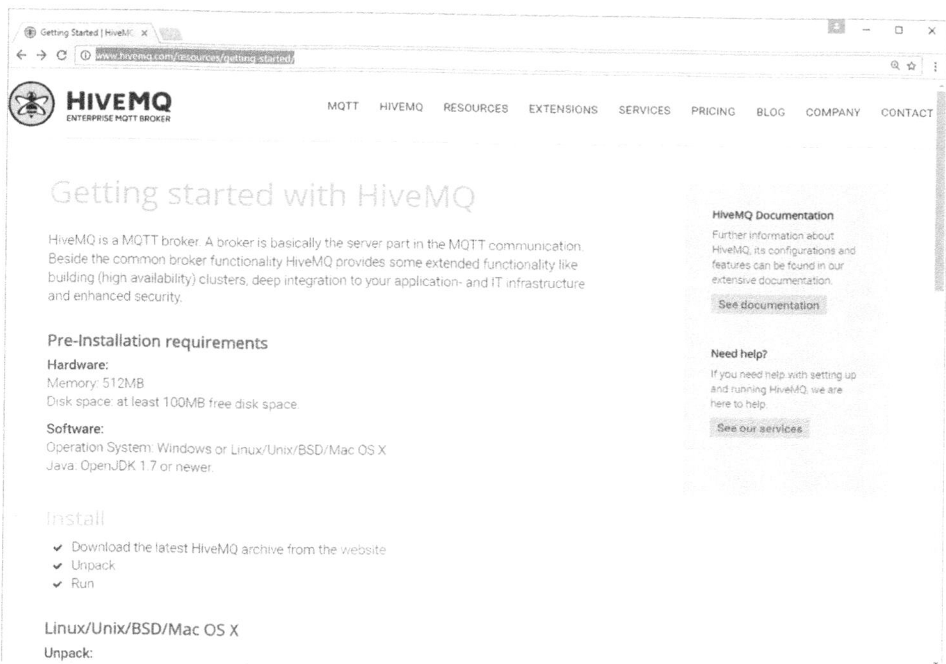

Figure B.1 The HiveMQ MQTT broker website.

Designing Embedded Systems and the Internet of Things (IoT) with the ARM® Mbed™, First Edition. Perry Xiao

Companion website: www.wiley.com\go\xiao\designingembeddedsystemandIoTwitharmmbed

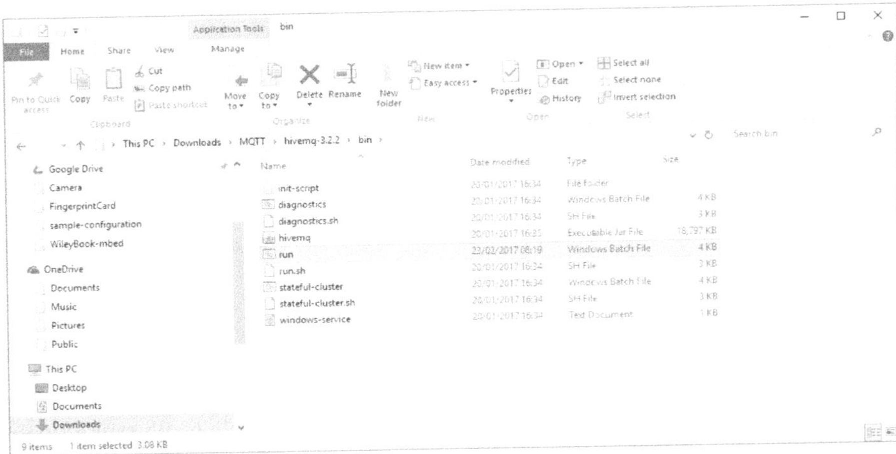

Figure B.2 The HiveMQ MQTT broker software folder.

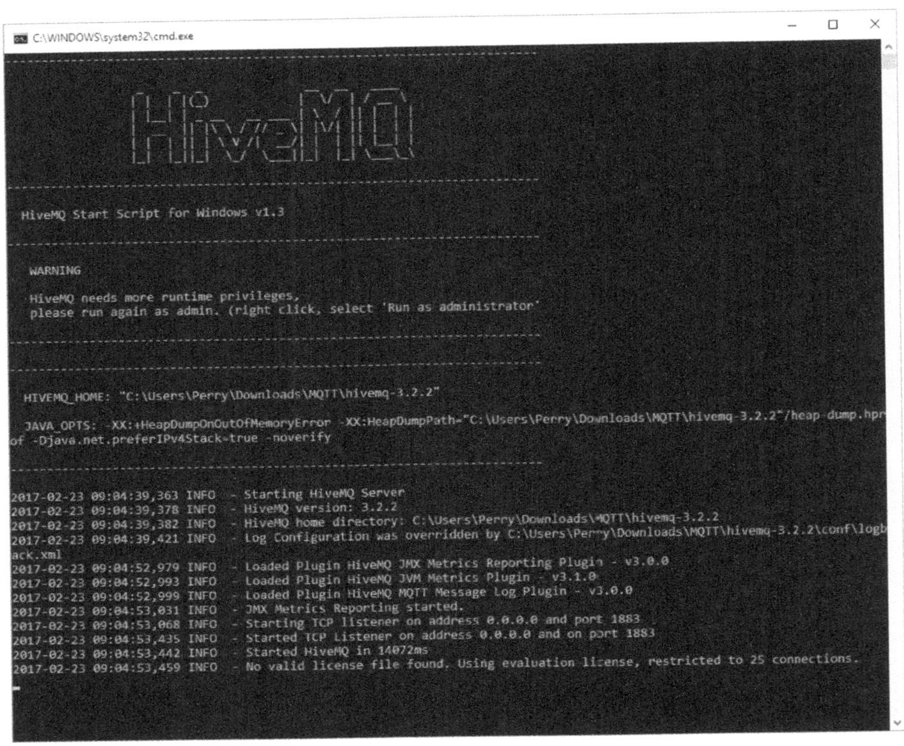

Figure B.3 The HiveMQ MQTT broker software screen outputs.

HiveMQ MQTT software builds on Java, so if you have not got the Java development kit software installed on your computer, you will need to download and install Java JDK (not JRE) software from its Oracle website:

http://www.oracle.com/technetwork/java/javase/downloads/index.html

To test the HiveMQ MQTT broker, you will need MQTT client software. Eclipse Paho is one of the most popular MQTT client software options. To use it, just go to the Eclipse Paho website and follow the instructions to download and install the software, as shown in Figure B.4.

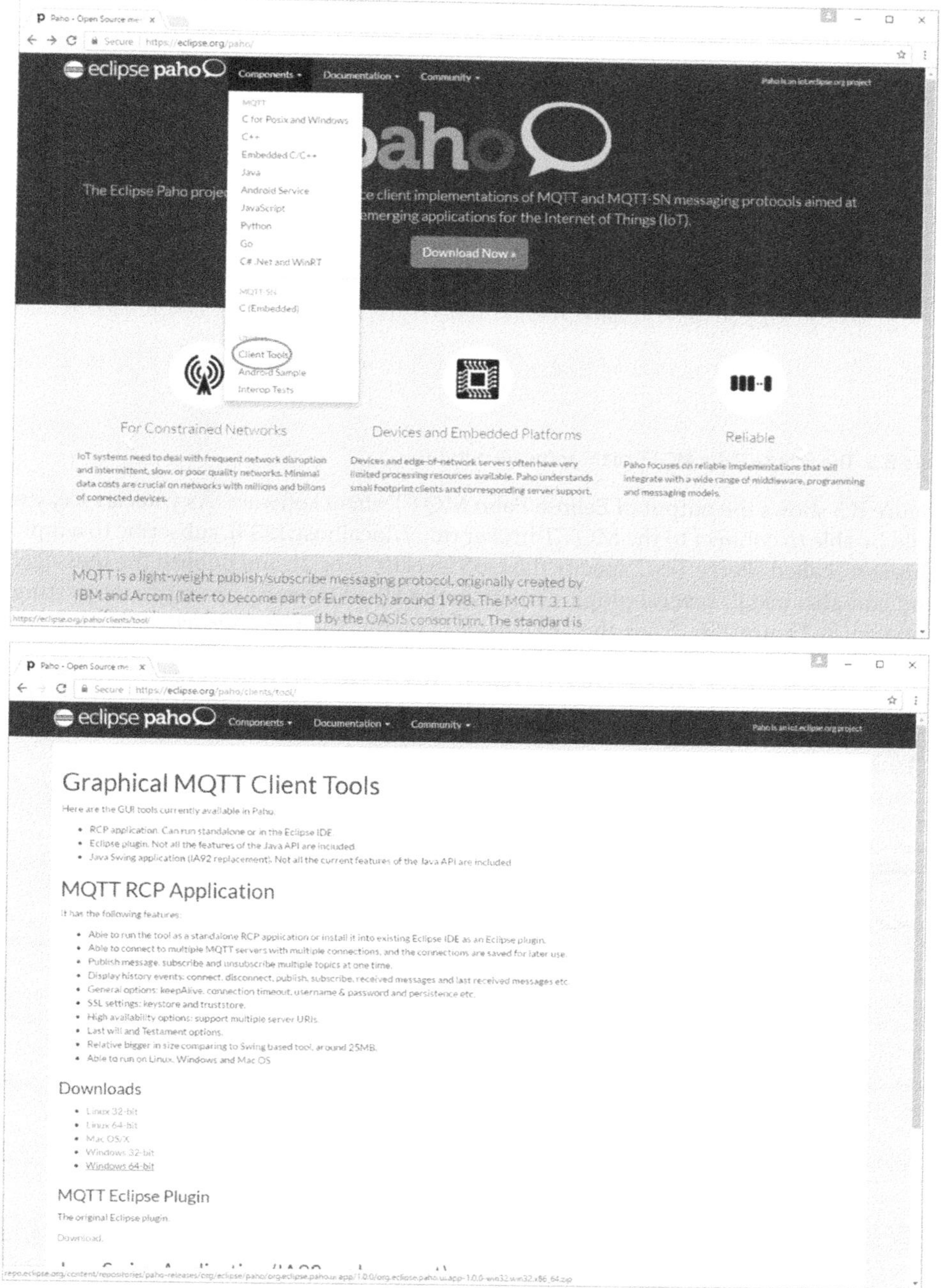

Figure B.4 The Eclipse Paho website (top) and the corresponding download page (bottom).

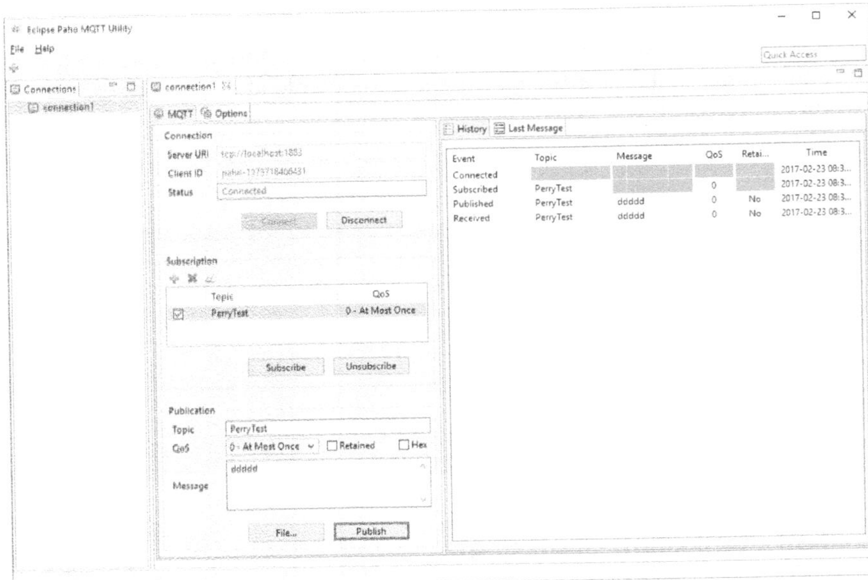

Figure B.5 The Eclipse Paho MQTT client software outputs.

Figure B.5 shows the output of Eclipse Paho MQTT client software. As you can see, you should be able to connect to the MQTT broker (tcp://localhost:1883), subscribe to a topic, in this case, called "Perry Test," specify the QoS level (0, 1, or 2), and publish the message.

You can also install several plugins to make the HiveMQ software more interesting. As shown in Figure B.6, from the "*Extensions*" menu, such as "*Security Plugins*" and

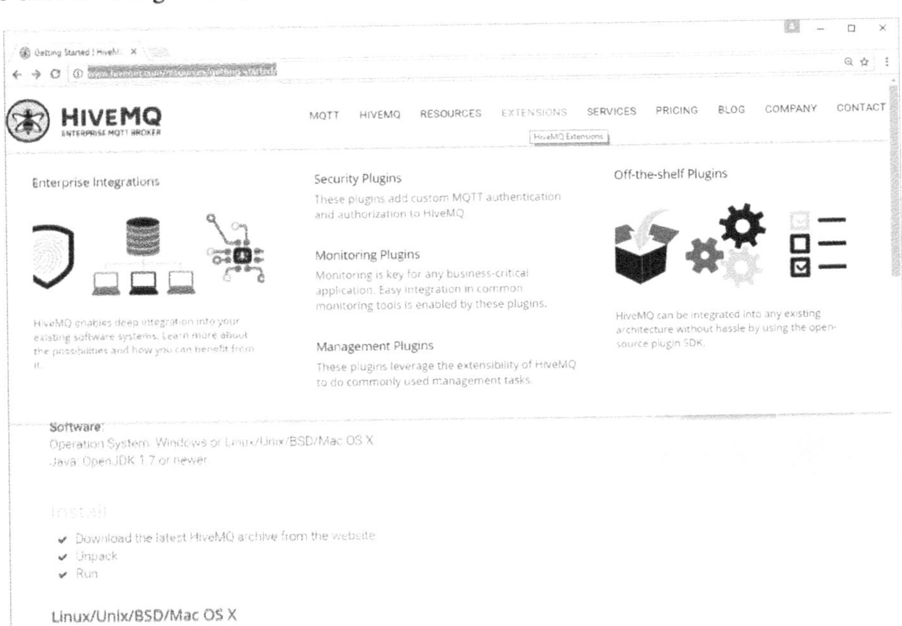

Figure B.6 The HiveMQ MQTT broker software plugins.

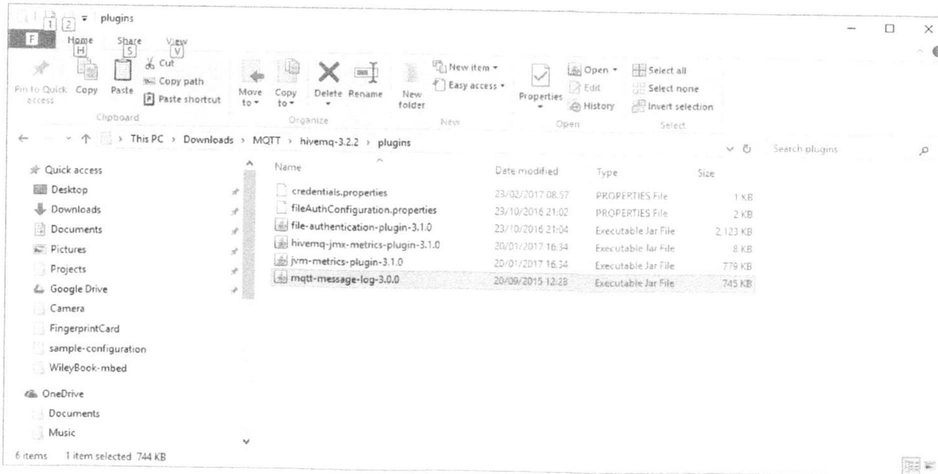

Figure B.7 The HiveMQ MQTT broker software plugins folder.

"*MQTT Message Log*" (inside the Monitoring Plugins). Just download and unzip the corresponding files to HiveMQ software's plugin folder, as shown in Figure B.7.

To implement security plugins, you need to modify the file "*credentials.properties*" in the plugins folder, as shown in Figure B.8. In this example, we added a new user called "perry" and password is "1111."

Alternatively, you can also use the "*file-authentication-plugin-utility-1.1.jar*" program in the plugins folder to manually add, configure, list, and remove users, as shown in Figure B.9. Just open the command prompt terminal program and from the plugins folder, run the following Java command:

```
java -jar utility/file-authentication-plugin-utility-1.1.jar
```

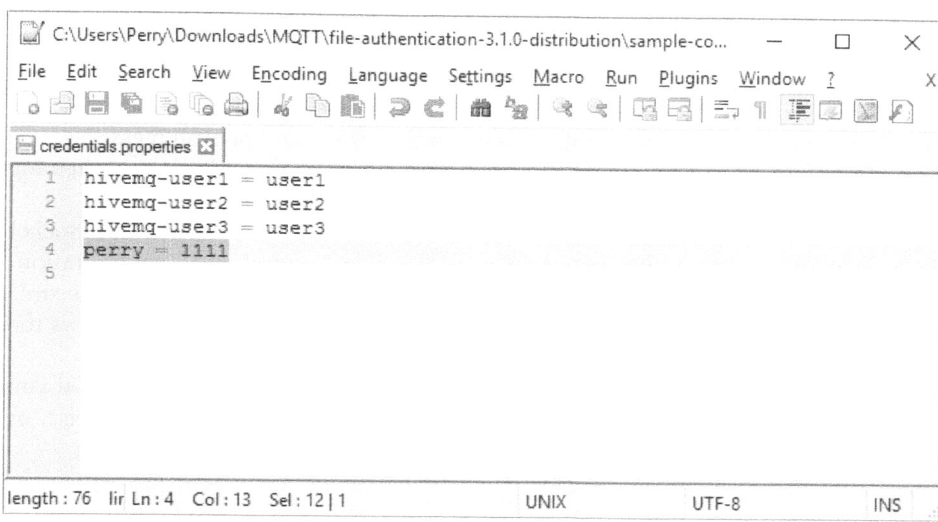

Figure B.8 Modification of the "credentials.properties" file.

Figure B.9 User management using file-authentication-plugin-utility-1.1.jar program.

To make the changes take effect, you need to re-run the HiveQM MQTT broker software, as we did before, shown in Figures B.2 and B.3.

Now, if you go back to the Eclipse Paho MQTT client software as shown Figure B.5, you will get an error message when you are trying to connect the HiveMQ MQTT broker. To get authenticated, from the "Options" tab in Paho software, tick the "Enable Login" checkbox and type in the username (perry) and password (1111). Go back to the "MQTT" tab; you should now be able to connect the broker and publish messages.

The "*MQTT Message Log*" plugin logs all the activities, users, and messages in a log file under the HiveMQ "*log*" sub-folder.

HiveMQ software can also provide WebSocket services. To activate the WebSocket service, go to the HiveMQ software folder, go to "conf" -> "examples" -> "configuration" sub-folder (Figure B.10), copy the file called "config-sample-mqtt-and-websockets.xml," paste it back to the "conf" folder, and rename it to "config.xml". Figure B.11 shows the content of the file, where WebSocket is provided on port 8000.

The "config.xml" file is the main configuration file for the HiveMQ software. If you want to keep the original "config.xml" file, you should copy it to somewhere else, or rename it first.

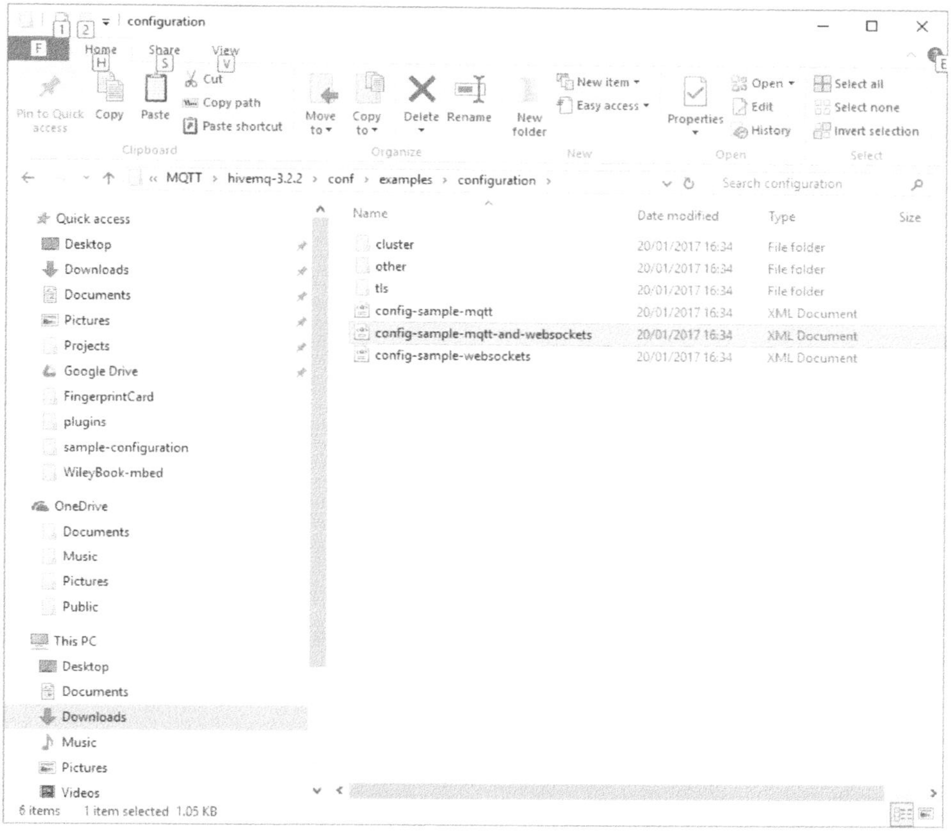

Figure B.10 The "config-sample-mqtt-and-websockets.xml" file in the "conf/examples/configuration" sub-folder.

Re-run the HiveQM MQTT broker software. It should now have the WebSocket service enabled, at port 8000 (Figure B.12).

To test the WebSocket service, you can use the HiveMQ WebSocket online client page (Figure B.13). You should be able to connect to the local host WebSocket service and send WebSocket messages. You can also view the WebSocket activities from the HiveQM terminal window, as shown in Figure B.14.

Further Information about HiveMQ MQTT:

http://www.hivemq.com/resources/getting-started/
http://www.hivemq.com/blog/hivemq-mqtt-websockets-support-message-log-plugin-2-min
http://www.hivemq.com/plugin/file-authentication/
http://www.hivemq.com/demos/websocket-client/

```
<?xml version="1.0"?>
<hivemq xmlns:xsi="http://www.w3.org/2001/XMLSchema-instance"
        xsi:noNamespaceSchemaLocation="hivemq-config.xsd">

    <listeners>
        <tcp-listener>
            <port>1883</port>
            <bind-address>0.0.0.0</bind-address>
        </tcp-listener>
        <websocket-listener>
            <port>8000</port>
            <bind-address>0.0.0.0</bind-address>
            <path>/mqtt</path>
            <subprotocols>
                <subprotocol>mqttv3.1</subprotocol>
            </subprotocols>
            <allow-extensions>true</allow-extensions>
        </websocket-listener>
    </listeners>
    <mqtt>
        <max-client-id-length>65535</max-client-id-length>
        <retry-interval>10</retry-interval>
    </mqtt>
    <throttling>
        <max-connections>-1</max-connections>
        <max-message-size>268435456</max-message-size>
        <outgoing-limit>0</outgoing-limit>
        <incoming-limit>0</incoming-limit>
    </throttling>
    <general>
        <update-check-enabled>true</update-check-enabled>
    </general>

</hivemq>
```

Figure B.11 The content of the "**config.xml**" file with WebSocket service (port 8000).

Figure B.12 The HiveMQ MQTT broker software screen outputs with WebSocket service enabled at port 8000.

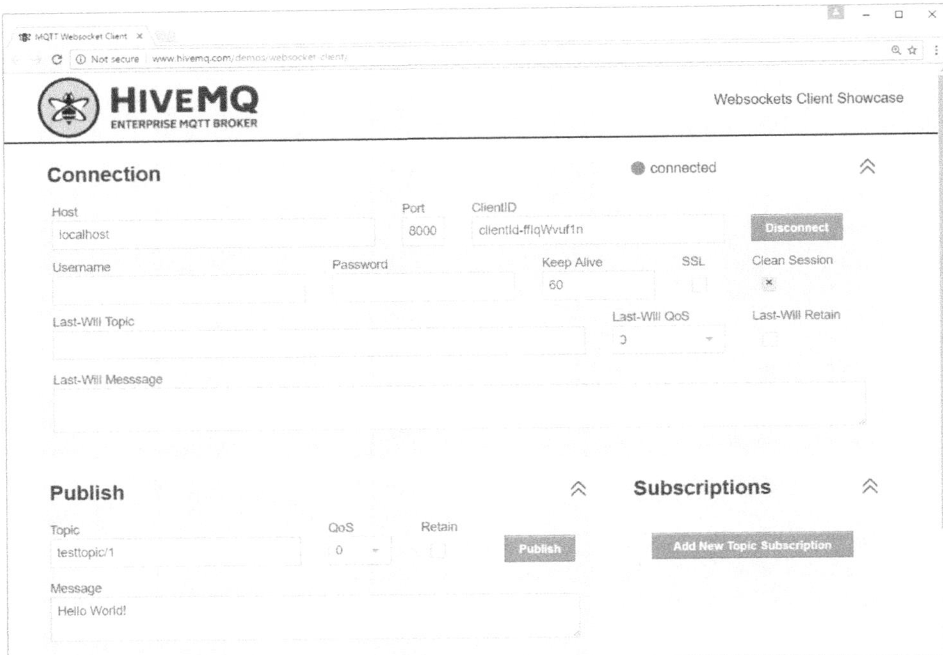

Figure B.13 The HiveMQ WebSocket online client.

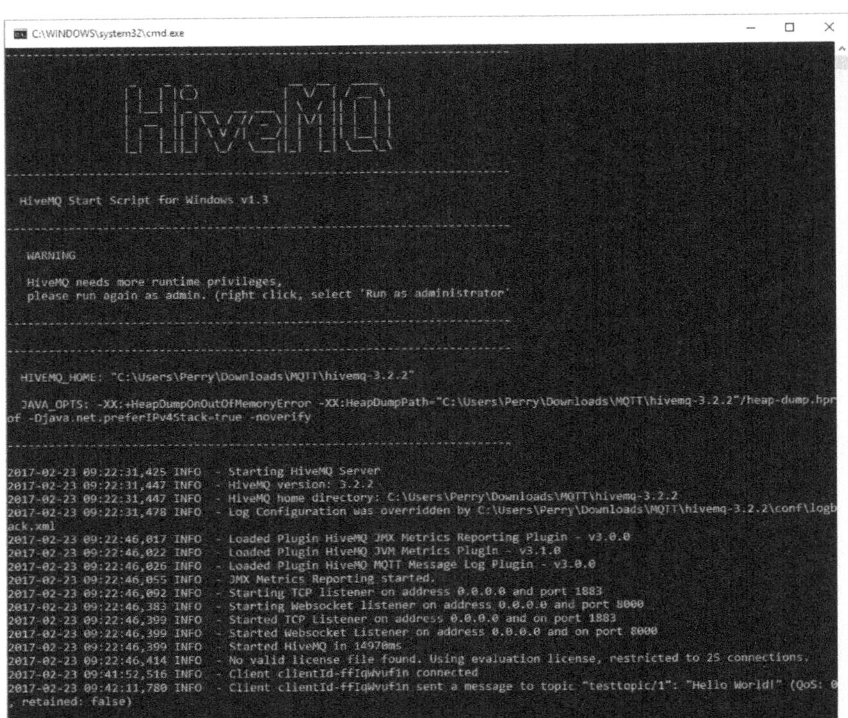

Figure B.14 The HiveMQ MQTT broker software screen outputs with WebSocket messages.

C

Node-RED on Raspberry Pi

Raspberry Pi is an excellent kit for learning. So if you have a one, you can also use Node-RED using Raspberry Pi, as the latest Raspberry Pi operating system Raspbian Jessie has Node-RED installed by default.

If you do not have the latest Raspberry Pi operating system, you can download it from the following website, as shown in Figure C.1.

https://www.raspberrypi.org/downloads/

You can also follow the instructions to install Node-RED by yourself, if it has not been installed.

https://howtonode.org/how-to-install-nodejs

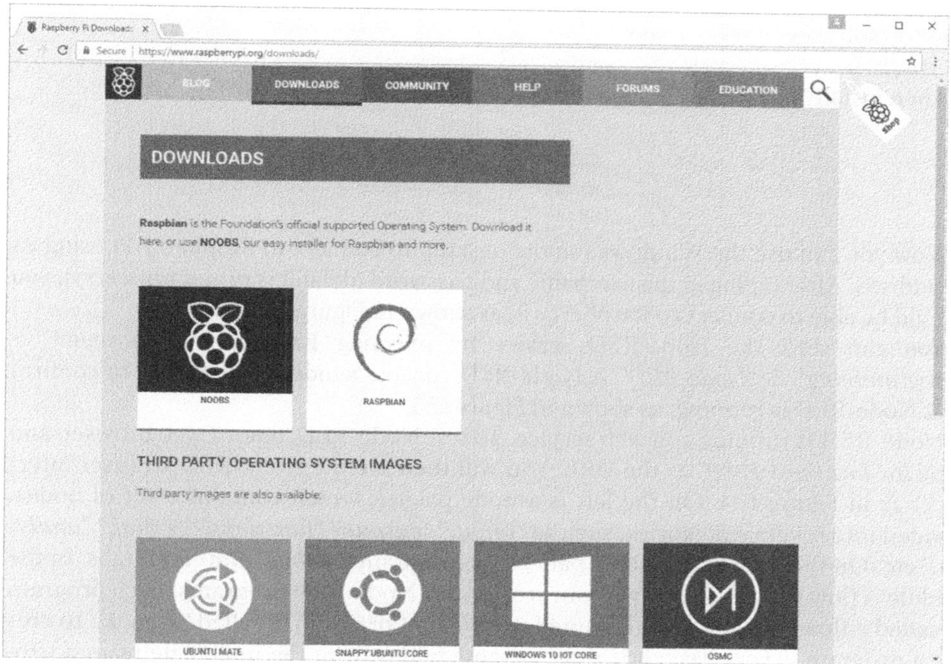

Figure C.1 Raspberry Pi operating system download page.

Designing Embedded Systems and the Internet of Things (IoT) with the ARM® Mbed™, First Edition. Perry Xiao

Companion website: www.wiley.com\go\xiao\designingembeddedsystemandIoTwitharmmbed

To install Node-RED, you need to install "**Node.js**" and "**npm**" packages first. From the Raspberry Pi desktop, just open a terminal software and type the following commands:

```
$ sudo apt-get update
$ sudo apt-get install nodejs
$ sudo apt-get install npm
```

Once you have **Node.js** installed, just use the following commands to install and run Node-RED:

```
$ sudo npm install -g node-red
$ node-red
```

When you are working with many devices, it is easier to access Raspberry Pi using SSH or using a remote desktop, as this saves the hassle of having an extra monitor.

Raspberry Pi disabled the SSH by default. To enable it, go from main menu -> "**Preferences**" -> "**Raspberry Pi Configuration**". A configuration window will appear, and from the "Interfaces" tab, make sure SSH is enabled, as shown in Figure C.2. Now you should be able to remote login to Raspberry Pi using SSH with any terminal software such as, putty.exe, Tera Term, etc.

To connect to Raspberry Pi using a remote desktop, you will need to install "*xrdp*" and "*tightvncserver*" packages first. To start from scratch, first remove the existing remote desktop software, if there is any:

```
$ sudo apt-get remove xrdp vnc4server tightvncserver
```

Then install "*tightvncserver*" and "*xrdp*" software,

```
$ sudo apt-get install tightvncserver
$ sudo apt-get install xrdp
```

Now you can use the Windows remote desktop to connect to Raspberry Pi using its IP address. After typing in the username and password (default is *pi* and *raspberry*), you should be able to connect to Raspberry Pi as shown in Figure C.3.

You can start the Node-RED service by selecting Raspberry main menu -> "*Programming*" -> "*Node-RED*". A Node-RED console window will appear to confirm that Node-RED is running, as shown in Figure C.3.

Node-RED is running as a web service. To test Node-RED, open a web browser, and type in "*localhost:1880*" as the URL. You will then see the Node-RED UI (user interface) as in Figure C.4. On the left is a node palette, which contains a set of nodes, divided into several categories, such as "*input,*" "*output,*" "*functions,*" "*social,*" "*analysis,*" etc. On the right is an output pane, which contains "*debug*" and "*info*" tabs. In the middle is flow canvas, where you can create your Node-RED programs. Each program is called a flow, and the default name for your first program is called "*Flow 1*". To create a program is very easy; just drag in some nodes from the node palette, and wire them up.

Let's use a simple MQTT example program to illustrate how to program in Node-RED. From node palette, drag in a "*mqtt*" node from the "*input*" category, and a "*debug*"

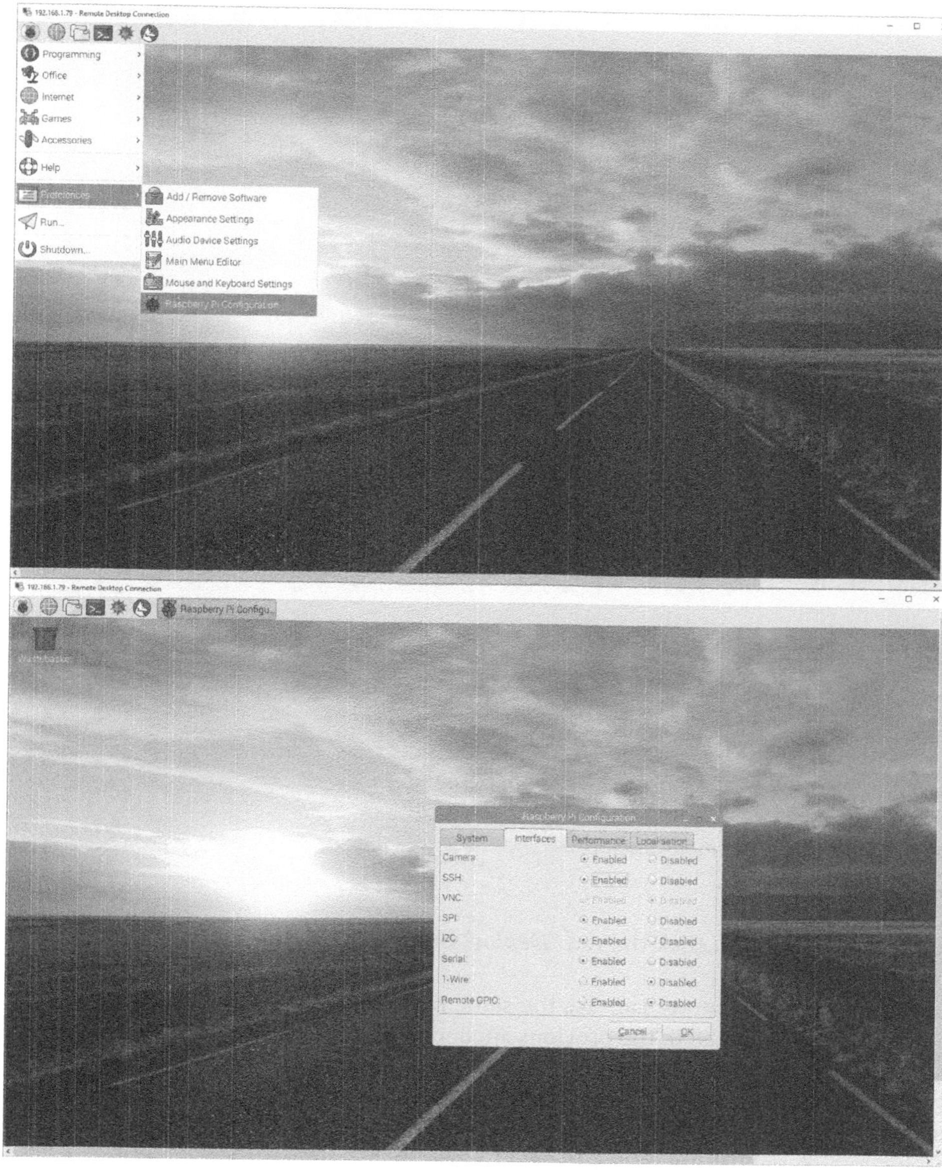

Figure C.2 Raspberry Pi Configuration.

node from the "*output*" category, wire them up as shown in Figure C.4 (top). Double-click the "*mqtt*" node to entre the configuration information as shown in Figure C.4 (bottom). This will connect to the HiveMQ MQTT broker we did in the previous appendix, running on your computer at port number 1883. The IP address is your computer IP address. It will subscribe to the topic "*PerryTest*." In this example, it will pick up any messages published to the topic "*PerryTest*," and display it on the "*debug*"

Figure C.3 Raspberry Pi remote desktop.

tab on the right. Now, click the red "*Deploy*" button on the top right to activate your program.

To test your program, we can use the same Eclipse Paho MQTT client software on your computer, as we did in the previous appendix. In this example, it published a message "*tttttt*" to the topic "*PerryTest.*" The message appeared both on the Eclipse Paho MQTT client, Figure C.5 (top), and Node-RED program "*debug*" tab, Figure C.5 (bottom).

You can also publish messages to MQTT broker by adding the second flow to the program. From the node palette, drag in an "*injet*" node from the "*input*" category, and a "*mqtt*" node from the "*output*" category, and wire them up as shown in Figure C.6. Please note that the "*inject*" changed its name to "*timestamp*" when it is placed on the flow canvas. This node will send out a time stamp message. Double-click the "*mqtt*" node and enter the configuration. Again, click the red "*Deploy*" button on the top right to activate your program.

Now, each time you press the "*timestamp*" node, it will publish a time stamp message to the topic "*PerryTest,*" and the first flow will receive the message, as shown in the "*debug*" tab in Figure C.7.

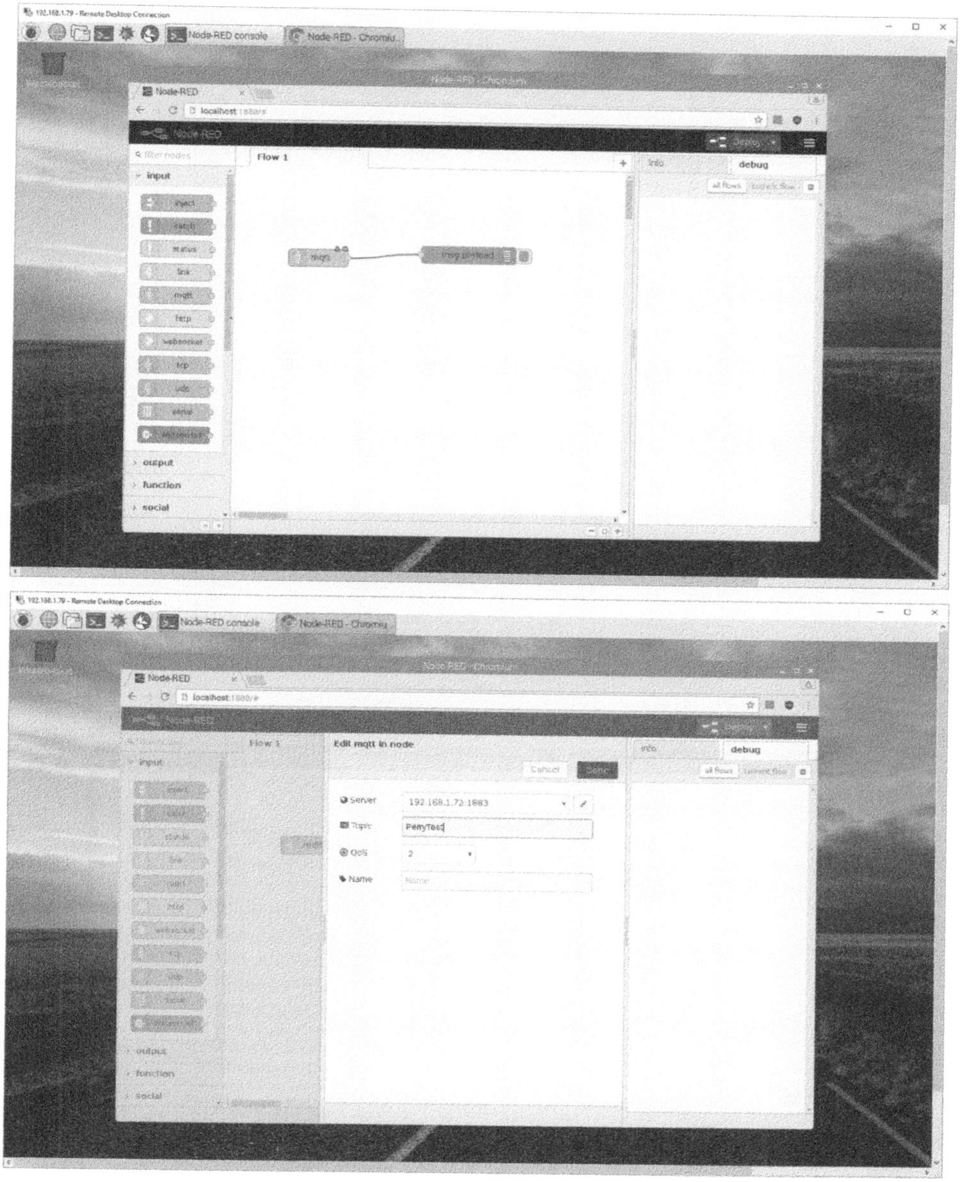

Figure C.4 Node-RED user interface (top) and Configuration of the "*mqtt*" node (bottom).

You may also want to install "*Palette Manager,*" which will allow you to install new modules. Following are commands:

```
$ sudo apt-get update
$ sudo apt-get install npm
$ sudo npm i -g npm@2.x
```

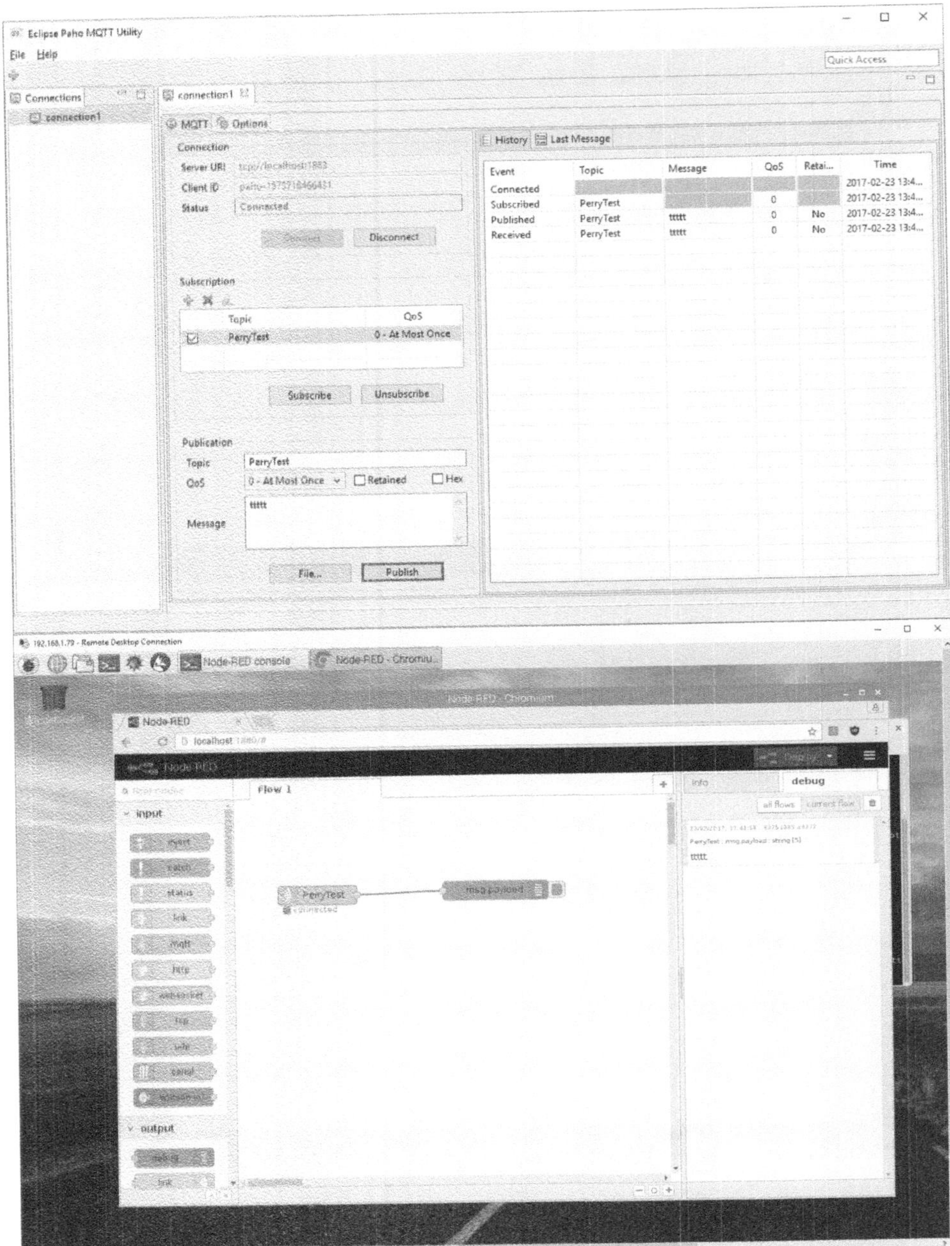

Figure C.5 Eclipse Paho MQTT client (top) and Node-RED output (bottom).

After the installation, restart the Node-RED. Then from the top right corner menu, you will see the item called "*Manage Palette*" (Figure C.8), from which you would be able to install modules such as Node-RED Dashboard, as shown in Chapter 12, section 12.5.3.

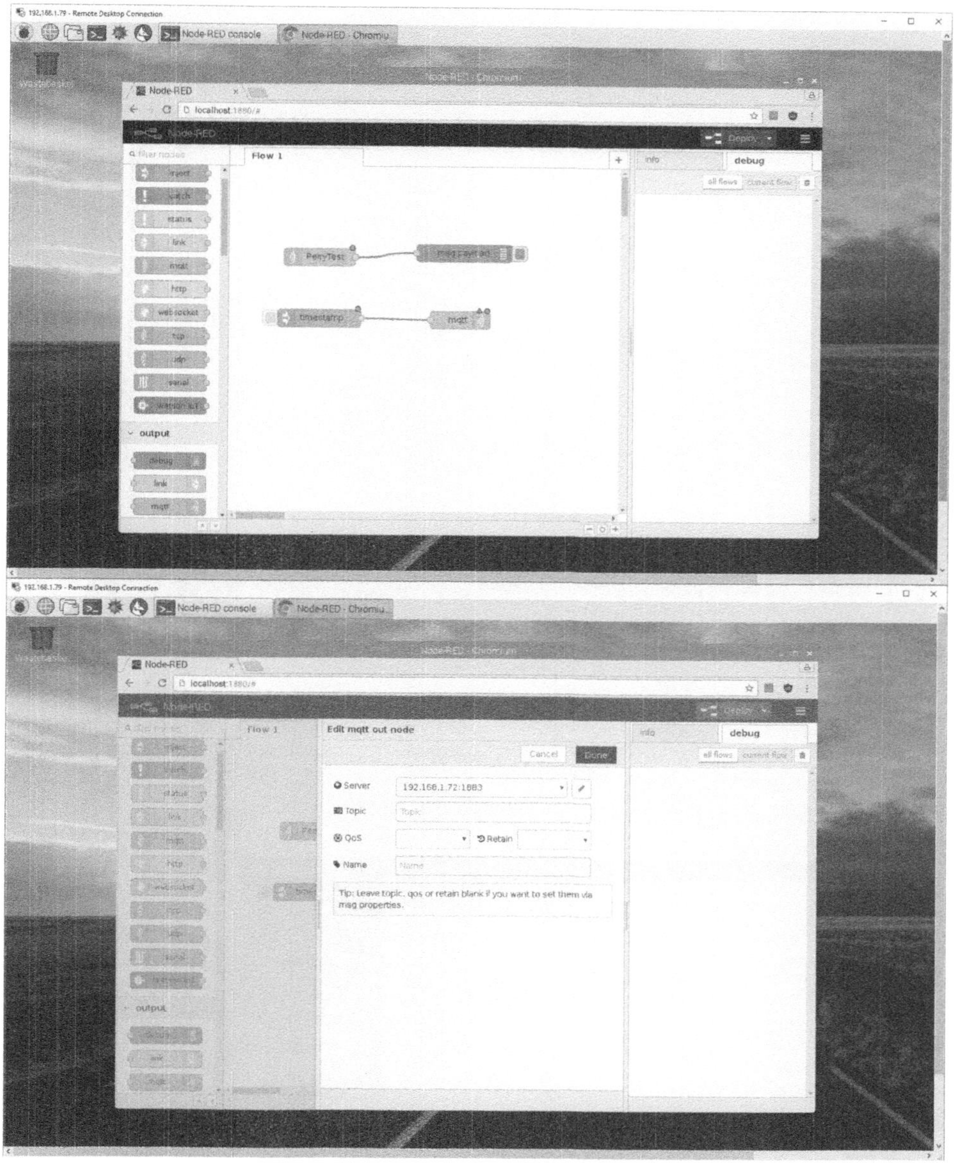

Figure C.6 Add a second flow to the program (top) and the configuration of "*mqtt*" output node (bottom).

There are many interesting examples online, which you can try by importing them using JSON (JavaScript Object Notation).

http://flows.nodered.org/
http://noderedguide.com/tag/example/
https://hub.jazz.net/project/sportshack/node-red-examples/overview

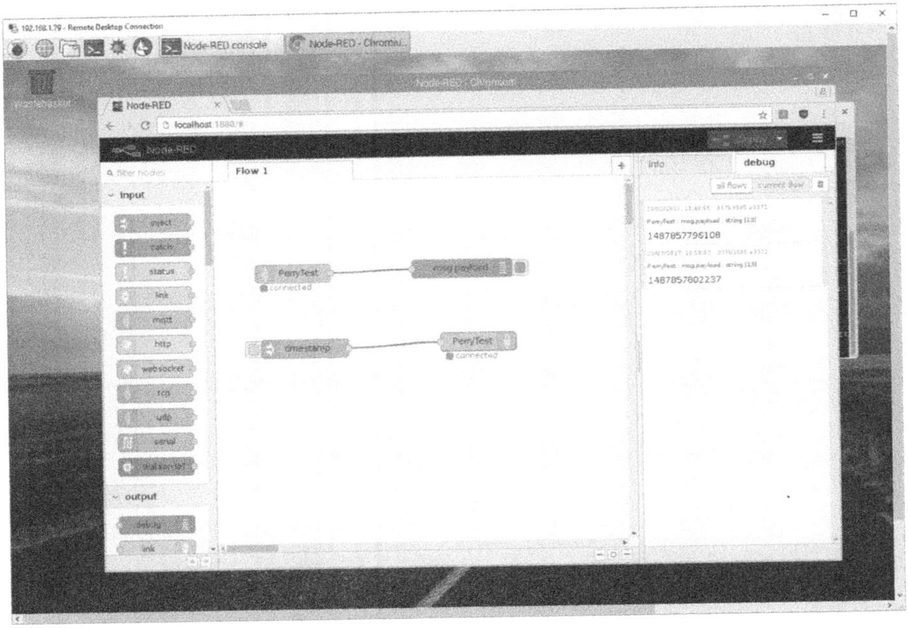

Figure C.7 The "*debug*" tab output of the program.

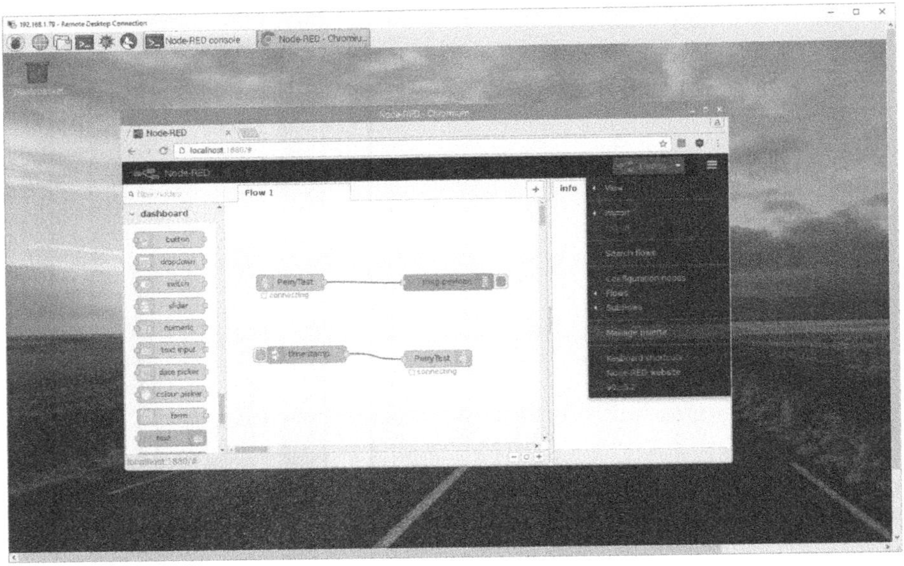

Figure C.8 The "*Manage Palette*" in the dropdown menu.

Further Information about Node-RED:

http://noderedguide.com/
http://nodered.org/docs/hardware/raspberrypi.html

D

String and Array Operations

String and/or array operations have always been an issue for many programmers, so in this appendix, we provide a list of example codes for commonly used string and array operations.

Define Strings

```
char str[256] ;
strcpy (str,"hello world");      //str = "hello world";
```

or

```
char str[60] = "Hello World";
```

or

```
char *str = "Hello World";
```

Concatenate Strings (Merge Strings)

```
char str[80] = "hello ";
strcat (str,"world ");      //merge "hello" and "world" into str
```

Sub-Strings

```
char *buff = "Hello World";
char* substr;
strncpy(substr, buff+7, 5);  //begin index: 7 substring length: 5
```

Designing Embedded Systems and the Internet of Things (IoT) with the ARM® Mbed™, First Edition. Perry Xiao

Companion website: www.wiley.com\go\xiao\designingembeddedsystemandIoTwitharmmbed

String Length

```
char str[32] = "hello, world";
int n = strlen (str)
```

String to Float/Double (sscanf)

```
char *buff  = "51.432";
double x;
sscanf(buff, "%f", &x);
printf("%f\n\r",x);
```

String to Float/Double (atof)

```
#include <cstdlib>
void main()
{
    float x;
    x = atof("3.14");
}
```

Read Numbers from a String

```
#include <stdio.h>

int main ()
{
  char sentence []="Tony   34   5.6";
  char str [20]; int x; float y;

  sscanf (sentence,"%s %*d %f",str,&x,&y);
  printf ("%s %d %f\n",str,x,y);

  return 0;
}
```

Combine Strings and Numbers

```
char buffer[256];
float x = 3.5;
int y =20;
n=sprintf (buffer, "%s %f %d", "Tom", x, y);
```

Split String into Two

```
#include <string.h>

char *token;
char line[] = "Hello World";
char *search = " ";

// Token will point to "Hello".
token = strtok(line, search);

// Token will point to "World".
token = strtok(NULL, search);
```

Split String into Many

```
#include <stdio.h>
#include <string.h>

int main ()
{
    char str[] ="- This, a sample string.";
    char * pch;
    printf ("Splitting string \"%s\" into tokens:\n",str);
    pch = strtok (str," ,.-");
    while (pch != NULL)
    {
        printf ("%s\n\r",pch);
        pch = strtok (NULL, " ,.-");
    }
    return 0;
}
```

Float to String

```
char buffer[50];
float x = 3.5;
n=sprintf (buffer, "%f", x);
```

Strings Array

```
char * const str[] = {
     "Hello",
     "World",
};
printf ("%s\t%s\n\r",str[0],str[1]);
```

Strings to Uppercase

```
#include<string.h>

int main() {
   char *str = "Hello World";
   strupr(str);
   printf("%s\n\r", str);

   return (0);
}

#include <stdio.h>
#include <ctype.h>

int main()
{
   int i = 0;
   char c;
   char str[] = "Hello World";

   while(str[i])
   {
      str[i]=toupper(str[i]);
      i++;
   }

   return(0);
}
```

Strings to Lowercase

```
#include <string.h>
#include <ctype.h>

char *strlwr(char *str)
{
  unsigned char *p = (unsigned char *)str;

  while (*p) {
     *p = tolower((unsigned char)*p);
      p++;
  }

  return str;
}
```

Strings to Lowercase 2

```
#include <stdio.h>
#include <ctype.h>

int main()
{
   int i = 0;
   char c;
   char str[] = "Hello World";

   while( str[i] )
   {
      str[i]=tolower(str[i]);
      i++;
   }

   return(0);
}
```

Compare Two Strings

```
char *c1   = "hello";
char c2 []   = "hello";

if (strcmp(c1,c2)==0){
     //do something
}
```

Compare Two Characters

```
#include "mbed.h"

Serial pc(USBTX, USBRX); // tx, rx

int main() {
    char c = pc.getc();
    if(c == 'a') {
        //do something
    }
}
```

Integer Array

```
int age[4];

age[0]=14;
age[1]=13;
age[2]=15;
age[3]=16;

or

int arr [5] = {1,2,3,4,5};

for (int i=0; i<5;i++){
   printf("%d", arr[i]);
}
```

Integer 2D Array

```
int age[4][3];

age[0][0]=14;
age[1][0]=13;
age[2][0]=15;
age[3][0]=16;

age[0][1]=14;
age[1][1]=13;
age[2][1]=15;
age[3][1]=16;

age[0][2]=14;
age[1][2]=13;
age[2][2]=15;
age[3][2]=16;
```

```
or

int ages [3][4] = {
    {1, 2, 3, 4},
    {5, 6, 7, 8},
    {9, 10, 11, 12}
};
```

Float Array

```
#include <stdio.h>

int main()
    float data[4096];
    for (int i=0;i<4096;i++)
    {
        data[i]=i*0.001;
        printf("%f\n\r", data[i]);
    }
    return 0;
}
```

Float 2D Array

```
#include <stdio.h>

int main()
    float data[4096][3];
    for (int i=0;i<4096;i++)
    {
        data[i][0]=i*0.001;
        data[i][1]=sin(i*0.001);
        data[i][2]=cos(i*0.001);
        printf("%10.2f \t %10.2f \t %10.2f \n\r", data[i][0],
data[i][1], data[i][2]);
    }
    return 0;
}
```

E

Useful Online Resources

Arm® Mbed™

https://www.arm.com/products/processors/instruction-set-architectures/index.php
https://en.wikipedia.org/wiki/ARM_architecture

---

mbed YouTube Playlist

https://www.youtube.com/channel/UCNcxd73dSceKtU77XWMOg8A/playlists

---

C/C++ Reference

http://www.cplusplus.com/reference/
http://en.cppreference.com/w/
https://www.gnu.org/software/gnu-c-manual/gnu-c-manual.html

C/C++ Tutorial

http://www.cplusplus.com/doc/tutorial/
https://www.tutorialspoint.com/cplusplus/
http://www.cprogramming.com/begin.html

WebSocket Tutorial

http://tutorialspoint.com/websockets/index.htm
https://developer.mbed.org/cookbook/Websockets-Server
https://www.fullstackpython.com/websockets.html

Designing Embedded Systems and the Internet of Things (IoT) with the ARM® Mbed™, First Edition. Perry Xiao

Companion website: www.wiley.com\go\xiao\designingembeddedsystemandIoTwitharmmbed

---

Python Tutorial

https://docs.python.org/3/tutorial/
https://www.tutorialspoint.com/python/

Java Tutorial

https://docs.oracle.com/javase/tutorial/
http://www.javatpoint.com/java-tutorial
http://javabeginnerstutorial.com/core-java/

MQTT Tutorial

http://mqtt.org/documentation
http://www.hivemq.com/blog/how-to-get-started-with-mqtt
http://www.ev3dev.org/docs/tutorials/sending-and-receiving-messages-with-mqtt/
https://learn.adafruit.com/mqtt-adafruit-io-and-you/overview

Node-RED Tutorial

https://nodered.org/docs/getting-started/first-flow
http://noderedguide.com/
https://developer.ibm.com/recipes/tutorials/getting-started-with-watson-iot-platform-using-node-red/

JSON Tutorial

https://www.w3schools.com/js/js_json_intro.asp
http://www.tutorialspoint.com/json/
https://www.javatpoint.com/json-tutorial

Node.js Tutorial

https://nodejs.org/en/
http://www.tutorialspoint.com/nodejs/
http://www.nodebeginner.org/

Index

Designing Embedded Systems and the Internet of Things (IoT) with the ARM® Mbed™, First Edition. Perry Xiao

Companion website: www.wiley.com\go\xiao\designingembeddedsystemandIoTwitharmmbed

Printed and bound by CPI Group (UK) Ltd, Croydon, CR0 4YY
17/09/2026
14964300-0002